MANAGING ISSUES IN GEOGRAPHY

Bob Hordern, Phil Lamb, Steve Milner

Hodder & Stoughton
www.hodderheadline.co.uk

Orders: please contact Bookpoint Ltd, 130 Milton Park, Abingdon, Oxon OX14 4SB. Telephone: (44) 01235 827720. Fax: (44) 01235 400454. Lines are open from 9.00–6.00, Monday to Saturday, with a 24 hour message answering service. You can also order through our website www.hodderheadline.co.uk.

British Library Cataloguing in Publication Data
A catalogue record for this title is available from the British Library

ISBN 0 340 80216 2

First Published 2003
Impression number   10 9 8 7 6 5 4 3 2 1
Year   2009 2008 2007 2006 2005 2004 2003

Copyright © 2003 Bob Hordern, Phil Lamb and Steve Milner

All rights reserved. No part of this publication may be reproduced or transmitted in any form or by any means, electronic or mechanical, including photocopy, recording, or any information storage and retrieval system, without permission in writing from the publisher or under licence from the Copyright Licensing Agency Limited. Further details of such licences (for reprographic reproduction) may be obtained from the Copyright Licensing Agency Limited, of 90 Tottenham Court Road, London W1T 4LP.

Typeset by Pantek Arts Ltd, Maidstone, Kent.
Printed in Italy for Hodder & Stoughton Educational, a division of Hodder Headline, 338 Euston Road, London NW1 3BH.

Contents

Section 1: Managing Change in the Human Environment

Chapter 1:	Population change	1
Chapter 2:	Rural–urban migration in LEDCs	7
Chapter 3:	The changing town and city centre	14
Chapter 4:	Pressure at the rural–urban fringe	24

Section 2: Managing the Physical Environment

Chapter 5:	Earthquakes and volcanoes	33
Chapter 6:	Weather hazards	49
Chapter 7:	Water and food supply	65
Chapter 8:	Pressures on the physical environment	89

Section 3: Managing Economic Development

Chapter 9:	Contrasting levels of development	105
Chapter 10:	Resource depletion	125
Chapter 11:	Economic development and the global environment	139
Chapter 12:	Tourism and the economy	156

Coursework advice	169
Key skills guidance	176
Appendix	180
Glossary	182
Index	190

Acknowledgements

The publishers would like to thank the following individuals, institutions and companies for permission to reproduce copyright illustrations in this book:

© Shaen Adey; Gallo Images/Corbis, 7.18; © Paul Almasy/Corbis, 7.23; AP Photo/Luis Elvir, 6.2; AP Photo/John McConnico, 5.14; AP Photo/Pavel Rahmen, 6.1; AP Photo/Yomiuri Shimbun, 5.1; AP Photo/Fabrizio Villa, 5.11; © AQA, 5.26; © Yann Arthus-Betrand/Corbis, 6.8; © Barry Aymes/Life File, 7.33 (first photo); © Tiziana and Gianni Baldizzone/Corbis, 7.33 (fourth photo); © Michael Boys/Corbis, 3.6b; © Peter Brookes/*The Times*, 11/04/00, Centre for the Study of Cartoons and Caricature, University of Kent, 9.8; © Dave Brown/The *Independent*, 6/11/98, University of Kent Centre for the Study of Cartoons and Caricature, 9.17; © Romano Cagnoni/Still Pictures, 7.33 (third photo); © CIMSS/Corbis Sygma, 6.5; CNES, 1992 Distribution Spot Image/Science Photo Library, 5.2; © Corbis/Sygma, p180 (Figure 2); © CWDE, 7.28; © John Dakers/Life File, 7.30 (top left); The Derby Telegraph, page 16; © Nigel Dickinson/Still Pictures, 10.5; The *Ecologist*, 11.1; © Ecoscene/Stephen Coyne, 7.26; © Ecoscene/Michael Cuthbert, 7.33 (second photo); © Ric Ergenbright/Corbis, page 11 (bottom); © Mike Evans/Life File, 3.6a; © Tony Garrett/*The Times* 10/07/01, 11.17; © Robert Garvey/Corbis, 10.7; © Global Water, p82 (top right); © Jeff Greenberg/Life File, p180 (Figure 1); © Paul Harrison/Still Pictures, 7.25, 7.27; © Adam Hart-Davis/ Science Photo Library, 8.3 (left); © Bob Hordern, 6.23, 8.2, 8.8 (bottom left), 8.8 (top right), 8.19 (top), 8.19 (bottom), p103 (Figure 1); © Emma Lee/Life File, 4.9, 4.12, 4.14, 11.7 (top); © Richard List/Corbis, 7.30 (top right); © London Aerial Photo Library/Trackair aerial surveys, 8.8 (bottom right); © Jenny Matthews/Network/Action Aid, 7.31; © Angela Maynard/Life File, 7.30 (bottom right); Manchester Evening News, pages 19 (right), 25 (top); © Steve Milner, 12.6; © Oxfam, 5.25 (left); PA Photo, 7.33; PA Photo/John Giles, 6.19; © Panos Pictures, page 10 (top); © Trevor Perry/Photofusion, 3.13; © Fiona Potter/Life File, 2.8; © Jose Fuste Raga/Corbis, 3.3; Science Photo Library, 11.11; © Tearfund, 5.25 (right); © Flora Torrance/Life File, 2.10, 7.30 (bottom left); © University of Dundee/Science Photo Library, 6.11; © Vincent Dedet/Still Pictures, 7.17; © Andrew Ward/Life File, 3.11; © Diana Whitfield, 3.7, 4.6, 4.10.

The publishers would also like to thank the following for permission to reproduce material in this book: © Manchester Evening News 1996 for extracts from 'Traders counter shopping city threat' (4.5); permission to redraw 7.14, granted by Nelson Thornes (original diagram from *Key Geography for GCSE 2 Book 2 Teacher's Resource Guide* published by Nelson Thornes); maps reproduced from Ordnance Survey mapping with the permission of the Controller of Her Majesty's Sationery Office, © Crown copyright, Licence No. 100019872; © Copyright Oxford University Press 1985, reprinted by permission of Oxford University Press: diagram from *Investigating Physical Geography* by Neville Grenyer (OUP, 1985).

AQA examination material is reproduced by permission of the Assessment and Qualifications Alliance.

Every effort has been made to trace and acknowledge ownership of copyright. The publishers will be glad to make suitable arrangements with any copyright holders whom it has not been possible to contact.

SECTION 1   MANAGING CHANGE IN THE HUMAN ENVIRONMENT

Chapter 1   Population change

Population explosion

The population of the world passed the six billion mark in February 1999. In 2000, the annual world increase in the number of people in the world was 80.85 million. This is an extra 173 people per minute! The rapid growth in world population has been called a **population explosion**. Such a rapidly growing population places great pressure on all the world's resources.

The population of the world has not always been growing so quickly.

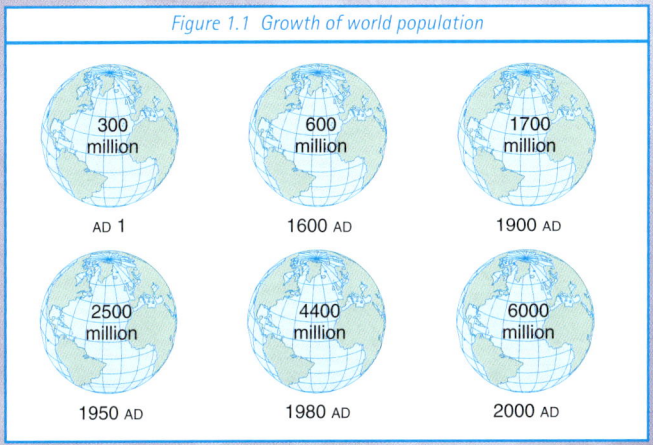

Figure 1.1 Growth of world population

- 300 million — AD 1
- 600 million — 1600 AD
- 1700 million — 1900 AD
- 2500 million — 1950 AD
- 4400 million — 1980 AD
- 6000 million — 2000 AD

As Figure 1.1 shows, it is thought that the population of the world 2000 years ago was about 300 million. For a long time there was very little growth. It took until AD 1600 for the population to double. The growth in population up to AD 1900 took place mainly in Europe and North and South America. The growth of world population speeded up after 1900, increasing by 53% in 50 years. The most rapid growth, or explosion, took off in 1950, especially in the LEDCs (Less Economically Developed Countries.) Here improvements in medicine helped more babies survive and enabled people to live longer. World population in the year 2000 was two and a half times that of 1950.

What could happen in the future?

The **United Nations (UN)** make long-range projections to try to predict what may happen to the world's population over the next 150 years.

These show a number of possible changes. The **constant** rate assumes that the world's population will continue to grow at its present level. The **medium** rate assumes that growth will level out. The **high** rate assumes an increase; the **low** rate assumes a decrease.

Figure 1.2 shows that unless recent population growth rates change, the world's population could reach 256 million (25.6 billion) by 2150. This is nearly four and a half times the population in 2000. The world would not be able to support this number of people. By the late 20th century it was clear that something had to be done to slow down population growth and it has now started to fall. If it continues to fall to the medium growth rate, the world's population will be 10 billion in 2150.

ACTIVITIES

1. Why is the term 'population explosion' used to describe changes in the world's population? (C1.2)
2. a How many years did it take the world's population to double from 300 to 600 million people? (N1.2)
   b After which date did the world's population increase most rapidly? (N1.2, N1.3) Suggest some reasons for this.
3. Why is rapid population growth an issue for concern? (C1.2)
4. Using Figure 1.2, describe the possible changes to world population predicted by the United Nations. (N1.1)
5. Look at the population figures below:

Region	% of total 1995	% of total 2150
Africa	12	24
China	22	14
Europe	13	5

Compare the changes to the percentage of the total world population in the regions shown. (C2.2, N1.3)

Research activity

6. Suggest reasons for the changes in two of these regions. (C1.2, IT1.1)

Section 1: Managing Change in the Human Environment

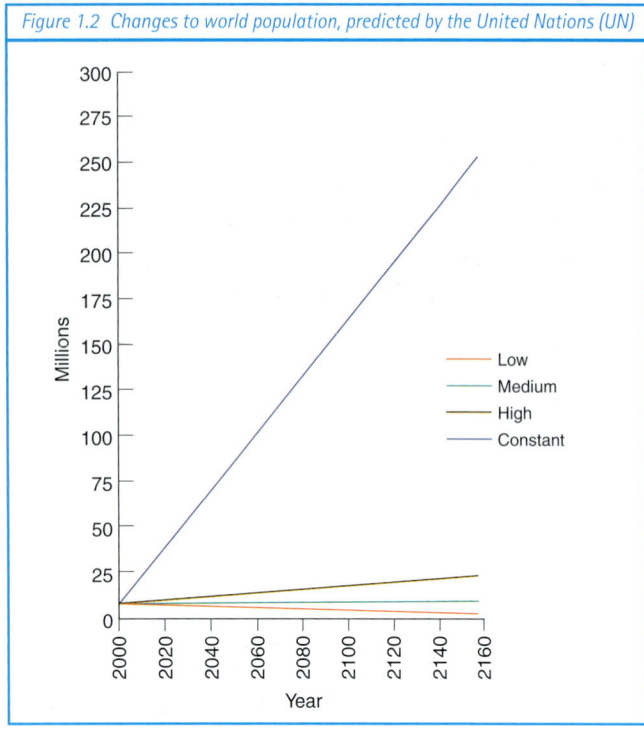

Figure 1.2 Changes to world population, predicted by the United Nations (UN)

The population cycle (demographic transition)

This is a model of population change. It shows that as a country develops economically, its population characteristics (birth rate and death rate) pass through a number of stages.

Any country or region can be placed at some stage of the population cycle. A few isolated groups of hunter-gatherers, e.g. in tropical rainforests, would be at Stage 1. The world's poorer countries such as Sudan, Somalia or Bangladesh would be at Stage 2. Many of the Newly Industrialising Countries (NICs), e.g. Brazil, Mexico, India, would be at Stage 3. Countries such as Canada, USA and Japan would be at Stage 4.

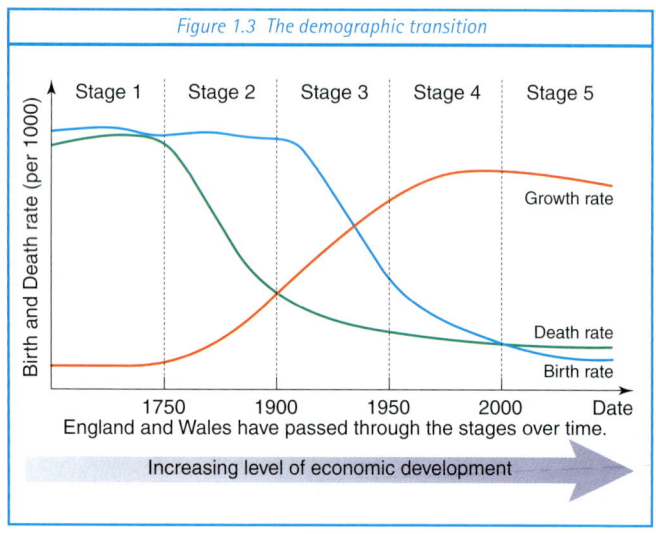

Figure 1.3 The demographic transition

England and Wales have passed through the stages over time.

Increasing level of economic development

ACTIVITIES

1. Make a copy of Figure 1.4 below.
2. Mark and label on to the graph the five stages. Stage 1 has been done for you. (N1.1)

   Stage 1: Birth rate and death rate high.
   Stage 2: Birth rate high, and death rate falling.
   Stage 3: Birth rate falling, death rate falling.
   Stage 4: Birth rate and death rate low.
   Stage 5: Birth rate declining, death rate low.

3. Describe the population growth rate at each stage. Choose your answers from the following list: rapid growth; slow growth; declining population; continued growth; steady population. (N1.1)

4. Shade in blue the area of the graph that shows a natural increase in population. Shade in green the area of the graph that shows a natural decrease in population. (N1.1)

Extension activity

5. Use an atlas to look up the birth rate and death rate of ten countries. Decide which stage (1 to 5) each country is at and add it to Figure 1.4. (N1.1)

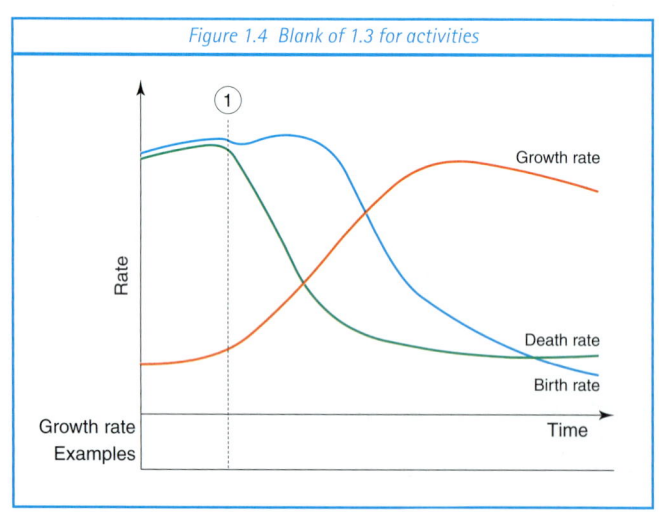

Figure 1.4 Blank of 1.3 for activities

Chapter 1: Population change

During Stage 1, population growth is slow, as both birth rate and death rate are high. Growth starts to take off in Stage 2, as birth rate remains high, but death rate starts to fall. Population growth starts to slow down in Stage 3 as the birth rate also starts to fall. By Stage 4, population growth is once again slow as both birth rate and death rate are low. We may be entering a fifth stage where population growth is negative, as birth rate is lower than death rate.

As countries develop economically, death rates will fall as medical facilities improve and there is an increased access to clean water. People will have a higher **life expectancy** and there will be a lower **infant mortality rate**. When there is less of a need to have large families, then the birth rate starts to fall.

ACTIVITIES

1. How does the information in Figure 1.5 help to explain changes in population growth rates. Set your answer out under two headings:
   - Effect on birth rate.
   - Effect on death rate. (N1.1)

Figure 1.5 Reasons for population change

Section 1: Managing Change in the Human Environment

Population structure

The way in which the population of a country or region is made up is called its population structure. This is best demonstrated using a **population pyramid**. The male population is shown on the left of the pyramid and the female population on the right. The age of each group is then shown in five-year intervals, e.g. 0 to 4 years or 75 to 79 years.

Population pyramids can be used to compare levels of economic development in different countries or regions. The shape of the pyramid will differ for a country at an early stage of economic development from one at a later stage.

The shape of the pyramid will change as the country passes through the population cycle, and **birth rates** and **death rates** start to change. Figure 1.6 shows that an LEDC (Less Economically Developed Country) has a wide base, showing large numbers in the 0 to 14 age group. The pyramid then quickly narrows off towards the top, as numbers in the older age bands fall off quickly. There are very few in the uppermost age bands. An MEDC (More Economically Developed Country) has a much narrower base, as birth rate is lower and the pyramid does not narrow as quickly, as the death rate is lower.

The dependent population in a country are those under working age and those above working age (see Figure 1.6). These two groups are dependent on the economically active population, i.e. those of working age.

ACTIVITIES

1. Look at Figure 1.6.
   a. What percentage of the population of the country are males aged 0–4 years? (N1.1)
   b. What percentage of the population of the country are aged between 30–34 years? (N1.1)
2. Look at Figure 1.7 below.
   a. Which pyramid, A or B, shows an MEDC? Give reasons for your answer. (N1.3)
   b. What is meant by:
      i. dependent population
      ii. economically active population? (C1.2)
   c. Which of the areas, X, Y or Z, represents the economically active population? (N1.3)

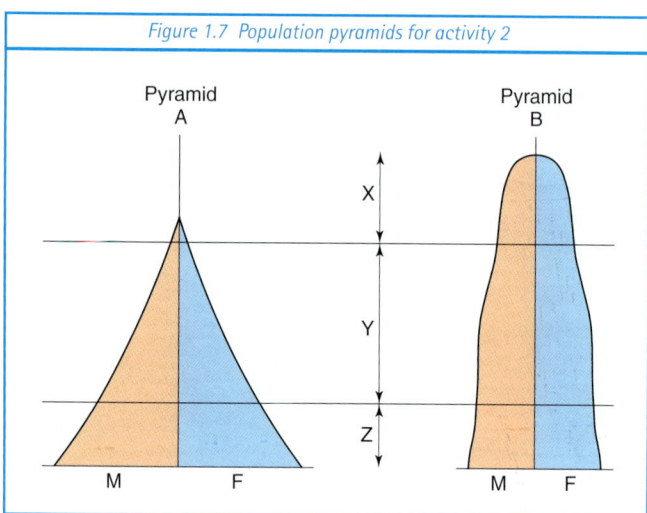

Figure 1.7 Population pyramids for activity 2

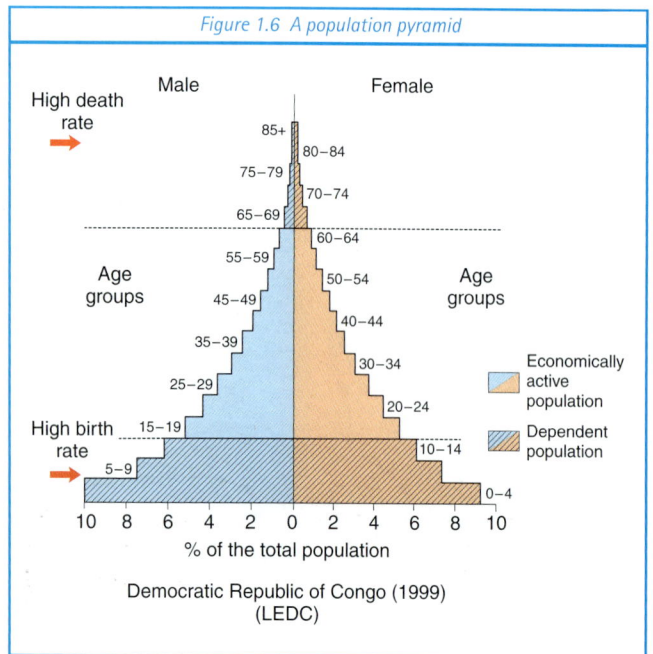

Figure 1.6 A population pyramid

Population structure and levels of dependency can be shown by comparing population pyramids for the UK (an MEDC) and Bangladesh (an LEDC) Figures 1.8 and 1.9 show the population structure of the countries in 2000 and how these are predicted to change up to 2050. This information can be used by governments to plan for the provision of services in the future.

The population pyramid for the UK in 2000 (Figure 1.8), shows that the widest part of the pyramid is the 30–59 age group. Above 60 years, the numbers in each age group gradually start to fall away. However, there are more females aged 60–85+. The numbers in the 0–30 age group are also lower.

By 2025 in the UK, the 'bulge' has moved up to the 55–64 age group. There are also larger numbers of people in the 65–85+ age group, especially women. The numbers in the younger age groups go down slightly.

Chapter 1: Population change

By 2050, the largest single age group of people in the UK is women aged over 85 years old. Approximately 16% of the population will be over 65 years old. The numbers in the younger age groups continue to fall and the base of the pyramid gets narrower. This is known as an **ageing population**.

The population pyramid for Bangladesh in 2000 (Figure 1.9), shows a **youthful population**. The widest part of the pyramid is at the 10–19 age group. Above this age, the numbers fall off quickly, showing a high death rate. The percentage in the 0–9 age group is lower than the 10–19 age group, showing a slight fall in the birth rate.

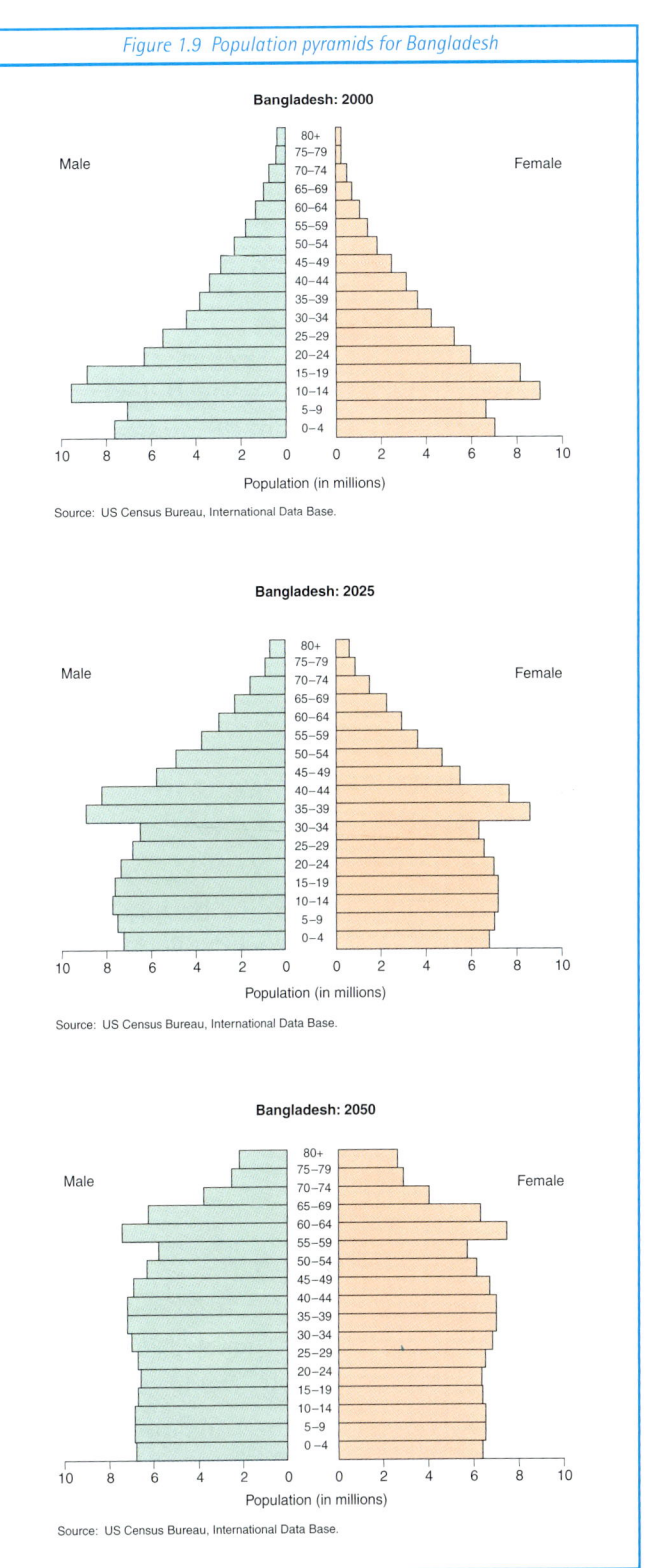

Figure 1.8 Population pyramids for the United Kingdom

Figure 1.9 Population pyramids for Bangladesh

Section 1: Managing Change in the Human Environment

By 2025, the 'bulge' has moved up to the 35–44 age group and the number of people aged over 65 has begun to increase. By 2050, the shape of the pyramid has become more like the UK in 2025. The pyramid has become narrower at its base and wider at the top as the number of people aged over 70 increases.

Managing problems caused by population change

The projected changes to population structure raise some serious planning issues for governments.

1. Problems caused by an ageing population

- Pressure on the National Health Service (hospitals, doctors' practices, care homes and social services, such as meals on wheels).
- People whose friends and partners have died often feel lonely and isolated. This is especially true of women.
- Many younger family members may have to act as carers for elderly relatives.
- There is a smaller percentage of the population that is economically active. These have to support the rest of the population and in some countries this has caused taxes to rise.
- Prejudice (ageism) against the 'burden' of elderly people may set in.
- Will governments in the future be able to support state pension for such large numbers of people?
- In some areas, entertainment and recreational provision may be geared more towards the elderly than the young.
- More houses will be needed in the future (see chapter 4.)

Managing an ageing population

- France and Russia both suffer from declining populations. People are encouraged to have larger families through state benefits.
- Governments encourage firms to employ and retain older workers.
- Governments restrict opportunities for early retirement.
- Immigration laws may be relaxed for key workers, e.g. doctors, nurses.
- Many people now take out private pension schemes to add to their income after retirement.
- Increased housing demands are met by brownfield development (see chapter 4.)

2. Problems caused by a youthful population

- Pressure for more schools to be built.
- Increased pressure on existing health services for infants and mothers.
- Unemployment when available jobs are taken up.
- The migration of younger people away from the countryside to the cities (see chapter 2) leaving an unbalanced population.

Reducing birth rates

As Figure 1.9 showed, birth rates in many LEDCs are expected to drop, as the need for large families is reduced. (Some of the reasons are shown in Figure 1.5, page 3.) The governments of many LEDCs, with the help of aid programmes from the richer countries of the north, or non-government organisations, have been able to set up simple, inexpensive educational campaigns and programmes, often aimed at women. Local schemes are set up to help people understand that family planning can lead to economic and health benefits. Programmes of contraception are made available. Some countries have taken much stronger measures, such as involuntary contraceptive injections or sterilisation, but these are unusual. Governments have also used a system of incentives and/or punishments to encourage people to have smaller families. China had a one-child policy. Free education and jobs were guaranteed by the state, but withdrawn if the family had more than one child. China's **population growth rate** fell dramatically and the policy has been relaxed.

ACTIVITIES

1. What is meant by an ageing population? (C1.3)
2. What is meant by a youthful population? (C1.3)
3. At what date will you reach the age of 40? Describe the possible problems caused by the structure of the population at this date. (N1.2, N1.3)

Rural–urban migration in LEDCs Chapter 2

Urbanisation

The proportion of the world's population living in cities is increasing fast. This process is called **urbanisation**.

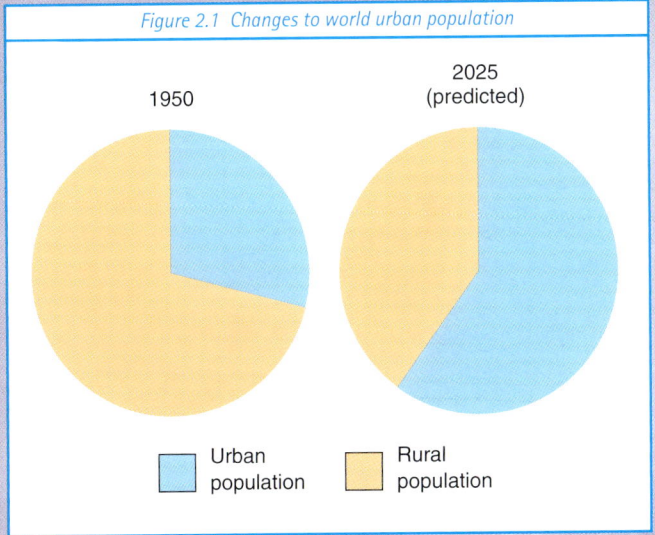

Figure 2.1 Changes to world urban population

Figure 2.1 shows that by 2025, it is expected that 60% of all the world's people will live in **urban** areas. This figure was only 29% in 1950. Figure 2.2 shows that urbanisation in MEDCs, such as the UK, mainly took place after 1750 and the industrial revolution. Urban growth in developed cities slowed down greatly from the mid-20th century. Today, most of the growth in urban population is taking place in cities in LEDCs. Cities in these countries started to grow rapidly in the second half of the 20th century and continue to do so. In 1950, 17% of people in LEDCs lived in urban areas; this is soon expected to reach 50%.

ACTIVITIES

1. What is meant by the term urbanisation? (C1.2)
2. Look at Figure 2.1.
   a. What percentage of world population lived in rural areas in: i 1950; ii 2025 (predicted)? (N1.2)
   b. Describe the changes to the world's urban population. (C1.2, N1.3)
3. Look at Figure 2.2.
   a. What was the percentage of urban population in MEDCs in: i 1970; ii 2000? (N1.3)
   b. What was the percentage of urban population in LEDCs in: i 1970; ii 2000? (N1.2)
   c. Compare the growth of urban population in LEDCs with that of MEDCs. (N1.3)
4. Look at Figure 2.3.
   a. In which continents is the most rapid growth expected? (C1.2)
   b. In which continents is slow growth expected? (C1.2)

Research activity

5. Find out why the urban population in MEDCs started to increase after 1750. (C1.2, IT2.1)

The world's largest cities

A millionaire city is a city of over one million people. In 2000, there were approximately 300 millionaire cities in the world. The continued growth of some urban areas has led to the development of megacities. A megacity is regarded by some geographers as being a city of over 5 million people, though more commonly it is said to be one of over 10 million people.

In 1940, only two cities – New York and London – had populations of over 5 million people. This had risen to 30 by 1990, ten of which were in Asia. The number in Asia is expected to continue rising to 19 by 2025.

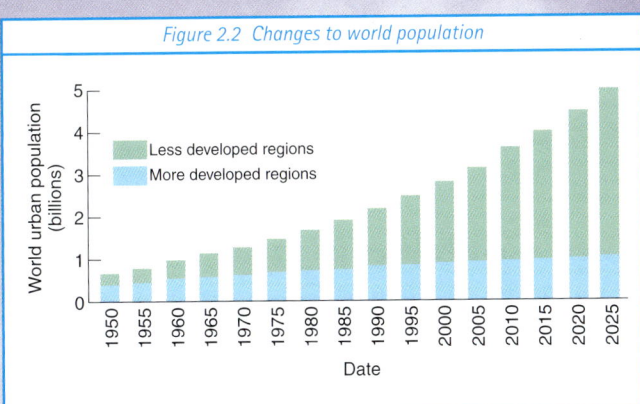

Figure 2.2 Changes to world population

7

Section 1: Managing Change in the Human Environment

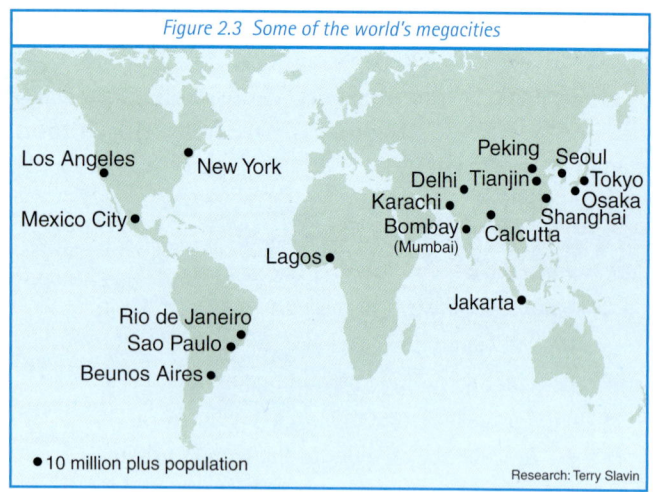

Figure 2.3 Some of the world's megacities

During the 1970s, there was great concern that the growth rates of some of the world's largest cities were getting out of control. However, the growth rates have slowed down and UN estimates were revised downwards by 1994. (Figure 2.4)

Figure 2.4 UN estimates for 2000 (millions)			
City	Estimate made in 1970	Estimate made in 1994	Actual population in 2000
Bombay	7.0	18.0	18.1
Cairo	16.4	10.7	10.6
Calcutta	19.7	12.7	12.9
Mexico City	31.0	16.4	18.1
Rio de Janeiro	19.4	10.6	10.6
Seoul	18.7	12.3	10.3

Factfile — Cities

- Only 15% of urban dwellers live in cities of over 5 million people. The UN predicts that this will only rise to 18% by 2025.
- 70% of urban dwellers live in cities of 2 million people or less.
- 60% of urban dwellers live in cities of 1 million people or less.
- Many of the smaller cities still suffer from the same problems as those found in the megacities.
- The number of countries which are rapidly urbanising is increasing.
- The number of cities that are growing rapidly is increasing.

Estimates cannot always be accurate, as it not always clear whether the population total is for the city itself, i.e. within the city boundary, or for the agglomeration. This is when the city sprawls outwards beyond its boundary, joining up with other, smaller urban areas. Figure 2.5 shows the populations of the world's top ten cities (the population within the city limits). How does it differ from Figure 2.4? Does it give any clues as to which cities are most crowded?

Figure 2.5 Largest cities of the world		
Rank	City	Population
1	Seoul, South Korea	10,229,262c
2	Mumbai (Bombay), India	9,925,891c
3	Mexico City, Mexico	9,815,795c
4	São Paulo, Brazil	9,811,776c
5	Jakarta, Indonesia	9,160,500c
6	Moscow, Russia	8,436,447e
7	Shanghai, China	8,205,598c
8	Tokyo, Japan	8,021,943e
9	Istanbul, Turkey	7,774,169e
10	New York City, USA	7,420,166e
10	Beijing, China	7,362,426c

'e' means estimated and 'c' means census figures.
© Global Statistics, 2000.

ACTIVITIES

1. What is a megacity? (C1.2)
2. Look at Figure 2.3. Name a megacity in each of these continents: Europe, North America, South America, Africa, Asia.
3. In which continent are most megacities expected to be by 2025?
4. Look at Figure 2.5. Which city has grown larger than its 1970 estimated population? (N1.1)
5. Describe one problem found in estimating the population of an urban area. (C1.2)

Three main causes of urban growth

1. Population growth within the city, caused by increasing birth rate and falling death rate.
2. The inclusion of 'new' urban areas, as changes take place as to what we classify as 'urban'.
3. **Migration**, i.e. the movement of population.

Chapter 2: Rural–urban migration in LEDCs

Migration

The size of the population of cities can change due to **immigration**, where people arrive from other countries. However, in cities in LEDCs, it is estimated that between 40 to 60% of urban population growth is due to **rural–urban migration**. This means that people are moving away from the countryside to go and live in the city. This process occurs due to negative factors, where people leave the countryside to get away from the difficult living conditions. It also occurs because of positive factors, where people are attracted to the urban area by the possible improvements to living standards offered by life in the city.

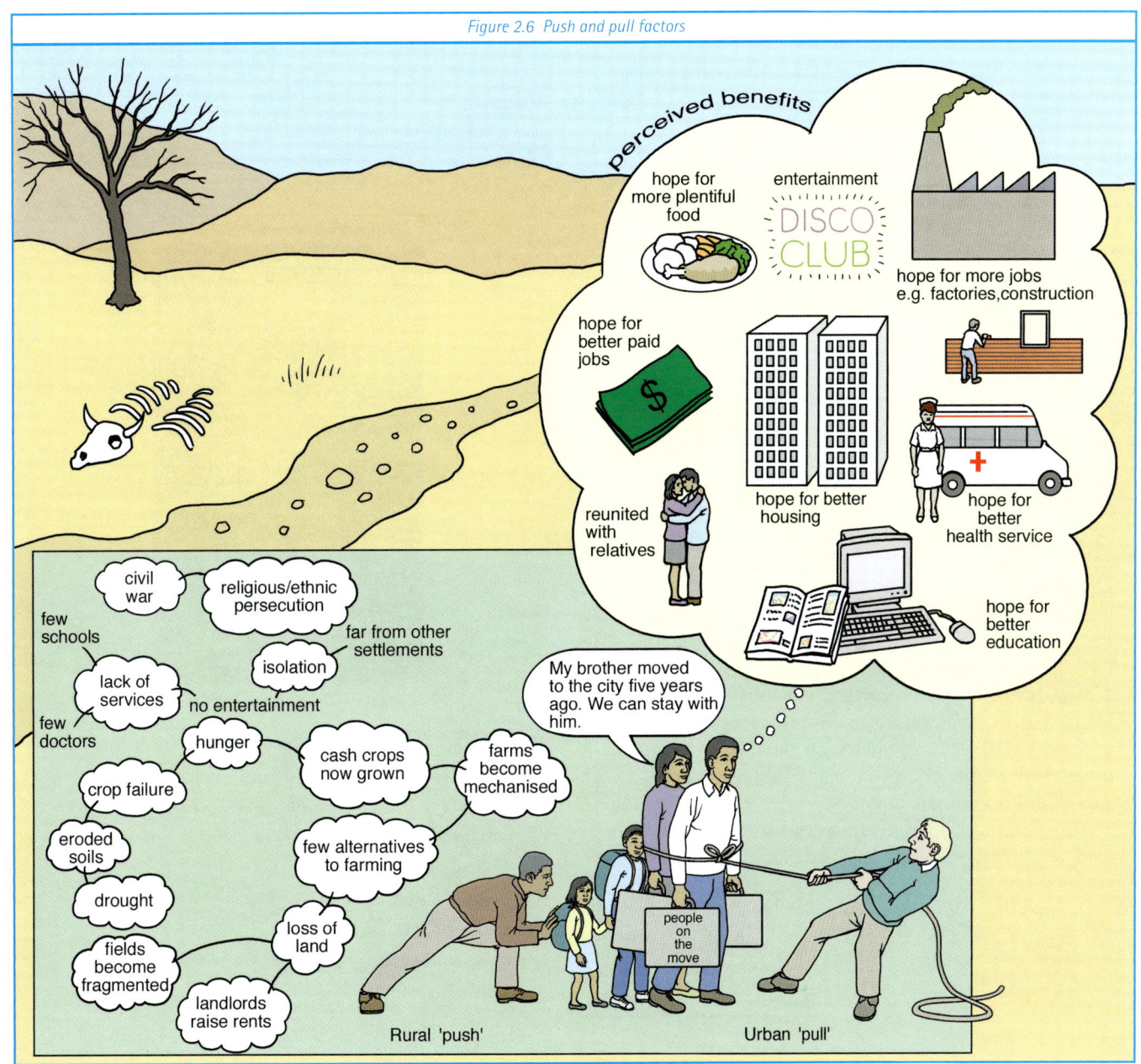

Figure 2.6 Push and pull factors

Section 1: Managing Change in the Human Environment

Shanty towns

As more and more migrants from rural areas arrive in cities in LEDCs, this creates many problems for the city authorities. The perceived benefits of life in the cities are not always to be found, especially when people first arrive. Jobs are hard to find, or are very poorly paid. People cannot afford to buy or rent houses, or they are just not available. Many migrants are forced to build their own temporary shacks out of any available materials. They become squatters as they are forced to occupy land which does not belong to them. As the number of people arriving each day is so great, large squatter settlements or **shanty towns** develop. These are often found on land that will not be built on by others, as it is polluted, on steep slopes or marshland, or next to rubbish tips. Increasingly they are built on unoccupied land at the rural–urban fringe. This causes the rapid outward growth of the city, i.e. **urban sprawl**.

Figure 2.7 Food stall in a shanty town

Figure 2.8 Shanty dwellings

shanty towns — Case Study

A description of a shanty town on the edge of Nairobi

As the number of water borne toilets increased, through the building of toilet blocks, the actual number of private toilets declined. More families have to share, increasing health hazards. Open sewers which run down the centre of many of the roads, overflow in the rain, infecting the water supply.

Many of the houses in the settlement are built from mud, boxes or plastic sheets, with corrugated iron roofs, and usually consist of just one room. Built close to each other, there is little privacy or chance to prevent the spread of disease.

Most forms of employment are temporary and low paid. Small stalls selling food, or scrap and other waste products salvaged from rubbish tips are common. City authorities are only able to dispose of 20% of the total refuse generated.

Global Eye, Autumn 2000

ACTIVITIES

1 Describe the features of a shanty town that you have studied. Set your answers out under these headings:
   - Housing
   - Services
   - Jobs
   - Other features

Managing the shanty town issue

Moving from conflict to co-operation

Demolition and eviction

This does take place in some cities. It usually involves armed police allowing workers to move in to bulldoze the shacks when the land is needed for other uses. Conflicts arise as shanty dwellers have no alternative housing and try to resist being moved on. This is a short term solution and does not solve the problem. It merely moves it elsewhere.

Figure 2.10 Government housing in a shanty town

Figure 2.9 Demolishing the shacks

Figure 2.11 Self-help groups

Authority provision

This involves local authorities building, often basic, low cost housing or improving existing housing. The houses improve people's living standards as they have facilities such as running water, a flushing toilet, connection to the sewerage system and an electricity supply. The houses are much safer than the shacks. There is less risk of fire and far greater protection from the weather. However, as many local authorities in LEDCs have limited funds, suffer from corruption or have other pressing problems, the number of basic houses that can be built is relatively small. It does not begin to solve the problem as demand for houses will be great. Also, even though the houses are low cost, many shanty dwellers will not have enough money for a deposit or have a regular income to keep up with the rent. Those with the higher incomes tend to get the homes.

Service programmes and self help schemes

Service programmes are when authorities plan out plots of land and install basic services, such as water and sanitation. Local people can then build their own houses (self build) on these plots of land. Self help schemes take this a step further. They are

ACTIVITIES

1 Look at Figure 2.12 (on page 12). Which stage of the model (1 to 5) describes each of these methods of managing problems in shanty towns?
   - Service programmes
   - Demolition
   - Authority provision
   - Self help programmes.

   Give reasons for your answers.

Section 1: Managing Change in the Human Environment

often run by community groups, who organise local people to help themselves by building not just houses, but other community facilities such as communal kitchens and medical centres. The authorities may provide loans of money, building materials and expert advice.

These different methods of managing the squatter problem can be shown as a model, passing through a number of stages, ranging from conflict to co-operation. A brief history of Villa El Salvador is shown in Figure 2.12. It is a *pueblo joven* (young town) set up in the desert on the south eastern edge of Lima in Peru. Figure 2.13 shows the location of Villa El Salvador.

Urban farming

City dwellers are encouraged to grow food and rear animals on any available plots of land. Produce can be sold at markets to provide an income or can be eaten. Often people are using traditional farming skills brought from the rural areas. Urban farming can help people break out of poverty, improve diet and raise health standards. It can also use up a city's waste once it has been turned into compost.

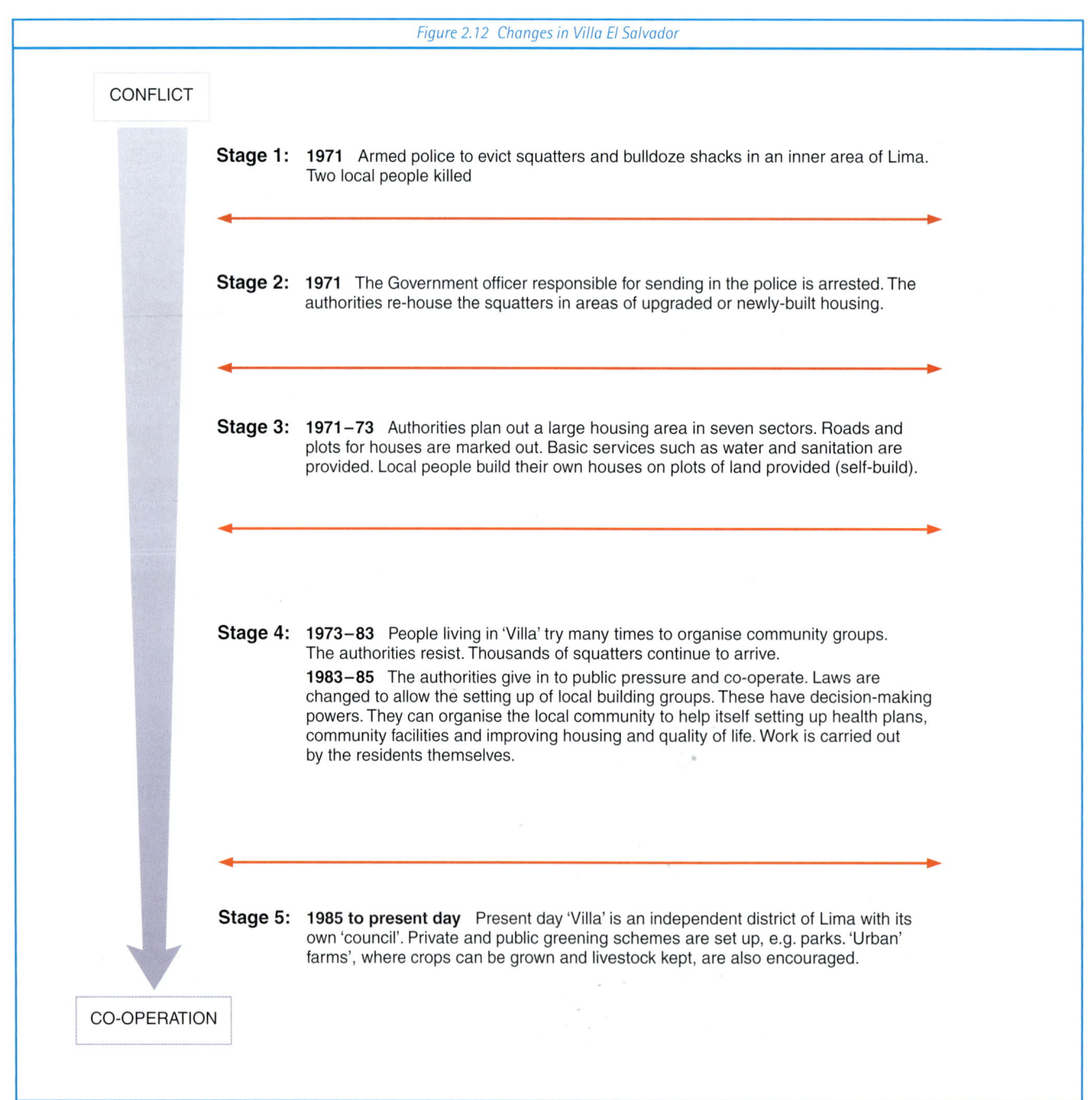

Figure 2.12 Changes in Villa El Salvador

CONFLICT

Stage 1: 1971 Armed police to evict squatters and bulldoze shacks in an inner area of Lima. Two local people killed

Stage 2: 1971 The Government officer responsible for sending in the police is arrested. The authorities re-house the squatters in areas of upgraded or newly-built housing.

Stage 3: 1971–73 Authorities plan out a large housing area in seven sectors. Roads and plots for houses are marked out. Basic services such as water and sanitation are provided. Local people build their own houses on plots of land provided (self-build).

Stage 4: 1973–83 People living in 'Villa' try many times to organise community groups. The authorities resist. Thousands of squatters continue to arrive.
1983–85 The authorities give in to public pressure and co-operate. Laws are changed to allow the setting up of local building groups. These have decision-making powers. They can organise the local community to help itself setting up health plans, community facilities and improving housing and quality of life. Work is carried out by the residents themselves.

Stage 5: 1985 to present day Present day 'Villa' is an independent district of Lima with its own 'council'. Private and public greening schemes are set up, e.g. parks. 'Urban' farms', where crops can be grown and livestock kept, are also encouraged.

CO-OPERATION

Chapter 2: Rural–urban migration in LEDCs

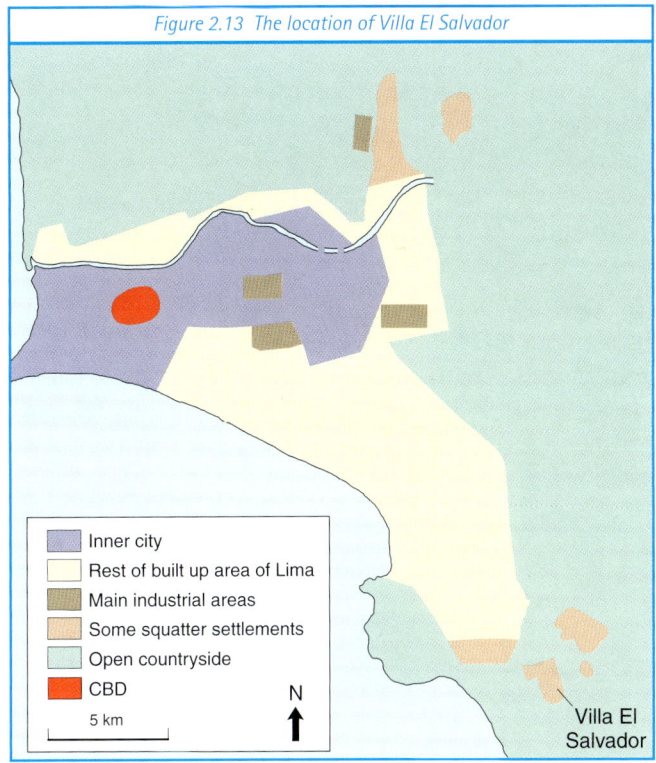

Figure 2.13 The location of Villa El Salvador

migrated to large cities such as Lima, because of the lack of facilities. It is important to tackle the issues raised by rural–urban migration in both urban *and* rural areas.

Slums of hope

Despite all of the environmental hazards, many migrants to the cities do see an improvement in their situation.

- A survey in New Dheli (India) showed that income could be twice as much as in the rural areas.
- Health of mothers and infants does improve when health and education programmes are run. Infant mortality is lower and life expectancy longer.
- In urban areas in LEDCs, family size has reduced from an average of six in 1950 to four today.

With continued management, these facts could help reduce world population growth.

In Bogota, Venezuela, you can see 'vertical agriculture'. Herbs and other plants are growing on balconies and rooftops. In Mexico City, sheep graze on the central reservations of dual carriageways.

Rural Enterprise Schemes

These schemes attempt to improve the lives of people in rural areas to stop them migrating to the cities. If people have paid jobs, they are less likely to leave in search of employment elsewhere. The traditional skills of people in rural areas can be used in craft industries such as weaving, pottery, wood, metal and leather working. Jamaica has set up small-scale sugar plants, using locally grown materials. These create 17 times more jobs than the sugar mills using modern machinery. Governments can support rural enterprise schemes through subsidies, training and advice. The Indian Goverment set up SIDO (Small Business Industrial Development Organisation) and passed laws to ensure that 600 products can be produced in rural areas only.

These schemes work alongside others set up to improve rural areas, such as land reform and improvements to health and education services. The Blue Peter appeal of 2000 raised money to help build simple field hospitals in remote parts of the Andes mountains. Here, many people have

Figure 2.14 Vertical agriculture

Chapter 3: The changing town and city centre

Land use in urban areas

In 1925, Ernest Watson Burgess, an American Sociologist, developed a model of city structure from studying Chicago and other American cities. A model is a simplified version of reality. The Burgess model helps us to understand urban land use by dividing a typical city up into a series of concentric zones, each with differing land use. There are many limitations to urban models, such as the concentric ring model. It was developed to study American cities and may not apply as much to the older European cities. It was also developed before large-scale car ownership made living at the edge of the city much easier for many people. It obviously does not take into account redevelopment in the inner city, or the movement of shops, industry and offices to areas on the edge of the city. It does, however, help us to locate the different areas of the city and to identify changes which may have taken place in urban areas. This chapter is concerned with the CBD and inner city areas only. Areas towards the edge of the city are considered in chapter 4.

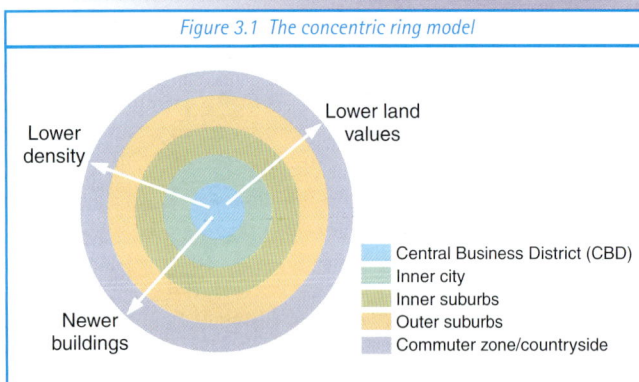

Figure 3.1 The concentric ring model

The Central Business District (**CBD**) was the old, historic core of the city. The pattern of land use developed as the city grew outwards. The age of the buildings becomes newer with distance from the CBD. Also, as you move away from the city centre, the amount of open space increases. This is because land values are highest in the CBD and lowest at the edge of the city. Land in the CBD is in greatest demand as it is the most **accessible** point from all areas of the city. Developers would be prepared to pay higher rents for a site in the central area. The density of buildings is, therefore, highest in the CBD. Developers build upwards rather than outwards and this is why there are so many tall buildings in city centres. The land use tends to be **commercial**, e.g. shops such as large department stores and office blocks, along with functional, such as local government, entertainment and transport.

The **inner city** is the area surrounding the CBD and here the **residential land use** has been high density, due to the high land values. There are closely packed rows of terraced houses, or blocks of high-rise flats.

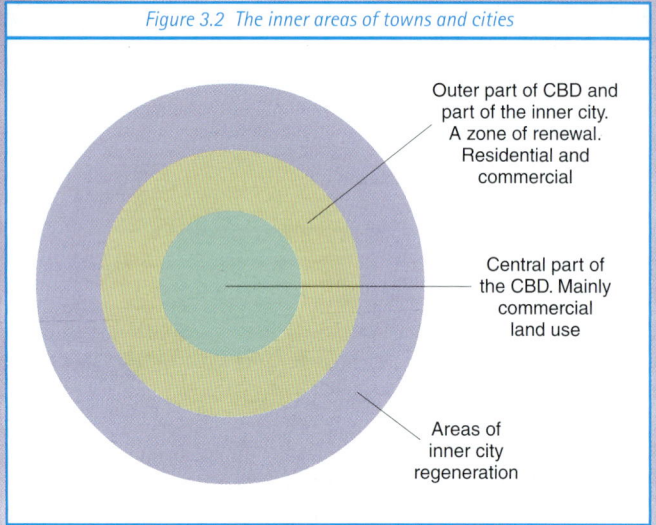

Figure 3.2 The inner areas of towns and cities

Land use changes

The Burgess model (Figure 3.1) is a simple way of understanding the pattern of land use in a city, but the function of the zones are now changing. Traffic congestion in inner areas of cities often means that the CBD is no longer the most accessible area. In some cities, land values in the CBD are starting to fall, and increase at some places on the edge of the city. This land is now under pressure from developers (see chapter 4.) Some of the old buildings in the CBD are being replaced as city centres are redeveloped as their functions change. New developments in the central areas of cities are now often at lower density, as the public demands more open space and pleasanter, 'greener' living, working and shopping environments.

Chapter 3: The changing town and city centre

Figure 3.3 Land use in the city centre

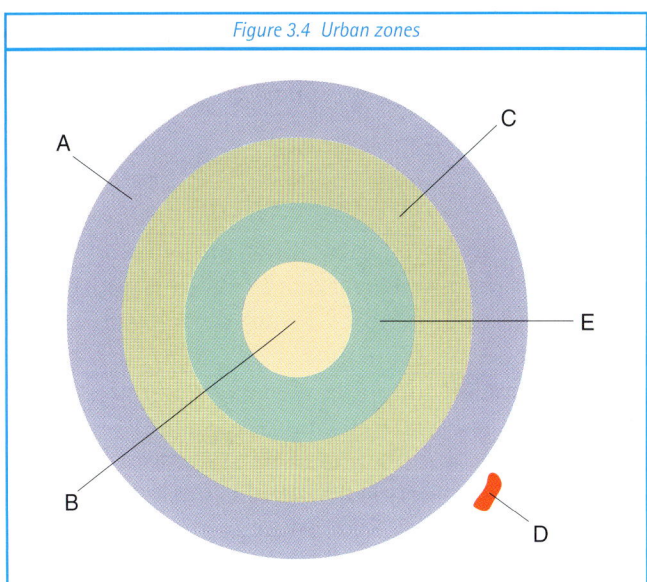

Figure 3.4 Urban zones

Changes in the city centre

The issue for many companies and **local authorities** is: will the way in which the city centre is used have to change in the 21st century?

Between 1980 and 1990, spending in city centre shops fell from 57% to 52% of the total. Many people now prefer the choice and convenience offered by out of centre, or out of town shopping centres (see chapter 4). These trends towards using the CBD less for shopping will be very difficult to stop. Some planners fear that the traditional town or city centre is in danger of becoming extinct. All councils in England and Wales now have to produce structure plans, to show how they will deal

ACTIVITIES

1 Match up the terms with their correct meanings.

Terms
- **a** Central Business District (CBD)
- **b** Urban model
- **c** Function
- **d** Inner city
- **e** Residential land use
- **f** Low density land use
- **g** Commercial land use

Meanings
- **a** The purpose for which the land is used.
- **b** A zone at the centre of the city occupied mainly by large shops, offices, entertainments and local government.
- **c** Land mainly used for housing, e.g. terraces, semi-detached, flats etc.
- **d** Buildings separated by areas of open space, e.g. large gardens, parks, undeveloped land.
- **e** Land mainly used for activities which make money. Businesses such as shops, offices etc.
- **f** A method of simplifying the structure of a city to make it easier to understand.
- **g** A zone just outside the city centre, which in the past consisted of terraces, old industry and high rise flats.

2 a Make a copy of Figure 3.4. Label the names of each urban zone (A–E).
  b Explain why zone B may no longer be the most accessible. (C1.3)

Section 1: Managing Change in the Human Environment

with planning issues. These follow guidelines laid down by government. The management of the town or city centre is an important part of the structure plan.

Changes to Derby city centre

In its structure plan, Derbyshire County Council shows that the number of empty shops is above average and that spending is slightly below average in the city centre. Now that more people own cars and often regard shopping trips as a day out for the family, they are prepared to travel as far as Manchester, Nottingham and Sheffield to shop. This is increasing competition for shops in Derby. There is now only one major foodstore in the city centre. The rest have either closed down or relocated away from the CBD. There are now seven out of centre supermarkets, built in the 1980s and 1990s. Due to these factors, the structure plan shows an attempt to **sustain** the city centre.

The plan includes:

- A five million pound refurbishment of the Eagle Centre market. Traditional 'market town' images are seen as attractive to shoppers, as are street markets. The redevelopment of a former supermarket in Exchange Street also provides extra space for new shops. These schemes hope to provide a pleasanter shopping environment.

- The Derby Promenade is an important environmental improvement and pedestrianisation scheme. As in many city centres, it aims to provide areas of public open space and traffic free areas. An image of a safe, clean environment is important in attracting shoppers or visitors.

- Town centres need to be more accessible if they are to be used by more people. In some centres, it may be appropriate to encourage car users. (In other centres it would be more appropriate to reduce the number of people using their cars, see page 20.) In Derby a 1 000-space, secure shoppers' car park has been built at Cock Pitt, on the edge of the CBD. Cycle parks are also considered. In all cities, it is important to promote the use of public transport, and in Derby permission has been given for the redevelopment of the bus station site. This will create a new bus station along with 12 000 square metres of retail and leisure facilities.

- Often in the CBD there is no room to build retail parks. If they are to be built, they should be just 'out of centre'. This would attract more people into the CBD as they would combine shopping trips with other purposes.

- In Derby, the council feel that with the development of the out of centre foodstores and retail parks, there is no demand for any more in the CBD. They will have to look to other services such as solicitors, estate agents, educational centres, doctors, dentists and leisure becoming more important. The council plans, along with the private sector, to diversify the uses of the CBD. In many town and city centres tourism is seen as a way of doing this.

- The council promotes leisure use and the 'evening economy'. Many cities are now granting permission for cafe bars, restaurants, hotels, conference centres, cinemas and bowling alleys as alternatives to shops.

Housing people in flats above shops helps with the upkeep of buildings and means that the centre is busier at night. Many councils are now seeing housing as a more important land use in the centre of towns and cities.

ACTIVITIES

1. Why are town and city centres used less by shoppers today? (C2.2)
2. Describe four ways in which town and city centres may be made more attractive to shoppers. (C2.2)
3. Suggest other activities, apart from shops, which may have to be encouraged to locate in town and city centres. (C2.2)
4. What is the 'evening economy' and why is it important? (C2.2)

Figure 3.5 Life in the town centre

Chapter 3: The changing town and city centre

Housing development in the CBD

There has always been some housing in the CBD, e.g. expensive penthouses at the top of office blocks. However, this has not been a major user of land. Now, especially in the outer areas of the CBD, housing development is becoming more common. There are many prestige developments taking place, often in converted buildings such as warehouses, which are turned into expensive flats. This process is called gentrification and is part of an attempt to attract people and money back into the central areas of cities.

The Government has said that 60% of all new houses needed up to 2020 should be built on **brownfield** sites. Many of these are to be found in the areas surrounding the commercial centres of cities. New urban village-style developments are being built on former warehouse and factory sites.

In the area just outside the outer edge of the CBD new housing developments are taking place, as these areas undergo a process of inner city **regeneration**.

ACTIVITIES

1. What is gentrification? (C2.2)
2. Suggest types of premises found in inner areas of cities, which may be converted into housing. (C2.2)
3. Describe the advantages of living in the city centre. (C2.2)

Figure 3.6a City housing development

Figure 3.6b Victorian semi

Swapping a Victorian semi for a trendy city apartment changed a family's life.

Section 1: Managing Change in the Human Environment

Hulme — Case Study

The Hulme district of Manchester lies immediately to the south of the CBD. It was one of the first areas outside of the city centre to be developed. In the 18th century, industries developed next to the nearby canal and River Irwell, and housing was built next to these industries. The building of the Manchester ship canal brought the development of more industry and housing. The houses were poor quality terraces, with a small yard and one outside toilet between two houses. By the first half of the 20th century, the condition of the houses had worsened to the point that they were regarded as the worst houses in Manchester. The houses were blackened by smoke from factories and coal fires, overcrowded and lacking in basic facilities. Many were infested with vermin.

The city council adopted a policy of **comprehensive redevelopment**. All the houses were to be knocked down and a totally new environment created. The oldest houses in the St George's area were **redeveloped** first in 1959. By the early 1970s, the whole of the housing area of Hulme had been completely rebuilt.

The terraces were replaced by high-rise developments. These were tall tower blocks and the 'crescents', deck-access blocks of flats. These flats were an improvement on the terraces in some ways. They had central heating and indoor toilets, and many offered panoramic views. However, they did create new problems. After the terraces were knocked down, communities were split up and people brought into Hulme from all parts of the city. Many, especially elderly people, felt isolated in the high-rise blocks and there was a lack of community spirit. Lift breakdowns made it difficult for parents with young children and the elderly to leave the flats. There were areas of open space, but few areas for children to play in and few community facilities or shops. By the late 1970s, many of the old industries had closed down and unemployment in the area was high. Crime levels increased as did problems such as drug abuse and truancy. The population of the area started to fall and shops and services started to close.

Throughout the 1980s the problems became worse.

Demolition of the crescents started in 1991 as a programme of **regeneration** was started. The city council entered into a partnership with housing associations and private companies in an attempt to 'breathe life back into the area'. This meant the improvement of some of the better high-rise blocks and also the building of a variety of styles of new houses that people would wish to live in and take pride in. The planning process involved Hulme residents themselves. Regeneration is not just about houses; schemes were put in place to try to attract investment from industry in order to create jobs. A new, large supermarket was built in the area. Community schemes and facilities, such as a doctor's surgery, were also included in the plan. There were also environmental improvements, such as the first new park in Manchester since Victorian times. The Hulme arch has been erected as a symbol of the rebirth of the district.

Figure 3.7 Inner city terraces

ACTIVITIES

1. Describe the early (18th and 19th century) developments in Hulme. (C2.2)
2. What is comprehensive redevelopment? (C2.2)
3. Describe the type of development undertaken in the 1960s and 1970s. (C2.2)
4. What problems were created by the 1960s/1970s developments? (C2.2)
5. What is inner city regeneration? (C2.2)
6. Describe ways in which the inner city area of Hulme is being regenerated. Set your answers out under these headings:
   - Housing developments
   - Economic developments (jobs)
   - Community developments (C2.2)

Chapter 3: The changing town and city centre

Figure 3.8 Hulme regeneration

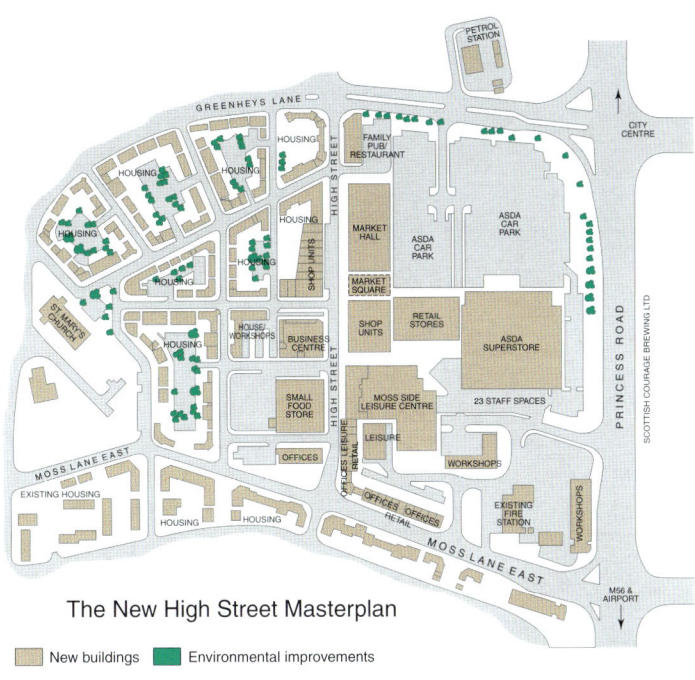

The New High Street Masterplan

New buildings ▪ Environmental improvements

Hulme in the 1970s

Hulme Arch

'Fear of crime is a major issue within Moss Side and Hulme, and its impact has a major influence on people's daily lives. It also influences potential investors such as retailers and new employers. It is important therefore that the area's determination to improve safety is supported and publicised. Local residents and agencies already meet to tackle community safety and have drawn up a strategy which aims to:
- reduce crime and the fear of crime;
- make the area safe and attractive; and
- counteract the negative images of the area.'

'As the new High Street develops it will bring many new job opportunities to the residents of Hulme and Moss Side.
  The building of ASDA has created dozens of new construction jobs and, once the store is open, ASDA will have a range of full and part-time jobs available for members of the local community. New jobs will also be created in other stores and smaller shops as well as the new offices and workshops.'

Section 1: Managing Change in the Human Environment

Changes to urban transport

As Figure 3.9 shows, car ownership in the UK has increased. Due to this and the improvement of road networks, people are now more mobile than in the past.

Figure 3.9 Car ownership in the UK

Year	Households with at least one car
1985	62%
1991	68%
1996	70%
2000	73%

The convenience and comfort of the private car has meant that fewer people use public transport, such as the railways. Many people now choose to live further away from their place of work and **commute** from the **suburbs** or from villages outside the city. The pattern of traffic flow along a main road into a city is shown in Figure 3.10.

This shows the rush hour traffic. Twice a day, when people are travelling to and from work, there are peaks of traffic at these periods. In addition to this, more freight deliveries are now made by lorries, whereas in the past more goods were delivered by train.

Problems caused by increasing traffic

In the inner areas of cities, traffic congestion has become a major issue. Traffic jams are now common in most towns and cities. Problems in inner areas are particularly bad, as it is here that roads from other parts of the city converge and traffic builds up. In many central areas, the original road network was not constructed to carry such large volumes of traffic, as roads are often narrow with many junctions. Congestion can have many damaging effects. Businesses claim that workers or deliveries being late due to traffic jams costs millions of pounds a year. 'Road rage' has become a recent problem as drivers' journey times are lengthened and they become frustrated.

Pollution of the air from traffic fumes is another problem, and is believed to be a cause of asthma. Cyclists wearing masks are now a common sight on city streets. In the past, lead in petrol was a cause of brain damage. Now we can use unleaded and lead replacement petrol (LRP) to prevent this health hazard. However, these fuels are non-renewable and the 647 million vehicles on the earth in 2000 are using up supplies at a rapid rate. Noise pollution caused by heavy traffic is also a problem for residents, as is the increased danger of accidents, damage to foundations from traffic vibration and blackening of brickwork from fumes. Often the value of houses near to a main road is reduced. Land in cities is used up as urban motorways are built or roads widened. The cost of building new roads and repairing existing ones is passed on to the taxpayer.

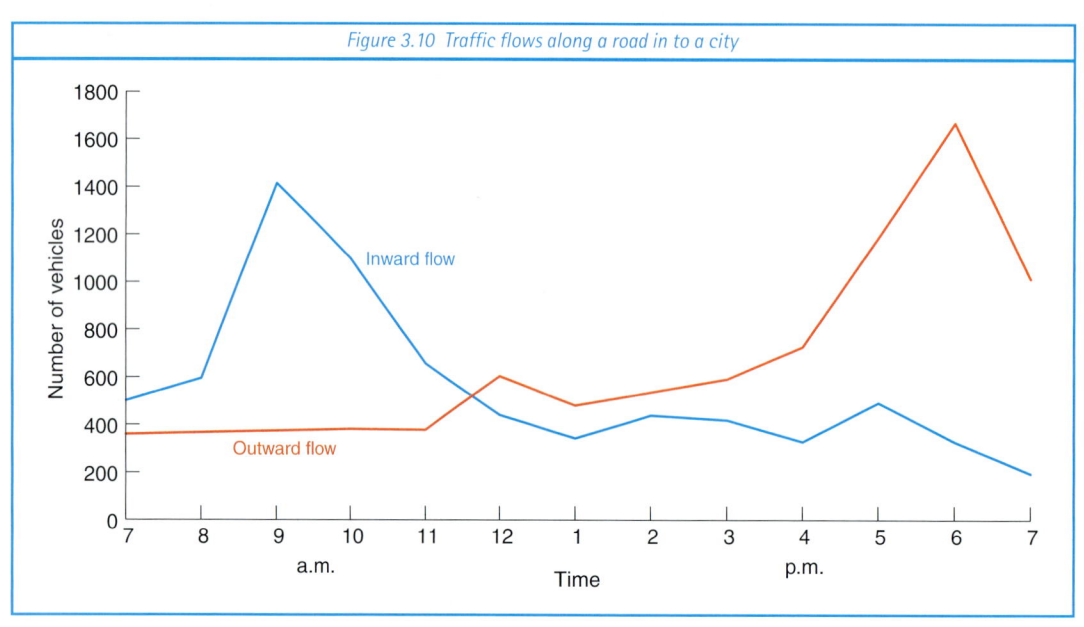

Figure 3.10 Traffic flows along a road in to a city

Chapter 3: The changing town and city centre

Solutions to urban traffic problems

Public transport is capable of carrying far more people than private cars. In an attempt to reduce the number of cars entering the central areas of cities, schemes have been put in place to encourage more people to use public transport. Cities such as Sheffield, Manchester and Newcastle now have Light Rapid Transit (LRT) systems in common with many cities in the EU (Figure 3.11) These are tramway systems which run from the edge of the city in to newly built trackways and stations in the CBD. People can travel quickly and in comfort on regular services, which go right into the heart of the city.

In some cities, bus fares are subsidised to encourage more people to use them. Most cities now have 'bus only' lanes which operate at peak times and enable buses to pass more quickly through slow moving traffic, speeding up journey times and encouraging more people to use the bus. Some cities, such as Leeds, have high occupancy lanes, which only cars with two or more passengers may use. This is an attempt to make more people 'car share'. Park and ride schemes are now a feature of many towns' and

Figure 3.11 Light rapid transit in Bonn, Germany

cities' transport system. Large, often free, car parks are provided on the edge of the city, or on the edge of the CBD. Commuters can leave their cars in the car parks and catch a free, or highly subsidised bus into the city centre.

In London, motorists have to pay to bring their cars into the city centre. Other local authorities are considering charging car users to enter the CBD. A system of tolls could be set up, with a pre-paid entry pass, where a bar code is read by a machine at the

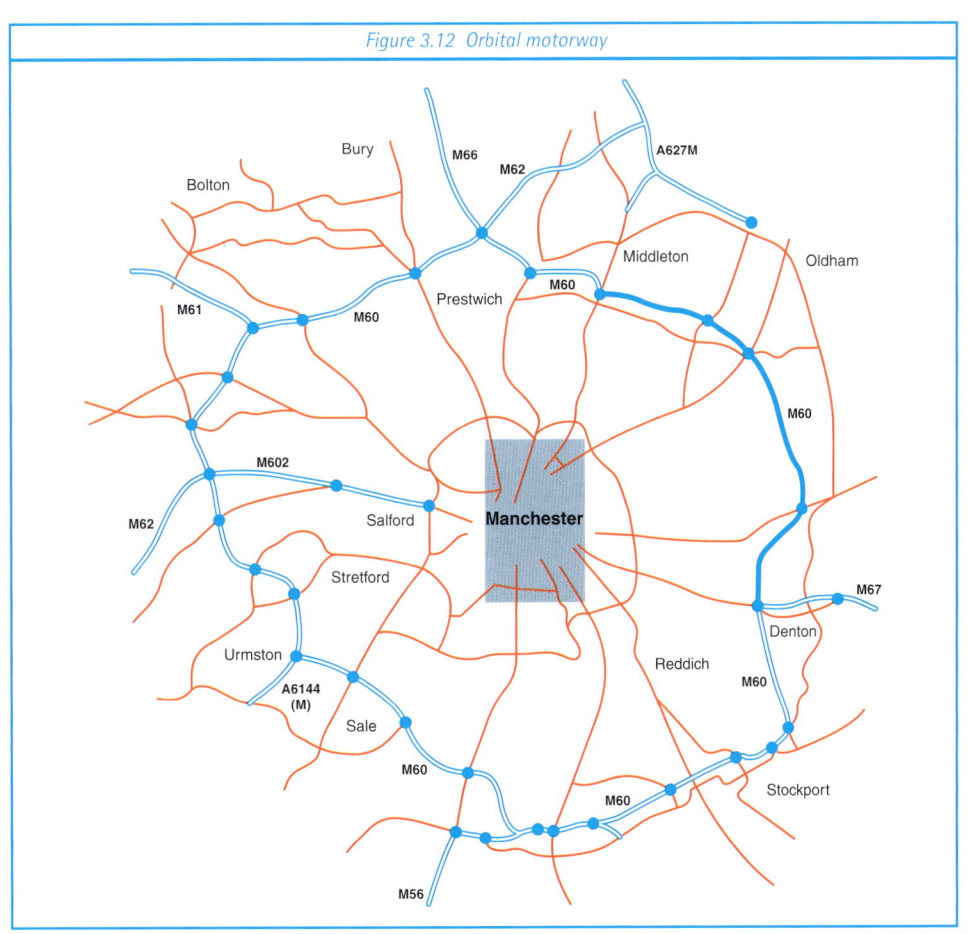

Figure 3.12 Orbital motorway

21

roadside. Other deterrents to stop people taking their cars in to city centres are schemes such as limiting parking spaces by use of double yellow lines, pedestrianising areas and using parking meters to allow for short stays only. These can be enforced by traffic wardens and wheel clamping teams.

Attempts are being made to redirect traffic around the cities, rather than through them. This can be done by building orbital motorways, such as the M25 around London or the M60 around Manchester (Figure 3.12). Outer ring roads serve the same purpose. Inner ring roads direct traffic around the edge of the CBD, as do clearly signposted 'through routes'. Contraflow or tidal flow systems are used to reduce congestion at peak times. They operate by having more lanes open to inward traffic than to outward traffic during the morning peak. This is reversed during the evening peak. The lanes open or closed to traffic are shown by lights suspended on a gantry above the road (Figure 3.13).

In some towns and cities, it is thought that up to 30% of the traffic at peak times is a result of the 'school run', where parents drop their children off at school. Many local authorities have issued leaflets to parents and pupils, encouraging them to walk or cycle to school. Many cities now have cycle lanes, marked with coloured tarmac. Some authorities are

Figure 3.13 A tidal flow system

considering an American-style, yellow school bus system, in an attempt to reduce traffic congestion.

The fuel crisis of 2000 showed our over-reliance on private vehicles and petrol. Developments in alternative, cleaner fuels such as electricity and gas could reduce air pollution in cities and some councils are now ensuring that any new council vehicles use these fuels. The government has reduced road tax for cars with smaller engine sizes as an incentive for people to use less polluting cars.

ACTIVITIES

1 Look at Figure 3.14 opposite.
   a What percentage of UK transport is by road? (N2.1)
   b Which three EU countries use road transport the most? (N2.1)
   c Which three EU countries use road transport the least? Use the information in Figure 3.14 to give reasons for your answer. (N2.1)
2 Look at Figure 3.10 (page 20).
   a Describe the pattern of traffic flow in the city throughout the day. (N2.1)
   b Give reasons for the pattern that you have described. (N2.1)
3 List five problems caused by traffic. (C1.2)
4 Describe three schemes to solve traffic problems. Choose one scheme which encourages people to use public transport; one scheme which deters people from taking their cars into the city and one which reduces pollution (air or noise). (C2.2)

Summary activity

5 Look at Figure 3.15 opposite. Which of the numbers (1–9) shows each of these developments in the inner area of a city:
- new urban village on former warehouse site
- inner ring road
- new urban park
- new tramway system
- new supermarket in inner city
- new small scale manufacturing units
- new estate of semi-detached houses
- improved terraced houses
- pedestrianised shopping area

6 a Choose any three of the new developments above. Discuss the likely impacts of the developments. List reasons for and against building each. (C1.1, C1.3, IT1.2)
   b For one of the developments, write a letter to the developers, either in support of, or against, the development going ahead. (C1.1, C1.3, IT1.2)

Chapter 3: The changing town and city centre

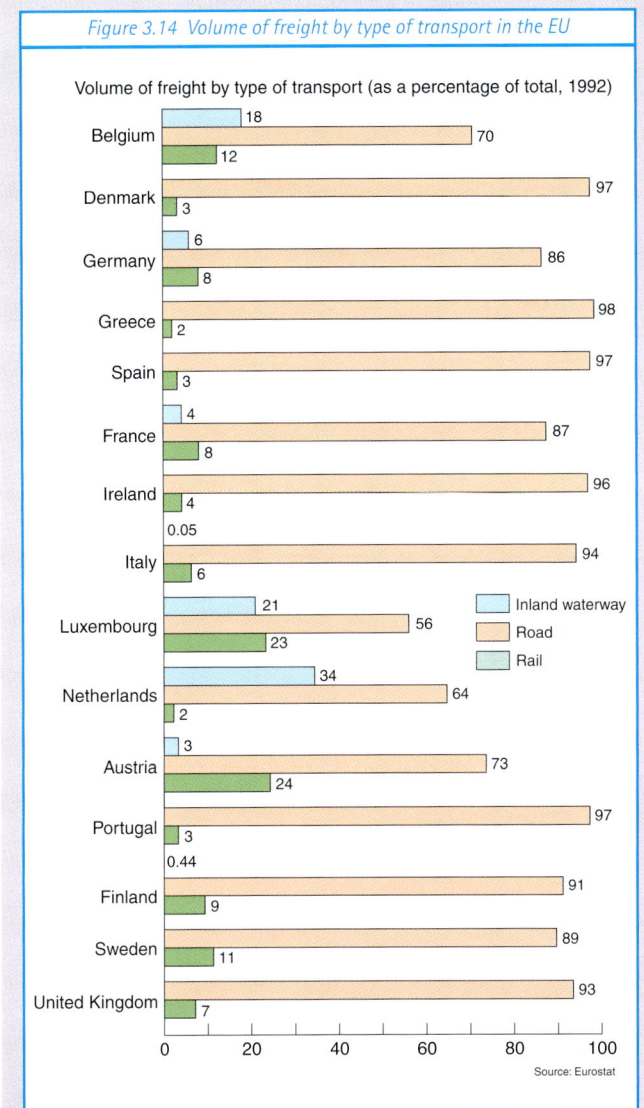

Figure 3.14 Volume of freight by type of transport in the EU

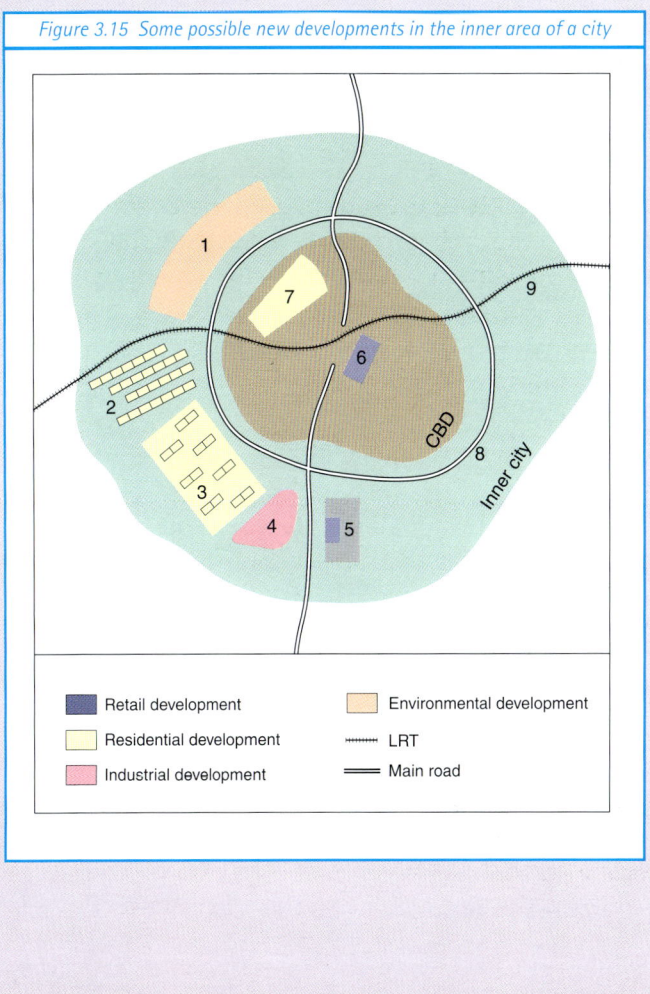

Figure 3.15 Some possible new developments in the inner area of a city

23

Chapter 4: Pressure at the rural–urban fringe

The **rural–urban fringe** is the area of mainly open land at the edge of the built-up area of a town or city. Land in the rural–urban fringe is under pressure from developers wanting to build upon it. Development at the edge of the built-up area can add to the problem of **urban sprawl**, the outward growth of the urban area.

A variety of users wish to occupy sites in the rural urban fringe (Figure 4.1). Many economic activities now see the area as being desirable, e.g. business parks, retail parks, science parks or large shopping centres. These land uses are in competition with recreational activities such as sports stadia, theme parks and leisure complexes. **Residential** and transport land uses add further competition for land in the rural–urban fringe.

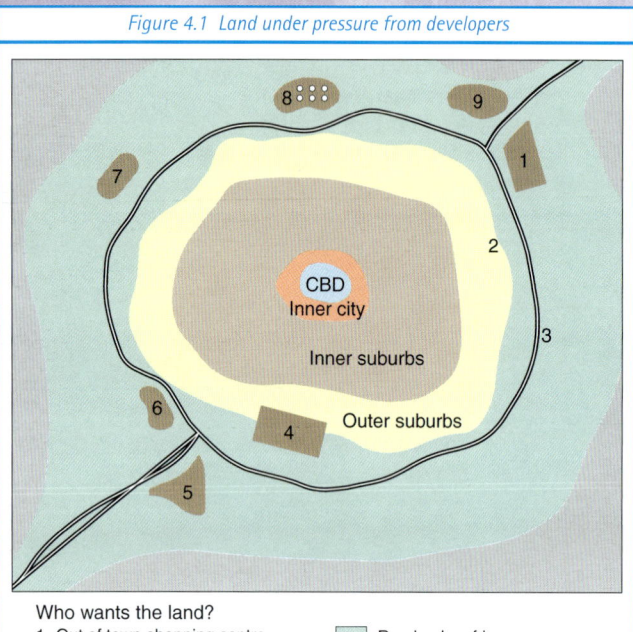

Figure 4.1 Land under pressure from developers

Who wants the land?
1. Out of town shopping centre
2. Suburban housing estates
3. Orbital motorway/ring road
4. Outer city council estate
5. Business Park/Science Park
6. Suburbanised village
7. Football ground/Sports stadium
8. Sewage works
9. Theme Park

Rural–urban fringe. Land under pressure from developers

Out of town shopping centres

Across the United Kingdom, large shopping centres have been developed on land in the rural–urban fringes of large towns and cities. The first was the Metrocentre in Gateshead, which opened in the mid 1980s.

What are the advantages of a location at the rural–urban fringe?

1. Large amounts of land are available for big buildings and possible future expansion. There is sufficient room for large car parks. This is important as many people now buy in bulk and use their car to shop.
2. Land values are cheaper than the inner areas of cities.
3. Close to orbital motorways or outer ring roads. This makes the site more accessible for customers and deliveries, both locally and from further away. Over 2 million people live within a half hour drive of the Trafford Centre in Greater Manchester.
4. A pleasant, 'green' or countryside image. This is favoured by modern service industries.
5. Suburban estates are nearby, giving a potential workforce and customers.
6. As there are few tall buildings, the site is prominent (can be clearly seen by customers).

ACTIVITIES

1. Give the meaning of the term 'rural–urban fringe'. (C1.2)
2. a Give examples of land uses which are in competition for land in the rural–urban fringe. (C1.2)
   b Give named examples of these land uses from your local area. (C1.2)
3. Copy and complete the passage below. Choose your answers from those in the brackets. (C1.2)

   The rural–urban fringe offers many [advantages/disadvantages] for modern [primary/service] industries. There are [small/large] amounts of [expensive/cheap] land. [Inner/outer] ring roads and [orbital/arterial] motorways make the location [inaccessible/accessible] for customers and workers. [Suburban/inner city] estates are nearby providing an available workforce.

Chapter 4: Pressure at the rural–urban fringe

The Trafford Centre — Case Study

The Trafford Centre (Figure 4.2) is a large out of town shopping and entertainment centre built on open land at Dumplington, Greater Manchester. The site is at the edge of the built-up area. The development is made up of over 280 shops and large department stores, along with a 20-screen cinema, bowling alley, restaurants and bars. There are large car parking areas surrounding the centre. Since the completion of the Trafford Centre, other tertiary (service) industries have been attracted to sites next to, or nearby it. These include retail parks, supermarkets, health clubs and restaurants.

ACTIVITIES

1. Using map evidence from Figure 4.4, give reasons why Dumplington was a good site on which to build a large shopping centre.
2. Describe how planners have made the Trafford Centre **prominent**. (C1.3)
3. Using map evidence from Figure 4.3, describe how **land use** has changed in the Dumplington area.

Extension activity

4. Suggest reasons why other tertiary industries have been attracted to a site near the Trafford Centre. (C2.3, IT2.1)

Figure 4.2 The Trafford Centre

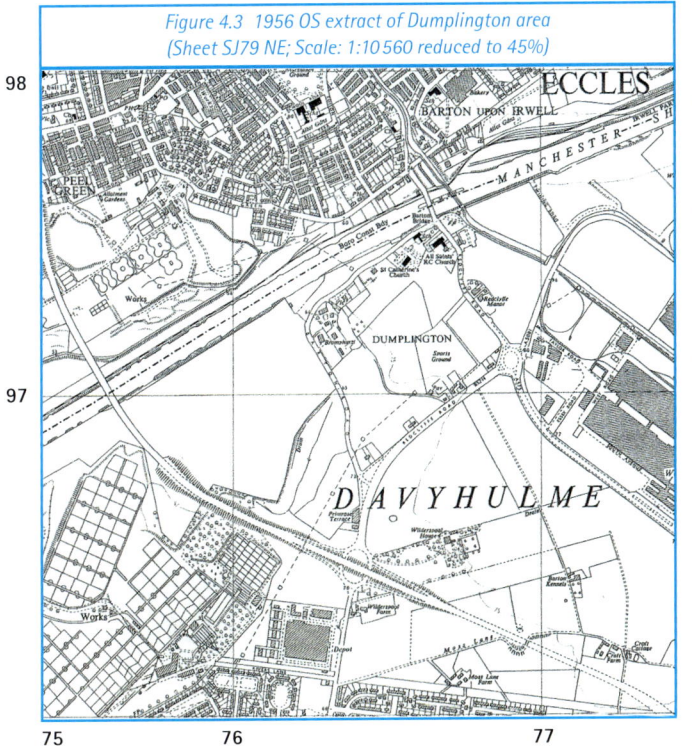

Figure 4.3 1956 OS extract of Dumplington area
(Sheet SJ79 NE; Scale: 1:10 560 reduced to 45%)

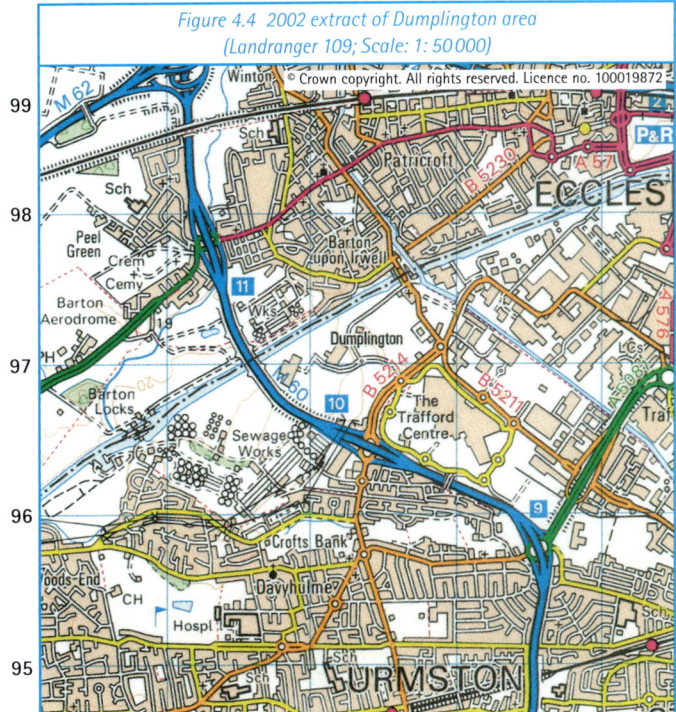

Figure 4.4 2002 extract of Dumplington area
(Landranger 109; Scale: 1: 50 000)

25

Section 1: Managing Change in the Human Environment

The impact of out of town developments on the CBD

Shop owners in the Central Business Districts of towns and cities are opposed to the development of out of town shopping centres. (Figure 4.5 and 4.6) They feel that the new centres will take their business away. Here are some reasons why:

Figure 4.5 Impact on the city centre

Traders counter shopping city threat

Traders and council chiefs in Bury have begun a multi-million pound fightback against the threat from the Trafford Centre shopping city.

Preparations for a £20m facelift of the Union Arcade in Millgate town centre shopping precinct are being finalised by developers.

A £2m scheme to modernise the town's famous market is planned and the search is on for cash backing.

And traders are due to hold talks with town hall chiefs over their proposals to create a Victorian-style shopping area with specialist stores.

© Manchester Evening News, 1996

Figure 4.6 Manchester city centre

A tale of two 'cities' in sales battle

Stores in Manchester city centre experienced a quiet start to the sales with many businesses reporting that trade was down on past years.

But it was a different story at the Trafford Centre shopping city which was experiencing one of its busiest times as crowds of bargain-hungry shoppers defied the wintry weather and packed the complex.

In Manchester itself there had been an increase in crowds after a slow start.

- Traffic congestion in, and on the way to, the CBD.
- Limited, expensive car parking.
- Some shops have relocated. Less choice.
- Some people prefer to shop 'all under one roof' and be protected from the weather.

Some town and city planners fear a vicious circle of decline (Figure 4.7.)

A study into the effects of the Meadowhall shopping centre (Sheffield) showed that the profits of nearby shopping areas declined after the new centre was opened (Figure 4.8.) These problems have led to schemes designed to modernise and upgrade city centre shopping areas (see chapter 3) in an attempt to attract shoppers in the area. One such scheme is shown in Figure 4.5.

Figure 4.8 The effects of Meadowhall			
Centre	Impact (First 36 weeks)	Centre	Impact (First 36 weeks)
Barnsley	−10.87%	Leeds	−3.00%
Bradford	−2.91%	Mansfield	−4.45%
Broomhill	−6.17%	Nottingham	−2.23%
Dewsbury	−6.03%	Rotherham	−7.79%
Doncaster	−3.47%	Sheffield	−11.54%
Halifax	−2.89%	The Moor	−16.52%
Hillsborough	−7.25%	Worksop	−2.20%

ACTIVITIES

1. **a** Which three shopping centres have suffered the most due to the building of Meadowhall? (N1.1)
2. Explain why a vicious circle of decline may occur in some town or city centres. (C1.2)
3. For *either* Bury (Figure 4.5), *or* a central shopping area that you know, describe what has been done to try and improve the environment for shoppers. (C1.2, IT1.1)

Residential land use

The growth of the suburbs

During the period between the First and Second World Wars (1920–30), cities in the United Kingdom began to grow outwards rapidly as what are now known as the 'outer suburbs' were built. These were estates of mainly low density housing such as detatched or semi-detached houses with space for gardens and garages (Figure 4.9). The lower land values at the edge of the city meant that this type of development could take

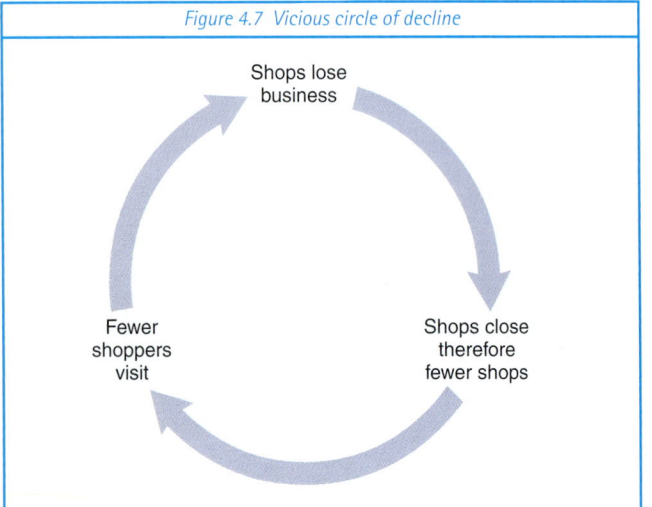

Figure 4.7 Vicious circle of decline

Shops lose business → Shops close therefore fewer shops → Fewer shoppers visit →

Figure 4.9 Suburban housing

Chapter 4: Pressure at the rural–urban fringe

place. People living in the suburbs were able to commute, perhaps to jobs in shops and offices in the CBD or in manufacturing industry in the inner city. This was possible due to improvements to public transport and later, due to increased ownership of private cars and the construction of arterial routes in to the city centre. Urban sprawl continued in the 1960s and 1970s, when comprehensive redevelopment of inner city areas was taking place (see chapter 3.) People were displaced as the old, slum terraces were knocked down and many large council estates were built at the edge of the city to house them (Figure 4.10).

Figure 4.10 An outer city estate

ACTIVITIES

1. Look at this list of urban features:
   - semi detached houses
   - gardens
   - garages
   - low density
   - small back yard
   - inter-war housing
   - cul-de-sacs
   - commute to workplace
   - high density
   - on-road parking
   - 19th-century housing
   - terraced
   - close to workplace
   - grid-iron pattern

   Put each feature into a copy of the table below, according to whether it is a feature of the suburbs or the inner city.

Feature of the suburbs	Feature of the inner city

2. Which of these groups of people are most likely to move to the outer suburbs? For each group chosen, explain why they would be more likely to move. (C1.2)
   - newly arrived migrants from an LEDC
   - people with increasing incomes
   - parents with young families
   - young, unmarried people
   - farm workers
   - highly skilled or qualified people

Counter-urbanisation

Counter-urbanisation has also placed pressure upon land at the rural–urban fringe. This process is the opposite of **urbanisation** where people move into towns and cities (see chapter 2.) It is when people are moving out of urban areas to live in villages just outside the city. People often choose to move out of large towns and cities as they become more affluent. Figure 4.11 shows that disposable income has increased.

As people have more money, they can afford to commute to work in the city and are able to afford the often more expensive houses. They wish for an improved quality of life. For them, this means a 'greener' environment, with more open space, fresh air and peace and quiet, away from the hustle and bustle of life in a large town or city. There is also a perception of lower crime rates.

This process has led to villages becoming suburbanised. This means that they become more like the suburbs and start to lose their distinct characteristics. New housing estates are added and the look of traditional, older buildings are changed as they are bought up and extended.

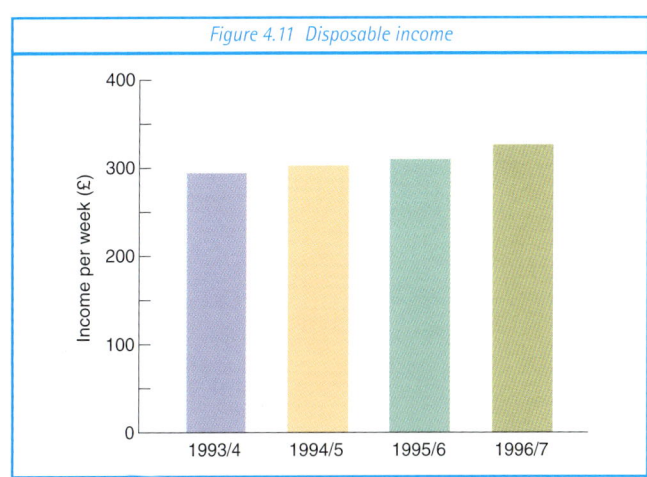

Figure 4.11 Disposable income

Section 1: Managing Change in the Human Environment

Figure 4.12 A new estate in a suburbanised village

Managing issues at the rural–urban fringe

Green belts

Green belts were set up by Act of Parliament in 1947. They are areas of mainly open countryside and small settlements surrounding urban areas (Figure 4.15.) The green belt is protected from development. The number and type of buildings that can be erected is restricted. Planning permission is not normally given to developers.

The purposes of green belts:

- To prevent urban sprawl by halting the outward growth of a city.
- To prevent cities or towns merging together to form a continuous urban area.
- To protect the open nature of the countryside and the character of the settlements in it.
- To preserve the land for farming and to provide access for recreation.

Figure 4.14 Will this green and pleasant land disappear under concrete?

ACTIVITIES

1. Add the labels below to Figure 4.13 to show some changes to a suburbanised village:
   - re-opened railway station
   - new estate of semi-detatched houses
   - new estate of detatched houses
   - newly widened road
   - improved, older houses in the core
   - new restaurant added to pub
   - barn converted to a house
   - row of terraced cottages knocked through

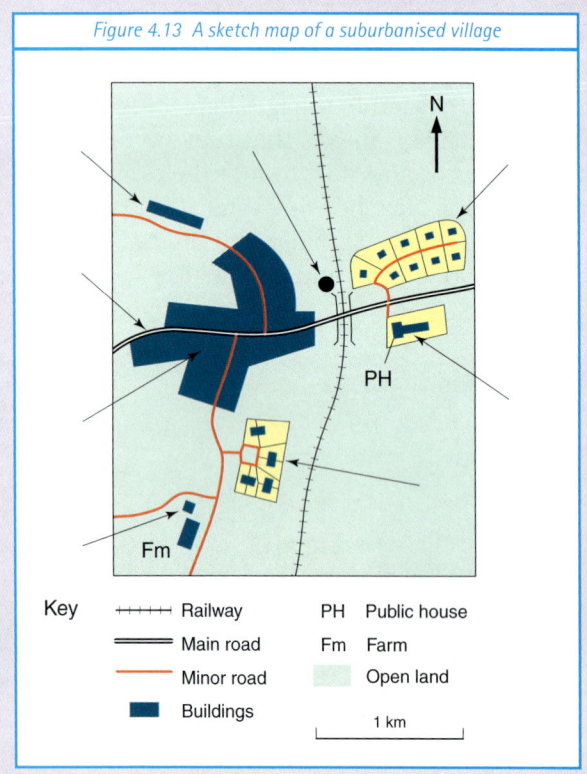

Figure 4.13 A sketch map of a suburbanised village

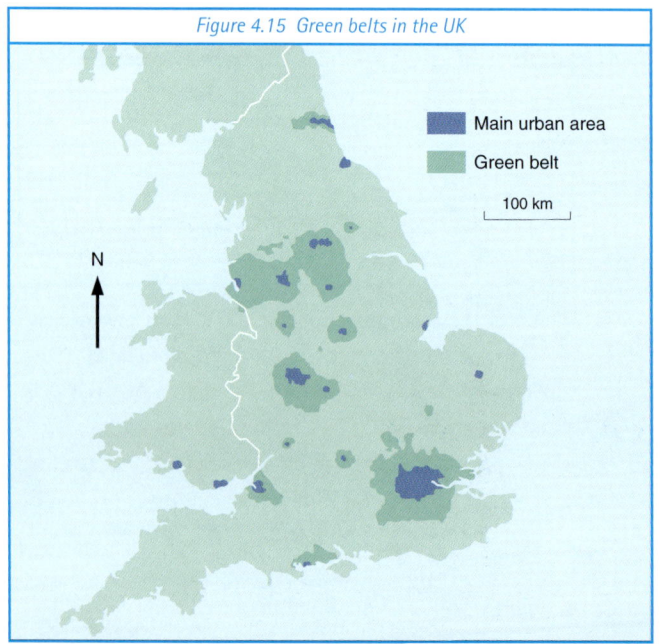

Figure 4.15 Green belts in the UK

Chapter 4: Pressure at the rural–urban fringe

THE ISSUE

Do we continue to protect green belt land or give in to pressure from developers who want to build there?

Some planners now feel that the idea of green belt is old fashioned. They claim that the rules were made when food was scarce, after the Second World War. Also at that time, many countryside dwellers were moving to the cities (rural–urban migration was taking place). Today the countryside is not just seen as a source of food from farms. Now more people want to live in and use the countryside. More people own cars and have access to the countryside (see page 20).

The Government has said that 'the issue would have to be widely debated before the rules are changed.'

Here are some arguements for and against changes which would allow development:

For
- Incomers bring money into the countryside.
- More money for the local council, raised from taxes.
- It is cheaper for developers to build on greenfield land.
- It creates jobs in the rural area.
- It stops people moving out of the countryside, as there are more shops, schools etc.
- The rules stop farmers from diversifying. This weakens the rural economy and therefore damages the rural environment.

Against
- Animals and their habitats are destroyed.
- The natural beauty of the area is spoilt, as new developments do not 'fit in' with the countryside.
- Villages become commuter settlements. This means more cars, therefore more air and noise pollution and more traffic congestion.
- Tourist income for the area may decline.
- It places demands on local councils, e.g. more roads, refuse disposal, schools etc.
- Incomers bring different values and ideas. This may lead to conflict.
- Incomers may work and shop elsewhere, creating dormitory settlements which have few benefits for the local economy.

The Government has said that up to 4 million new houses may be needed to be built by 2020 and that 40% of these would be built on **greenfield sites**, many in areas of green belt.

The region of the UK with the greatest demand for housing is the South East of England. Here it is expected that between 1991 and 2011, 850 000 new homes will be built.

Other Government policy

The Government has stated that 60% of the new houses needed by 2020 should be built on **brownfield** sites within the existing urban areas (see chapter 3).

The Government also recently restricted the building of any new out of town shopping centres. The last two schemes to be allowed to go ahead were Braehead (Glasgow) and the Trafford Centre (Greater Manchester) in 1992. However, in August 2000, permission was granted to extend the Metrocentre (Gateshead) by 10% and planned extensions to Merry Hill (West Midlands) are being considered.

ACTIVITIES

1. What is a green belt and what is its purpose? (C1.2)
2. Use an atlas and Figure 4.15 to name the urban areas of the UK that are surrounded by green belt.
3. Explain why green belts have been described as an 'old fashioned' idea. (C1.2)
4. Describe how use of the countryside has changed in recent years. (C1.2)
5. Look at Figures 4.16 and 4.17 on page 30.
   a. Explain why 4 million new houses may be needed by 2020. (C1.2)
   b. In which three counties of South East England will most new houses be needed? (C1.2)
6. Look at the arguments for and against building on greenbelt land. State your opinion on the issue, giving reasons as to why you feel that way about it. Try to include some reasons not already given. (C1.1, C1.2, C1.3, IT2.1)
7. List three groups of people who may be for building on greenbelt land and three groups who may be against it. (C1.1, C1.2, C1.3, IT2.1)

Section 1: Managing Change in the Human Environment

Figure 4.16 Why do we need more houses?

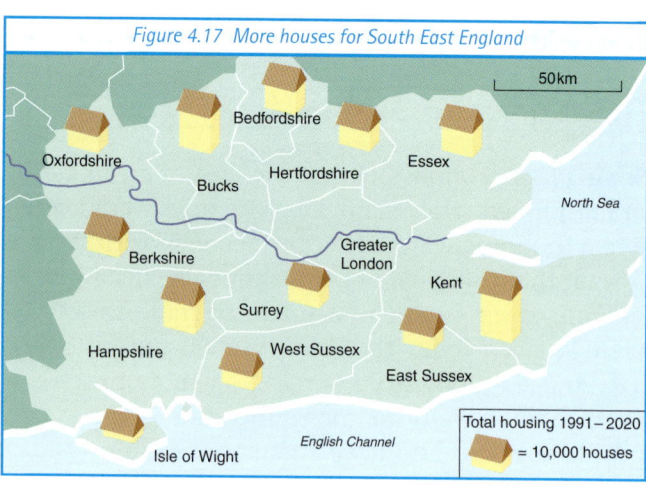

Figure 4.17 More houses for South East England

Environmental protection

Some planners argue that the benefits of development at the edge of the urban area are greater than the problems caused. What do you think? Are you in favour of short term economic gain, as opposed to long term environmental damage?

There will always be pressure on local councils to let developments on greenfield sites go ahead, especially if jobs are involved. It is argued that developments should go ahead, but should be carefully managed so that the damaging effects are kept to a minimum.

Here are some ways in which development may be managed:

- Car parks can be built underground.
- Developments can be screened by fast growing trees.
- Earth mounds can be created to reduce noise.
- Earth mounds and the sites of the developments themselves can be landscaped to give a greener appearance.
- Natural materials and colours may be used to reduce visual pollution, e.g. green glass does not stand out.
- Designers can use local stone and building styles which 'blend in' with the area.
- Sites of special scientific interest (SSSIs) can be relocated.

Some environmentalists argue that these schemes can never compensate for the damage that new developments cause. They see fast growing trees and plants used in landscaping as visual pollution in themselves (especially if they are not natural to the area). They do not feel that new, large, modern buildings will ever fit in with a rural or semi-rural environment. They argue that even when great care is taken, it is never possible to recreate habitats in their former glory.

ACTIVITIES

1 a Describe ways in which environmental damage can be reduced in the building of a new development that you have studied. (IT2.1, C1.3)

   b How successful do you think these measures will be? (IT2.1, C1.3)

Section 1: Exam-style questions

Exam-style questions for Section 1

Question 1 – F Tier
Look at the five sites marked as A to E on Figure 1.

Question 2 – H Tier
Look at the five sites marked as A to E on Figure 1.

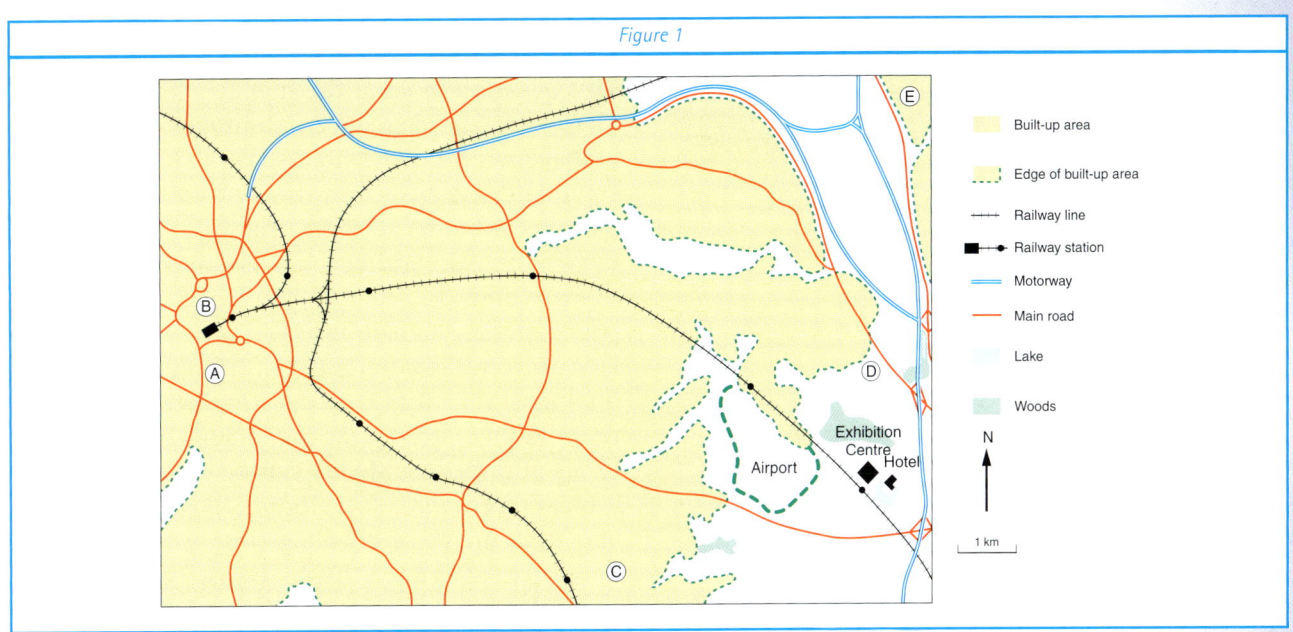

Figure 1

(a) (i) Site B is in the CBD. Give two pieces of evidence to show this. **(2)**

(ii) A developer wants to use the land at site D to build a large shopping centre. Give three advantages of a location such as this. **(6)**

(b) (i) Which of the terms (A to D) below, means the rapid outward growth of a city?

A Rural–urban migration.
B Urban sprawl.
C Urban regeneration.
D Urban zone. **(1)**

(ii) Here are three methods of managing the outward growth of a city.

- Greenbelt policy.
- Blending in developments with the environment.
- Development on brownfield sites.

Choose **two** methods and explain how they may help manage urban growth. **(4)**

(c) Many people from the village at E on Figure 1 travel daily into the city to work.

(i) What name is given to a person who travels daily into the city to work? **(1)**

(ii) Describe the problems that the daily movement of people into a city may cause. **(6)**

(20 marks)

(a) (i) Which of the sites, A to E, is in the CBD? **(1)**

(ii) Give **three** reasons for your answer to (a)(i), above. **(3)**

(iii) A developer wants a site on which to build a large shopping centre. Decide which site, A or D, the developer would choose.

Write the letter of the site that you have chosen. _____

Explain why the land at the site chosen is under pressure from developers. **(6)**

(b) New developments can cause the outward growth of a city. Describe ways in which this may be managed. **(6)**

(c) Describe the problems that may be caused by the daily movement of people from village E, towards the city centre. **(6)**

(22 marks)

Section 1: Exam-style questions

Commentary on exam-style questions for Section 1

F Tier (a)(i) and H Tier (a)(i) and (ii)

This question is settting the scene. At H tier, candidates have to make a decision as to which site is the CBD, whereas F tier candidates are told this information. Tier H candidates also have to find more pieces of evidence to support their answer. Candidates should refer to the map and notice that site B is surrounded by an inner ring road, contains a main railway station, has all the main roads converging upon it and is approximately 10 km from the edge of a built up area.

F Tier (a)(ii) and H Tier (a)(iii)

This question looks at the causes of the issue. Again, tier H candidates have to make a decision as to the best site, whereas tier F candidates are instructed to look only at site D. Tier H candidates could justify either choice of site, but must always ensure that they are able to develop their responses and, where possible, use case study examples to access all 6 marks, as longer answer questions such as this will be marked according to a level of response. Marks are not awarded for every point made, but a level (low, middle or high) is allocated for the quality and depth of response. Tier F candidates have to develop three responses in order to gain the maximum 6 marks. At both levels, the nature of the advantages *must* be clear. For example, 'There is plenty of open space ... for car parks and possible future expansion', 'there are large suburban estates nearby ... for a large customer base', 'it is next to a main road, which leads to a motorway ... so there is easy access for customers and deliveries'. Each of these developed statements would gain 2 marks. This is much better than merely stating 'It is near a motorway', which would only gain 1 mark.

F Tier (b)(i)

1 mark for stating B, urban sprawl.

F Tier (b)(ii) and Tier H(b)

This question looks at the management of the issue. Tier F candidates are given a number of management schemes to discuss, whereas tier H candidates are not given these prompts and would be expected to recall them. Tier F candidates would have to explain and develop a point about each of two methods, in order to gain a maximum marks. For example, 'A greenbelt is an area of mainly open land surrounding a city **(1)** ... It limits the outward growth of the city **(1)** as development is severely restricted by law **(1)**.' At the H tier, there would be levels of response. Merely stating schemes would be low level responses, developing these schemes, as above, would be middle level responses, continued development of the scheme, or use of case study examples, would access the higher levels.

F Tier (c)(i)

1 mark for stating 'commuter'.

F Tier (c)(ii) and H Tier (c)

This is common to both tiers. (Approximately 30% of questions are common.) It is marked according to levels of response and looks at the nature of the issue. Answers such as 'There will be congestion', would be low level. Development of more than one point would access the middle level, i.e. an explanation of why congestion may occur at a particular point. 'Congestion will occur at places where roads converge on entering the CBD.' Further development of the same point, or use of case study examples, would give access to the higher level mark. Two points, fully developed would gain a maximum mark.

SECTION 2 MANAGING THE PHYSICAL ENVIRONMENT

Chapter 5 Earthquakes and volcanoes

Kobe, Japan

'Rescue teams in Kobe fought fires and searched for the dead and missing after the disaster that Japan thought could never happen had apparently killed 1,800 and injured more than 6,330. The earthquake, which struck at 5.46 am measured 7.2 on the Richter Scale, and was Japan's worst for 50 years. Buildings, roads and railways designed to withstand severe tremors, buckled.'

The Times, Wednesday, January 18, 1995

Figure 5.1 *The Hanshin Expressway near Kobe was believed to be quake-proof*

Gujurat, India

'Damage from last month's Gujurat earthquake, which killed at least 30,000 people is so bad that the Indian government is considering abandoning some towns. The quake measured 7.9 on the Richter Scale and the worst damage was within 125 miles of the epicentre near Bhachau. UNICEF, the United Nations Children's charity, said that 1,700 primary schools in the region had been destroyed.'

Guardian, Friday, January 26, 2001

Figure 5.2 *Satellite photography picks out smoke from Mount Etna, Sicily*

Mount Etna, Italy

'Streams of lava heading down the slopes slowed yesterday, raising hopes that villages would escape destruction.' Mount Etna is Europe's tallest and most active volcano. Although its eruptions are far less violent than its famous neighbour Vesuvius which buried the Roman city of Pompeii in AD 79, it has regularly released lava every few years throughout the last two millennia.

Independent, Friday, July 27, 2001

The risks of living with earthquakes and volcanoes

California

What is the risk?

In this, the richest of America's states, running 10% of the US economy, some 34 million people live in towns and cities astride the San Andreas Fault. Linked to this major split in the earth's surface is a zone famous for its earthquake potential. Here an earthquake flattened the city of San Francisco in 1906, and again caused damage in 1989 and 1994. Scientists monitoring the region expect that a far larger earthquake, the so-called 'big one', will probably occur in the next 30 years. Expecting a quake is one thing, predicting when it will happen is another.

Is it worth it?

Despite these risks, clearly California is a popular place to live. There are many reasons for this:

- a sunny climate and good beaches
- nearby mountains with spectacular scenery
- National Parks with volcanic features (including Mount St Helens)
- fertile valleys with rich soils for agriculture (market gardening and fruit)
- an international reputation for wine production
- multi-million dollar industries (like Hollywood and its film companies)
- some of the world's most expensive real estate (homes and property)

Sicily

What is the risk?

Sicily's north-eastern corner is overshadowed by Mount Etna. The risk here is relatively small as lava flows slowly down its slopes posing little serious threat to the villagers below. Despite its active state the volcano holds few surprises – no massive explosions or gas avalanches here.

Is it worth it?

Local people remain close by because:

- volcanic soils weather down to produce fertile soils, supporting vineyards and small farms
- the steep slopes double as ski runs in the winter months
- volcanic features themselves attract large numbers of tourists to the area

> 'As the streams of molten lava swallowed up the ski station and threatened the souvenir shops and cafes, and evacuation plans were being drawn up, police were putting up warning signs and turning away sight-seers and disaster tourists.'
>
> *Independent,* Friday, July 27, 2001

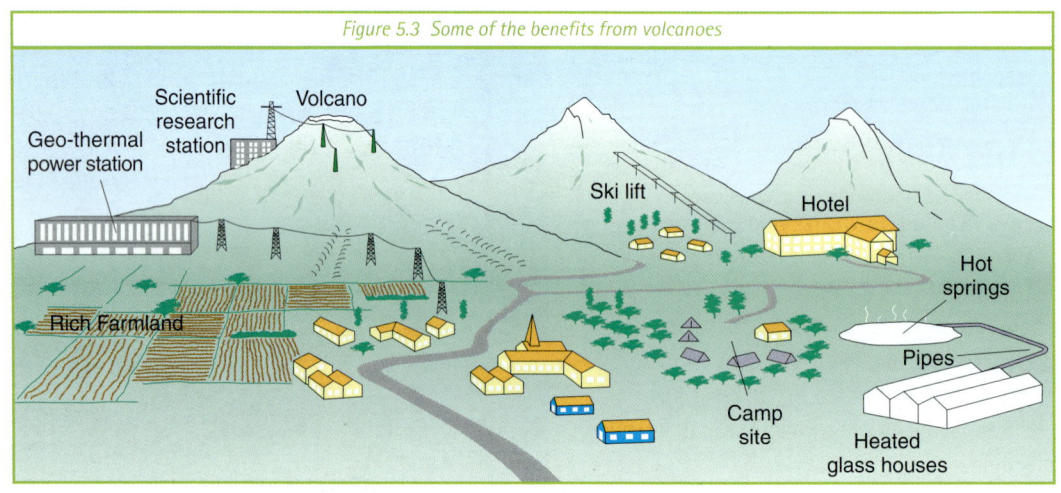

Figure 5.3 Some of the benefits from volcanoes

ACTIVITIES

1. Which do you feel was the worst disaster, Kobe or was Gujurat? Explain why. (C1.1)
2. Why do Californians risk living with the 'big one'? (C1.1)
3. Using Figure 5.3, list the benefits of living near to a volcano. (C1.3)

Earthquakes

Earthquakes are a fact of life in some parts of the world. When sudden movements take place within the Earth (at the **focus**) shock waves are sent to the surface. These tremors may damage buildings, destroy towns and even split open the ground. Most damage will occur around the **epicentre**, a point on the Earth's surface directly above the focus. From here the effects of the shaking will spread outwards, but decreasing in strength as they do. There are various types of shock waves that create different effects, but importantly they allow scientists to understand where quakes begin and how they develop. The instrument used to record and measure earthquakes is a **seismograph**.

Major earthquakes occur when there are already weaknesses in the Earth's crust, as these are along the plate boundaries (see Figure 5.8). Here large **fault**, like the San Andreas Fault in California, have the potential to produce massive earthquakes. In other parts of the world like the UK such events are rare and confined to minor tremors.

How strong are earthquakes?

The strength or power of earthquakes varies enormously and this is recognised in the way that they are measured on the **Richter Scale**. The scale runs from 0 to 9, but increases logarithmically. A scale 3 quake, for example, is 10 times as destructive as one recorded as scale 2.

An alternative to this is the **Modified Mercalli Scale**, which describes the effects rather than the power of an earthquake. This scale goes from 1, which is hardly noticeable, to 12 which causes total devastation. (see Figure 5.5)

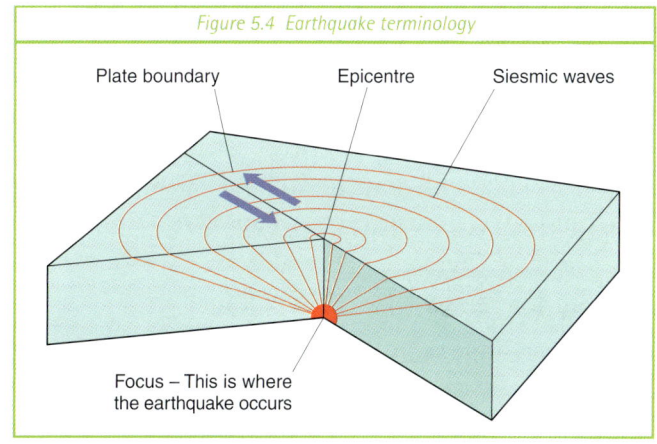

Figure 5.4 Earthquake terminology

ACTIVITIES

1. List and explain each of the six terms in **bold type** on this page. (C1.3)
2. Why do we need two earthquake scales – Richter and Mercalli? (C1.1)

Figure 5.5 A cartoon view of the Modified Mercalli Scale for earthquakes

Section 2: Managing the Physical Environment

What makes earthquakes worse?

The effects that result from earthquakes do not entirely depend upon their strength. Other factors are often more important:

- the population density of the area (the number of people at risk)
- the quality of the buildings (their ability to resist collapse)
- the rescue and emergency services (food and medical resources especially)
- the wealth of the region or country (cost of rebuilding)

In 1993 an earthquake with a magnitude of 6.4 killed 23 000 people in Maharastra (India).

A similar quake hit Los Angeles (USA) in 1994. The buildings rode the shock waves and only 60 people died (many indirectly from heart attacks).

Of the 100 000 killed in the Gujarat earthquake in 2001, most were crushed by collapsed buildings.

ACTIVITIES

1. Why are the strongest earthquakes not also the deadliest? (Figure 5.6) (C1.2)
2. Suggest which of the five factors listed above, were most important in:
   a Maharastra   b Los Angeles   c Gujarat
3. What is the main feature of the pattern of earthquakes shown in Figure 5.6? (C1.2)
4. Identify the earthquakes numbered 1–4 in Figure 5.7. (C1.2)

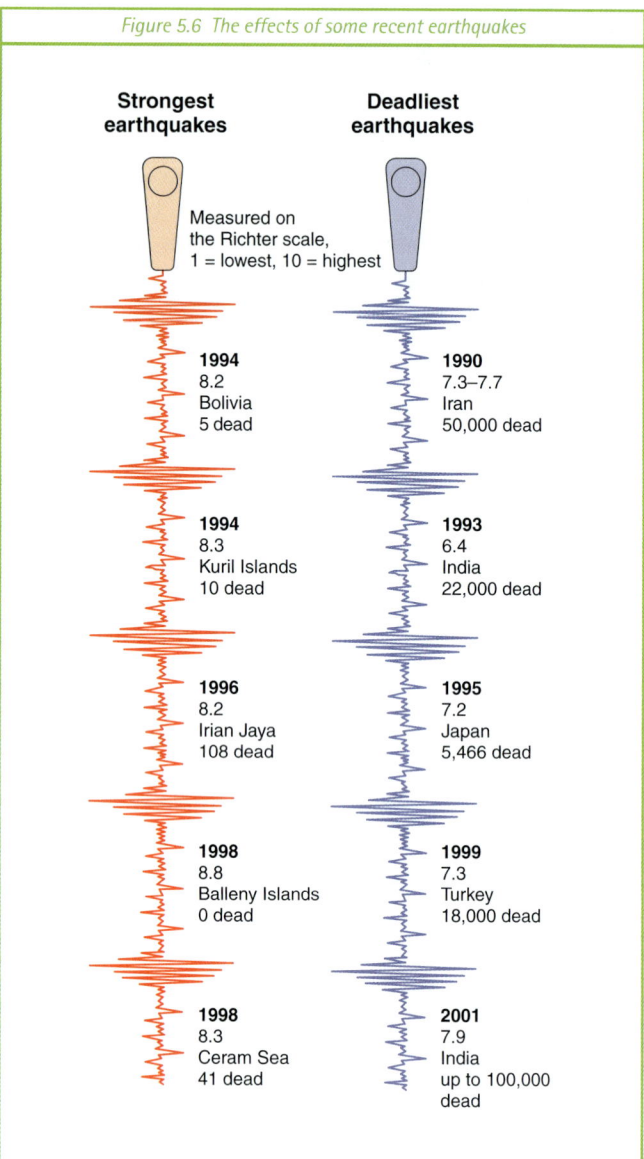

Figure 5.6 The effects of some recent earthquakes

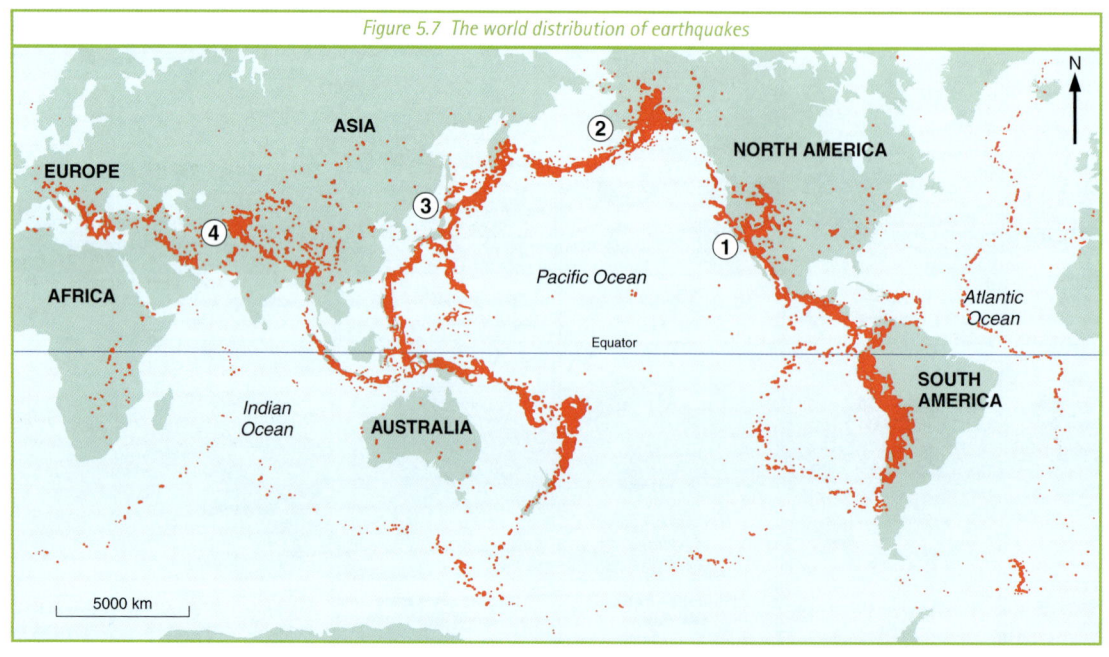

Figure 5.7 The world distribution of earthquakes

Chapter 5: Earthquakes and volcanoes

The KOBE Earthquake

'It lasted only 30 seconds but it was enough to wipe out 2,500 lives and destroy Japan's belief in its shock-proof structures and technology.'

'Japan is at the meeting point of three tectonic plates, and the city of Kobe itself sits close to a geological fault stretching from Kyoto to Osaka.'

'Experts put the destructive power of this quake down to the shallowness of the epicentre, which was only 16 km below the island of Awaji and 20 km from Kobe. They also warned of the likelihood of further quakes'.

Adapted from newspaper reports from
Daily Express, January 18, 1995

This earthquake that took place in Osaka Bay on January 17, 1995, killed 4 571 people and injured 15 000 others. Some effects were immediate and a direct consequence of the quake, whereas other effects happened subsequently. These two sorts of effect are said to be primary and secondary.

Primary effects

- 30 seconds of shaking and aftershocks
- large numbers of casualties (see above)
- 125 000 buildings collapsed
- 7 000 buildings burned out
- half a million people evacuated
- transport and communications disrupted
- water, gas and electricity supplies cut off

Secondary effects

- unable to use fires and medical services
- business in the city and port disrupted
- water and food shortages
- people afraid to return home
- effects on shares, banks and currency

ACTIVITIES

1. Why did an earthquake happen in Osaka Bay (Figure 5.8)? (C2.2)
2. Why were the effects so devastating? (C2.2)
3. Why was 'Japan's belief in its technology' destroyed? (C2.2)
4. Why do casualty figures in disasters seem to change in the days following an earthquake? (C2.2)

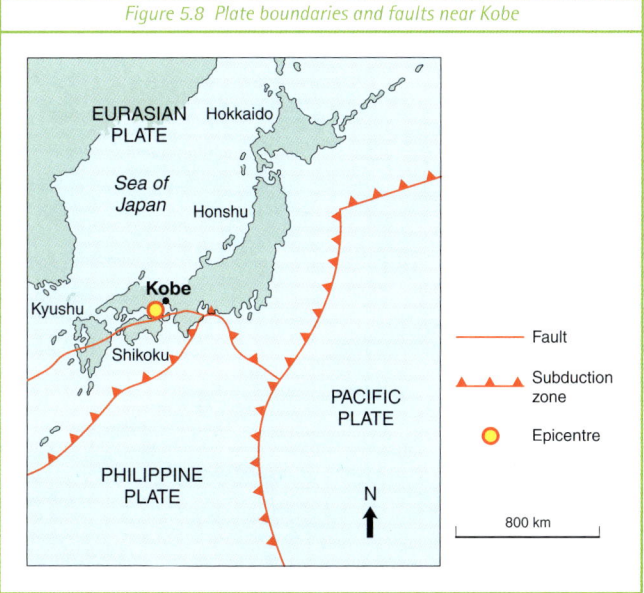

Figure 5.8 Plate boundaries and faults near Kobe

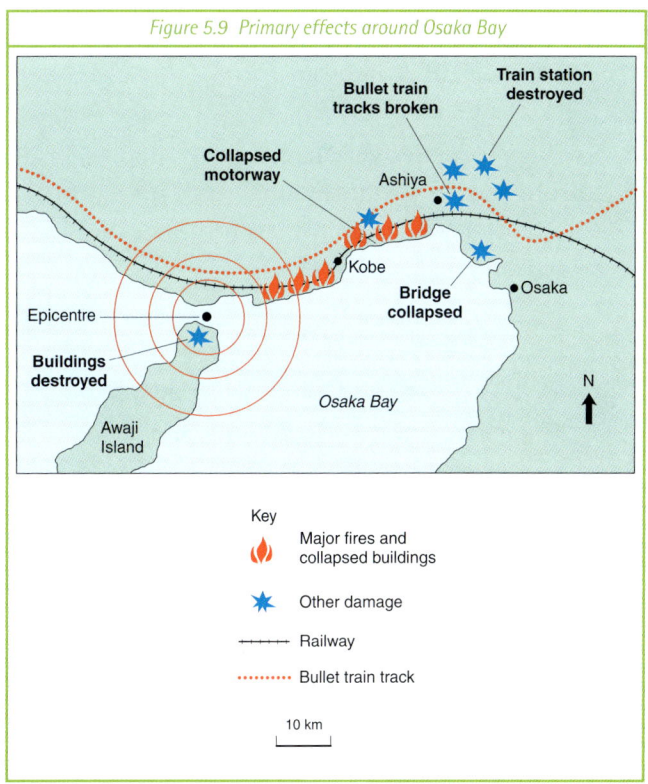

Figure 5.9 Primary effects around Osaka Bay

Section 2: Managing the Physical Environment

Volcanoes

Volcanoes form when material from the mantle escapes through cracks in the Earth's crust and reaches the surface. Molten rock, or **magma**, rises up and spills out to form rivers of lava. Ash, gases and even volcanic 'bombs' may also be ejected. Many of the world's 600 or so active volcanoes are found in the same weak areas of the crust as earthquakes. Whilst some volcanic events are slight, others involve major eruptions.

How often do they erupt?

- **active** volcanoes are likely to erupt at any moment
- **dormant** volcanoes have not erupted recently
- **extinct** ones are unlikely to do so again

What are volcanoes like?

Volcanoes vary a lot, but they can be put into different categories, based upon how often they erupt (see above), the way they erupt, or what they look like:

1. The best-known view of a volcano is that shown in Figure 5.10. This has the typical features of a **crater** and central **vent**, through which lava emerges. Successive lava flows and layers of ash build up to form a **cone**. Mount Etna is an example of this composite form.

2. Where the lava is more fluid (and basic), it flows easily and creates very large, gently sloping **shield** volcanoes like Mauna Loa. This sits on the floor of the Pacific Ocean forming the island of Hawaii.

3. Mount St Helens with its steeper **dome** shape shows how thicker (acid) lava solidifies more quickly. This also leads to more violent eruptions.

4. In other parts of the world **fissures** or large cracks in the Earth's crust allow lava, steam and gases to escape over a wide area forming a flat plateau. Much of India has been made in this way.

ACTIVITIES

1. Describe the features of a typical volcano like Mount Etna. (C1)
2. Why do many other volcanoes around the world look different to this? (C1)

Hazard facts

Volcanoes produce a wider range of hazards than simply streams of lava or volcanic ash. The lava itself in places like Hawaii can flow at 300 000 litres per minute and be over 1 000°C. In other places, such as the island of Montserrat in the Caribbean, the main danger comes from ash; in Montserrat it fell over half of the island. On the neighbouring island of Martinique, blasts of hot gases were able to vapourise everything in their path. In Colombia, heavy rain triggered by the heat caused mudflows (called lahars) and floods to fill river valleys.

Figure 5.11 Lava flowing from Mount Etna

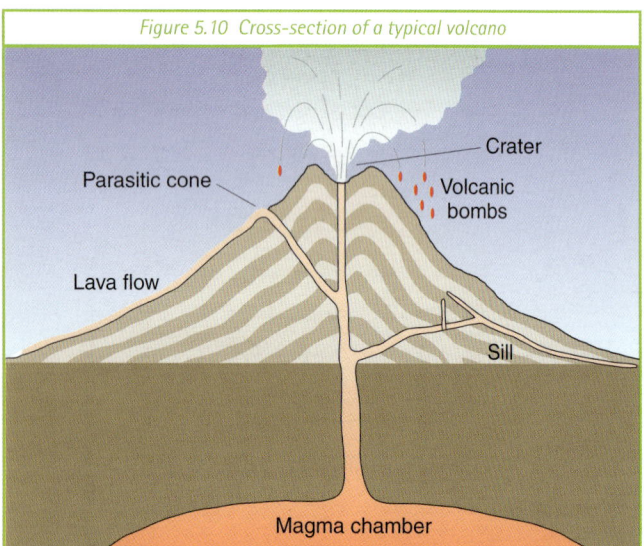

Figure 5.10 Cross-section of a typical volcano

Figure 5.12 Hazards produced by volcanic eruptions

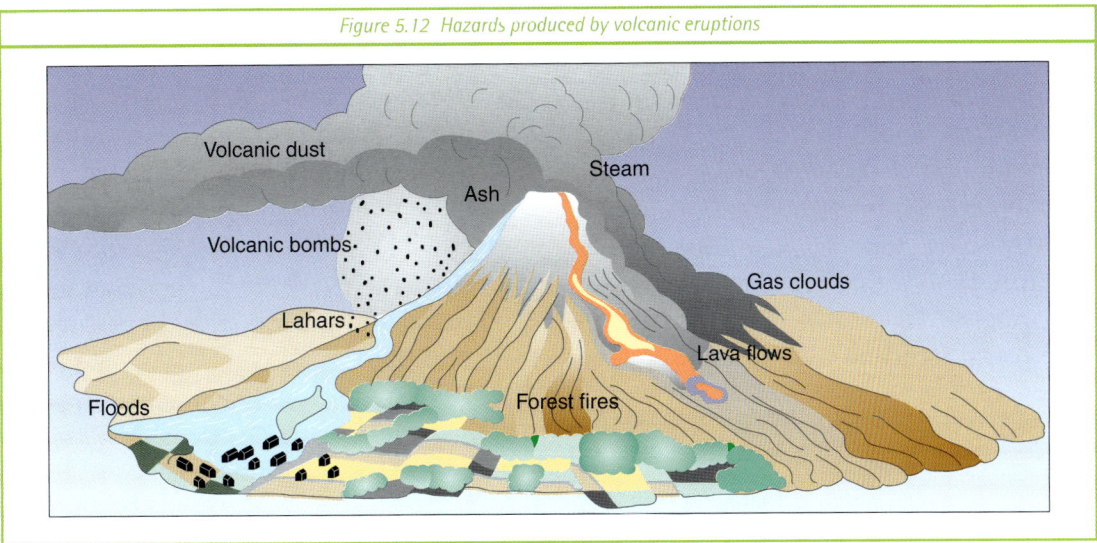

When major eruptions occur, vast amounts of debris can be thrown into the atmosphere leaving a crater in the volcano. At a later time this may collapse and create a huge hole or caldera within the volcano. Some volcanic effects can be global, as dust sent into the atmosphere may block out sunlight affecting temperatures around the world over a many of years. Mount Krakatoa in Indonesia was credited with having this effect in 1883. Earthquakes and tidal waves (**tsunami**) can also occur at the same time.

Where do volcanoes occur?

The distribution of volcanoes follows much the same pattern as earthquakes (see Figure 5.13).

ACTIVITIES

1. Use the text and Figure 5.12 to list the dangers people face living near volcanoes. (C2.2)
2. Which part of the world has most earthquakes? (C2.2)
3. Identify the volcanoes 1–4 in Figure 5.13. (C2.2)

Figure 5.13 The world pattern of volcanoes

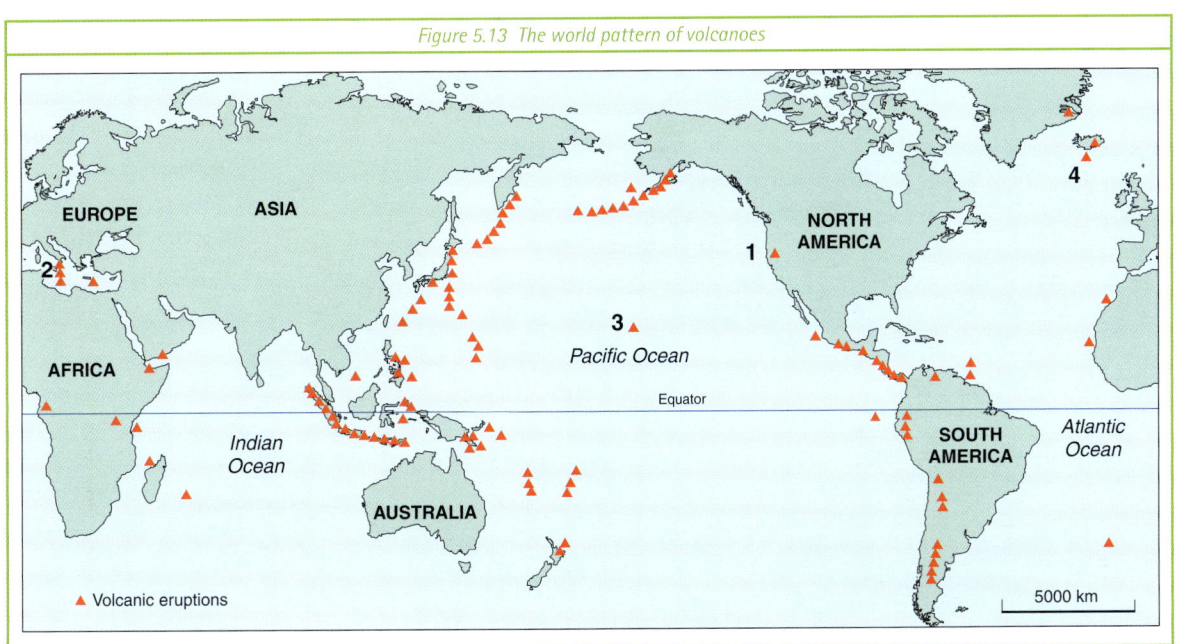

Section 2: Managing the Physical Environment

Montserrat

In the early 1990s this Caribbean island was seen as a tropical paradise occupied by 11 000 people. In the south of the island, the Soufriere Hills Volcano had caused a major eruption about 10 000 years ago but had been dormant for 400 years.

- In January 1995 eruptions began and the capital city, Plymouth, soon became covered in a layer of volcanic ash (see Figure 5.14). The natural environment and farmland were damaged by sulphur mixing with rainwater. Farmers and fisherman lost their livelihood.

- In January 1997 clouds of ash, gas and rocks flowed down the White river valley destroying the remaining cotton and cattle farms. By August more than half of the island was uninhabitable and 23 people had died (see Figure 5.15).

- Since then 7 000 of the population have left the island and the future of the rest remains uncertain.

This eruption is well documented in books. Visit the Montserrat Volcano Centre via the volcano Internet site: www.volcano.und.nodak.edu

Figure 5.14 Wasteland: a traditional British phone box lies half buried in volcanic ash in the deserted town of Plymouth, on the Caribbean Island of Montserrat. Residents were forced to abandon everything when volcanic ash from the Soufriere Hills volcano covered the town in 1966.

ACTIVITIES

1. How did the eruption of the Soufriere Hills volcano affect:
   a the environment
   b people living in the south
   c the island's economy? (C1.2)

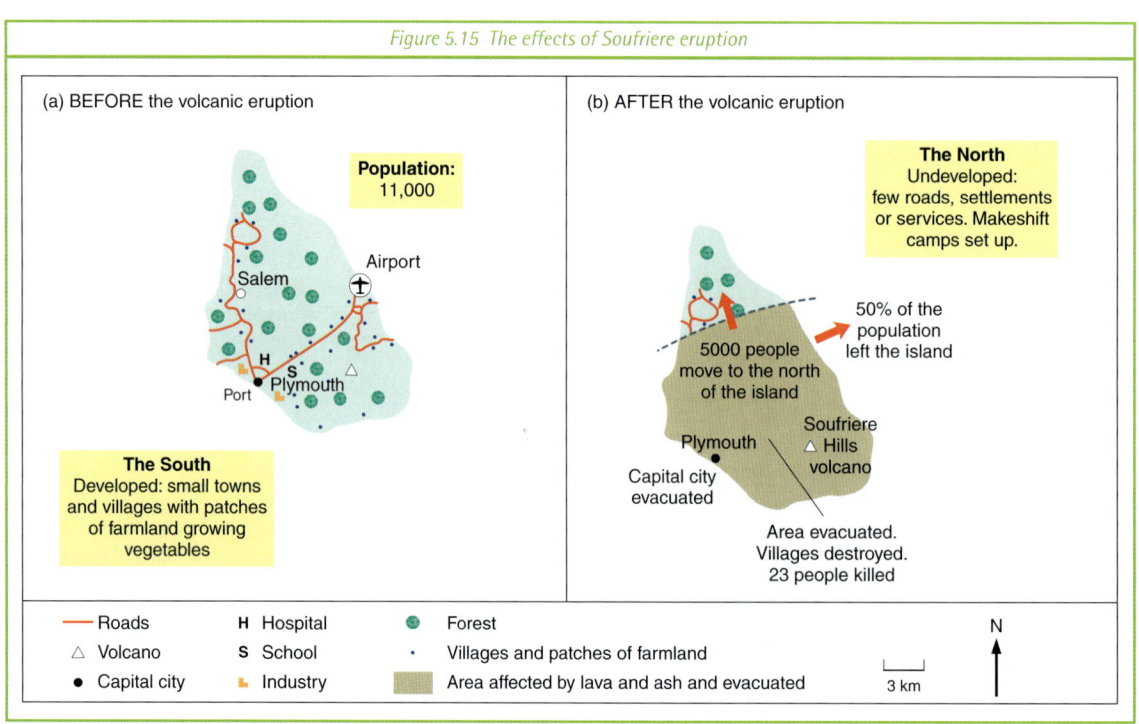

Figure 5.15 The effects of Soufriere eruption

Plate Tectonics

The global distribution of earthquakes and volcanoes shows that the Earth's crust has a pattern of weaker zones. These mark the boundaries of what scientists call **tectonic plates**. (see Figure 5.17)

This pattern of plates was discovered over a century ago, suggested by the way that the continents seem to fit together like a giant jigsaw. Only relatively recently have scientists been able to explain how **convection currents** in the Earth's mantle make these pieces of the continents '**drift**' or move. Proof of what was happening eventually came from geological and ecological evidence, some of which was gathered from the ocean floor. This research showed that the plates themselves are being recycled in zones of **construction** and **destruction**.

The Earth's present continents are what remain of a much larger super-continent that has gradually split up over hundreds of millions of years (see Figure 5.17).

Plate boundaries

There are four types of plate boundary:

type	description	examples
A Destructive	Plates meet, and the continental crust forces the thinner oceanic crust back down into the mantle	Much of the so-called 'ring of fire' that surrounds the Pacific Ocean
B Constructive	Plates move apart, and the mantle escapes upwards forming new oceanic crust	The mid-Atlantic ridge that runs north to south along this sea bed
C Conservative	Plates which slide sideways rather than lose or gain material	The California coastal area is a classic example
D Collision	Plates coming together and forcing both crusts upwards	The Himalayan zone which has some of the world's highest peaks

Figure 5.16 The four types of plate boundary

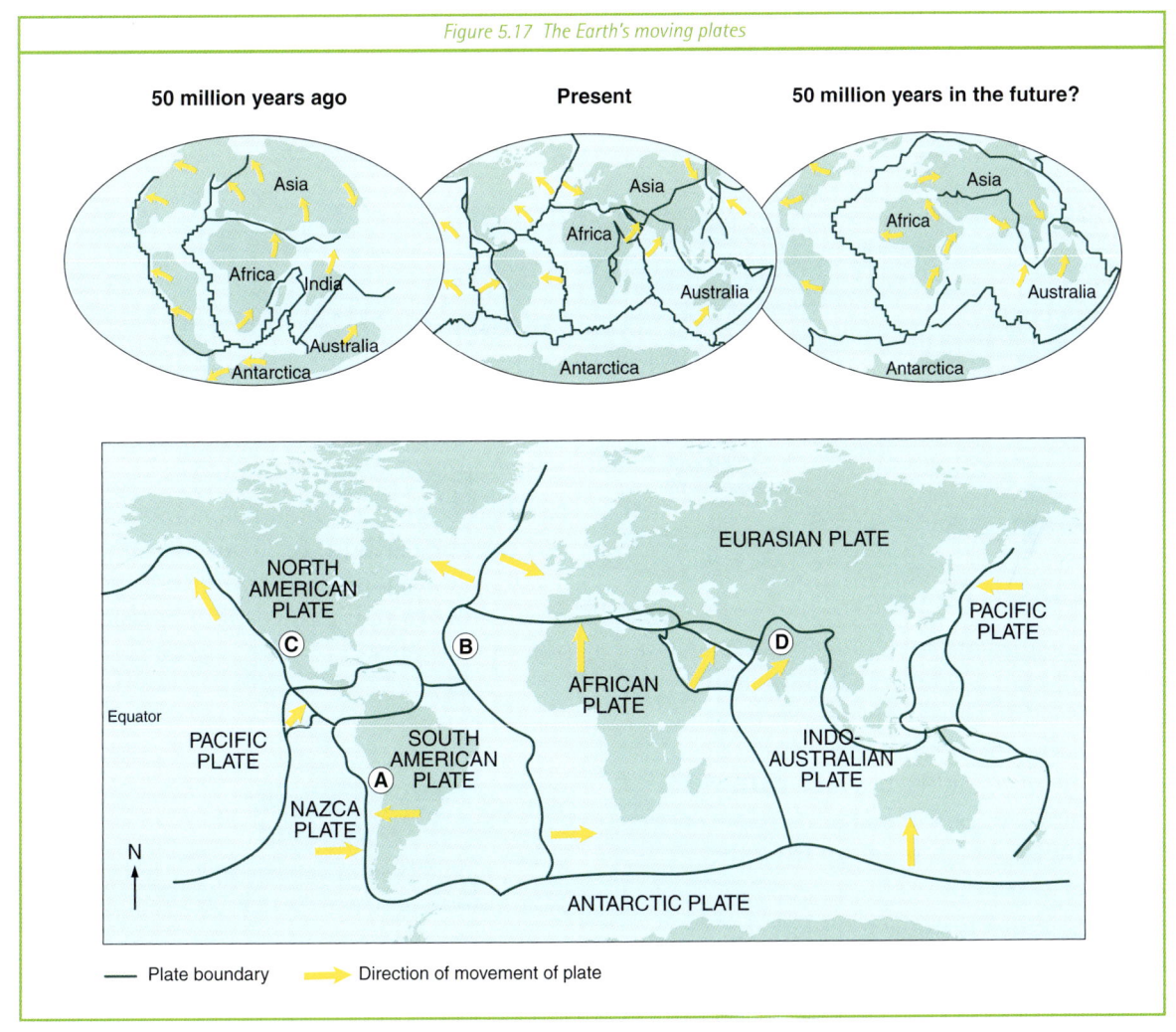

Figure 5.17 The Earth's moving plates

Section 2: Managing the Physical Environment

Destructive plate boundaries

This type of boundary occurs where plates move towards each other and oceanic crust is forced down under the thicker continental crust. This also pulls down the ocean floor creating deep **ocean trenches**, and buckles the nearby land into mountain ranges. Friction between the two plates can trigger sudden movements, which lead to earthquakes. Heat from the **subduction zone** below generates bubbles of magma, which escape upwards into faults and the vents of explosive volcanoes.

The plate boundary, which runs along the western coastline of South America, parallel with the Andes Mountains, illustrates all of these features. Montserrat and other Caribbean Islands have a similar origin but have formed offshore.

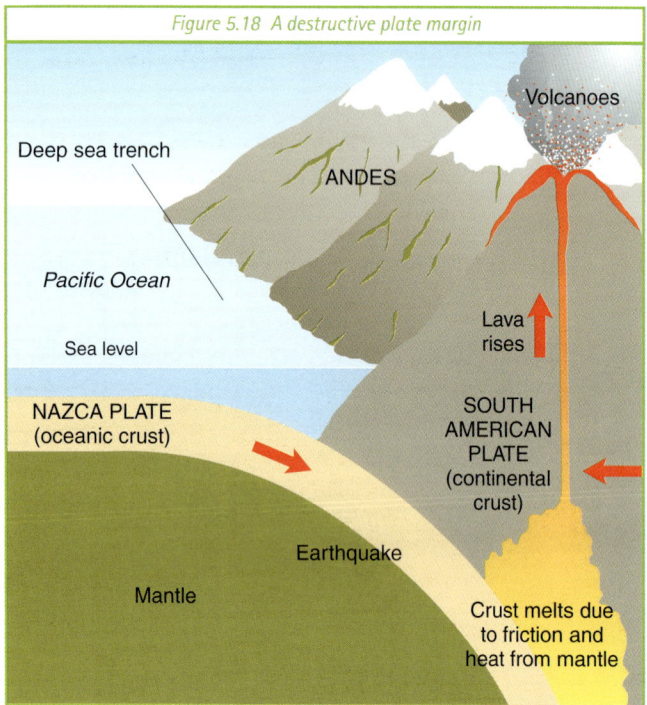

Figure 5.18 A destructive plate margin

Constructive plate boundaries

These plate boundaries are found on the ocean floor. Because the crust here is relatively thin, magma escapes upwards easily and forms new sections of sea floor. The plates are moving apart, like a giant conveyor belt, allowing faults and **rift valleys** to form. Within these ridges and faults, there are some very large volcanoes, which reach up from the seabed to appear as islands.

The Mid-Atlantic Ridge is the world's largest example showing these features. It includes volcanic islands like Iceland and Ascension Island.

Conservative plate boundaries

Conservative plate boundaries do not produce or recycle much material. Instead they slide sideways past each other. Scientists suggest that as two plates move past each other, locally areas of crust may temporarily 'stick' together. When these areas do finally move again, the sudden release of pressure causes even larger earthquakes.

In California, along the San Andreas Fault this situation seems about to occur again, just as it did in 1906.

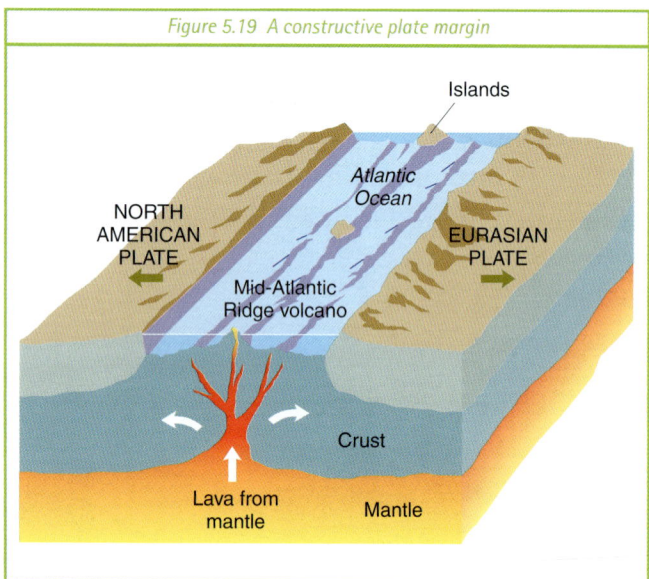

Figure 5.19 A constructive plate margin

Collision plate boundaries

Collision plate boundaries represent a sort of stalemate, where the edges of two continental plates are both forced upwards creating high mountain ranges and triggering earthquakes.

This is the process which is making the Himalayas and causing earthquakes from the Mediterranean through the Middle East and into India. There is even evidence that this is still increasing the height of Mount Everest.

ACTIVITIES

1 Using the words in **bold type** from the previous page, explain how and why the continents have 'drifted' apart. (C1.3)

2 Summarise the four types of plate margin. (C1.3)

Managing earthquakes and volcanoes

Coping with the power of earthquakes and volcanoes is not easy, but for a variety of reasons people cannot, or do not wish to, abandon areas where they occur (see Figure 5.3, page 34).

Can earthquakes and volcanoes be predicted?

Predicting the arrival of an earthquake accurately is not yet possible. Whilst plate tectonics allows scientists to say roughly where they might happen, it seems that we are quite unable to say when.

- Studies in California along the San Andreas Fault do suggest that it is possible to map areas where there has been little or no activity for some time; these 'seismic gaps' may be next to move.
- In California detailed **hazard maps** have been drawn up which estimate the risk of earthquakes occurring (see Figure 5.20). Building regulations, emergency plans and even insurance premiums use this information.
- Japanese scientists have also detected changes in the level of water in the sea, lakes and wells prior to earthquakes.
- Smaller tremors or foreshocks also often lead to larger quakes.
- Changes in animal behaviour have long been seen as indicators of disasters in many civilisations.

Predicting volcanic eruptions has been more successful, notably in Hawaii. It should not be forgotten, however, that 75% of all eruptions in the last 200 years have been from volcanoes thought to be dormant or extinct.

- The shape of a volcano does begin to change prior to erupting as the pressures beneath build up. 'Tiltmeters' recorded this prior to the Mount St Helens eruption.
- Changes in gravity, seismic readings and the chemicals in volcanic gases and dust are used successfully on Hawaii.
- **Exposure maps** for volcanic risks, similar to the hazard maps, are now in place on some Caribbean Islands (see Figure 5.21).
- Satellite heat images may help in future.

Figure 5.20 Earthquake hazard risks for San Francisco

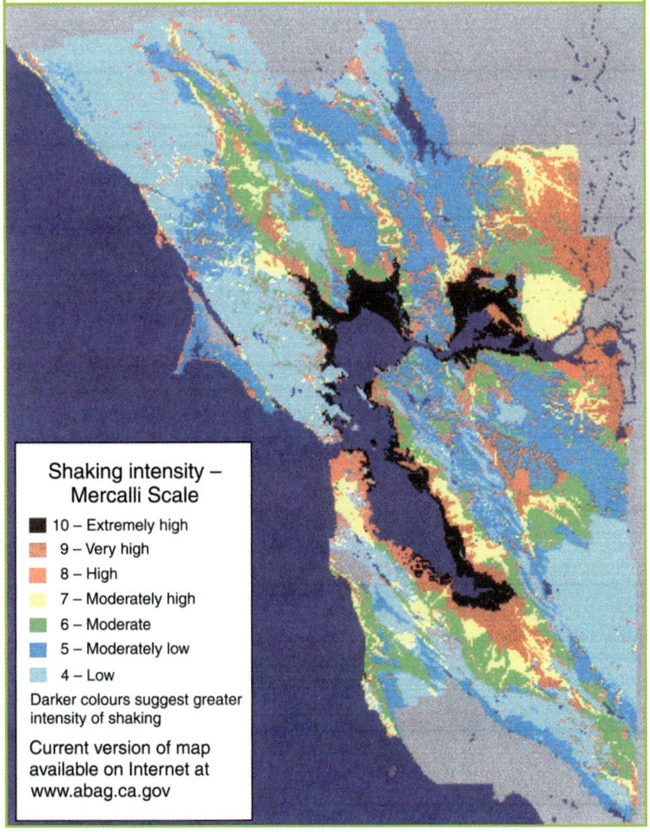

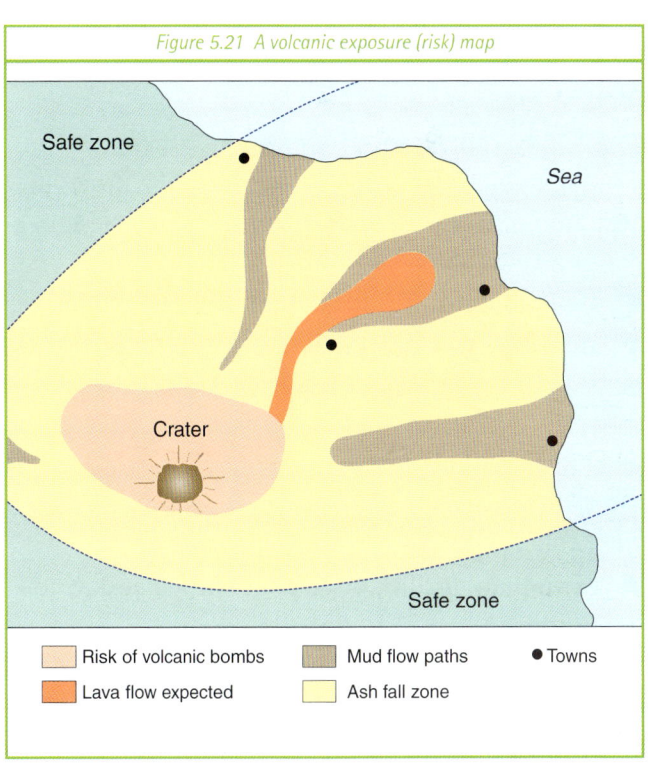

Figure 5.21 A volcanic exposure (risk) map

Section 2: Managing the Physical Environment

Planning for earthquakes

There are many ways to prepare for earthquakes: from trying to predict events, to organising emergency services, to long term planning.

In More Economically Developed Countries (MEDCs) technology and finance do allow elaborate precautions to be taken. Three areas in which much progress has been made are: building design, transport structures and emergency planning.

1. Building design

The success of building design is in its structure (see Figure 5.22) rather than its sheer strength, especially in major cities where many of buildings are 'skyscrapers'. Key points for designers to consider are:

- securely connecting walls, roofs and floors
- using timber or flexible materials to absorb the shock
- having deep and firm foundations
- 'X' structures to prevent twisting
- 'pyramid' design (Transamerica building in San Francisco)
- protection from dangerous materials like glass
- roof counter-weights to dampen shocks
- surrounding open spaces for emergency access, assembly and evacuation.
- use of canals to hold water for fire fighting
- automatic shut-offs for gas supplies

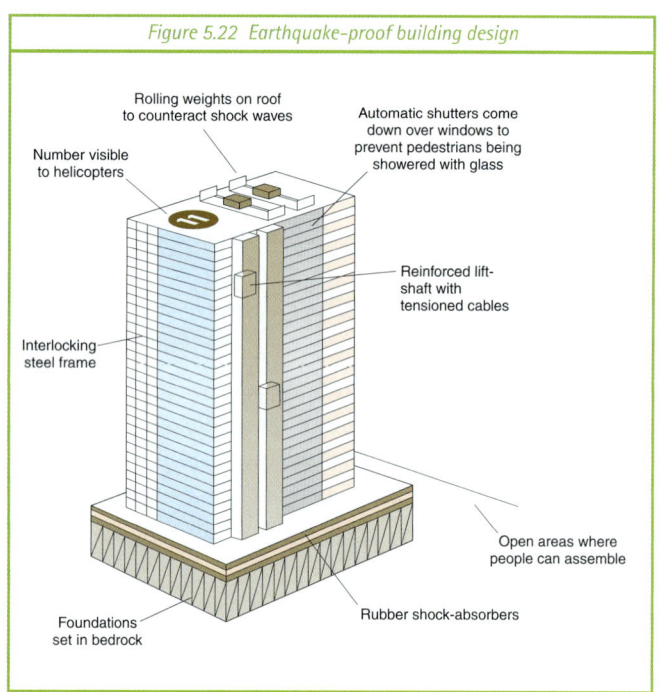

Figure 5.22 Earthquake-proof building design

2. Bridges, motorways and railway tracks

These are especially important as urban roads carry vast amounts of people during rush hours. Design features (see Figure 5.23) include:

- cable supports linking girders and columns
- steel coils to bind concrete supports
- concrete infils between motorway piers

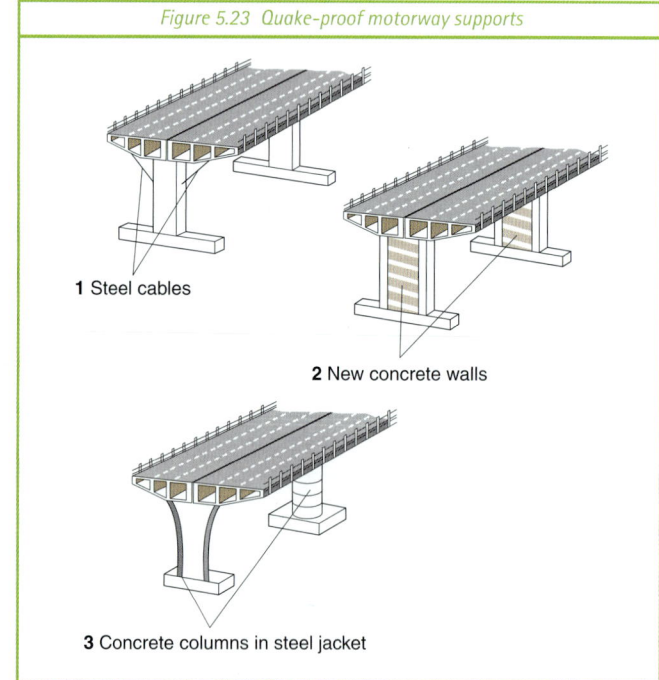

Figure 5.23 Quake-proof motorway supports

3. Emergency planning

These services are always stretched in disasters but their role is vital if lives are to be saved.

- earthquake 'drills', often on a particular day, to train people how to respond
- full range of emergency services ready
- evacuation procedures
- broadcasting information

ACTIVITIES

1. List ways to predict tectonic hazards. (C1.3)
2. List ways to make large structures earthquake-proof. (C2.2)
3. How many of the items in these two lists could be used in an LEDC? (C2.2)

Coping with disasters in LEDCs

It has been suggested that:

'there is no such thing as a natural disaster – only disasters made more or less serious by how much money countries are able to spend.'

In Less Economically Developed Countries (LEDCs) the effects of an earthquake are usually far worse than in richer countries, and the management options they can use are more limited. This is because there are:

- limited supplies of medicines and hospital facilities
- poor roads and links to remote places
- few if any emergency services
- few buildings that can withstand shocks
- shortages of food and water
- insufficient trained workers
- no funds to buy food, shelter and clothing

Foreign aid

For many such countries the only realistic answer is to ask for foreign aid or to rely on the work of international agencies like the Red Cross and charities like DEC (Disaster Emergency Committee) (see Figure 5.24).

Recent appeals of this sort bring home the awful message of what disaster really means. Two of the most recent examples seen in 2002 are the volcanic eruption at Goma (Democratic Republic of Congo) and the earthquake in Nahrin Province (Afghanistan).

- **Goma** – Mount Nyiragongo erupted on January 17, 2002, sending lava flows into the city of Goma. Homes and businesses were destroyed and 250 000 people fled into neighbouring Rwanda. Goma will be remembered as the place to which some 2 million refugees came in 1996, to avoid the civil war and genocide in Rwanda. DEC and Christian Aid launched an appeal to bring water, shelter and food to those forced to leave their homes in Goma.
- **Nahrin** – Up to 1 500 were feared dead following an earthquake measuring 6.2 on the Richter Scale which shook this part of Afghanistan on March 25, 2002. Some 150 km North of Kabul this was the recent scene of fighting between US and Taliban forces. Oxfam launched an immediate appeal for blankets and clothing, and the Afghan Red Crescent Society moved in emergency medical teams. The immediate need was for food, shelter and medical personnel.

Figure 5.24 An appeal for funds by DEC

THE INDEPENDENT

EARTHQUAKE APPEAL

Organised by

Disasters Emergency Committee

representing 15 major charities
Registered Charity No. 1062638

If you would like to help, please send a donation to:

**DEC India Earthquake Appeal
PO BOX 2710
IND3101
London W1A 5AD**

or telephone **0870 60 90 900**
http://www.dec.org.uk

ACTIVITIES

1. Why is it especially difficult to cope with disasters in LEDCs? (C2.2)
2. Why are Goma and Nahrin not entirely natural disasters? (C2.2)
3. Which of the DEC partners (see Figure 5.25) have you heard of before? (C2.2)

Figure 5.25 Two of the charities working for DEC overseas

Section 2: Managing the Physical Environment

Further tasks

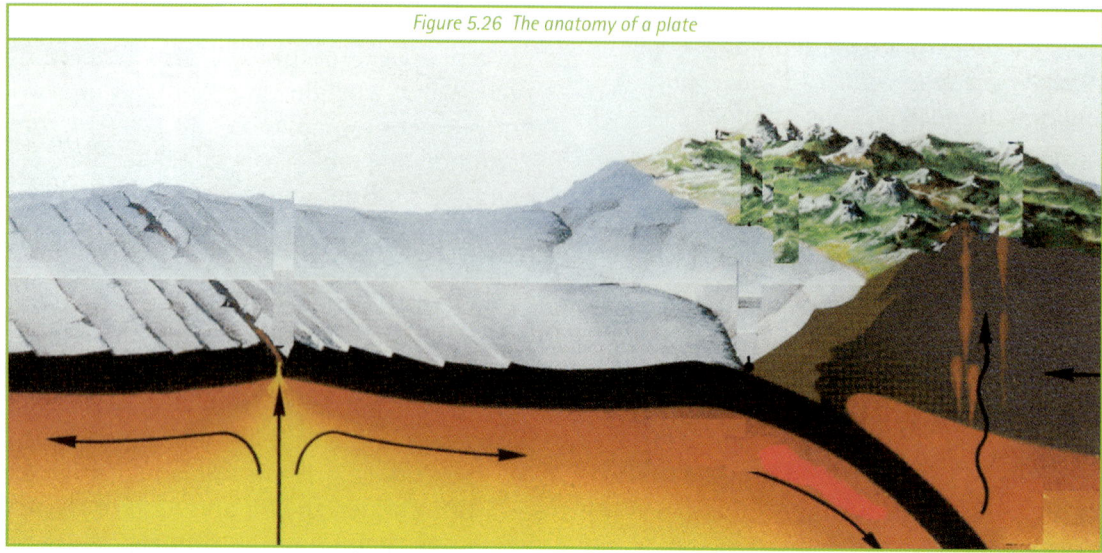

Figure 5.26 The anatomy of a plate

ACTIVITIES

1. Copy Figure 5.26 and label the diagram as carefully as you can. (C2.3)
2. Make notes beneath describing how the two plate boundaries re-cycle the Earth's crust. (C2.3)
3. Make a table to show the advantages and disadvantages of living close to a volcano. (C2.3)
4. Write a brief report (50 words) explaining how 'We can predict where an earthquake may occur but not when'. (C2.3)

Research

1. Research a case study of a volcanic eruption – either building on one looked at in this chapter or on a new one. For instance, there are lots of details about Mount St Helens in other books. (IT1.1)
2. Research an earthquake event. There are lots of facts and photographs printed about Kobe (see references list). (IT1.1)
3. Research and draw up a diagram of the structure of the Earth, showing the core, mantle and crust, and labelling the processes at work. (IT1.1)
4. Research the work of aid agencies like DEC (Disaster Emergency Committee), Oxfam and Christian Aid via the www.dec.org portal (IT1.1)

Decision-making

The most useful topics could centre on managing earthquakes, deciding which types of scheme might suit a particular situation or country. Two examples might be:
i How successful was Japan in coping with the Kobe earthquake? (PS, Wo)
ii What should be done for the people of Montserrat? (PS, Wo)

References

Most newspapers have a website that can get you to recent or even real-time disaster news. Just type in the name of the paper and the place or topic you want to investigate, e.g.: www.guardian.co.uk/natural disasters/

'Kobe – the anatomy of an earthquake', *Geographical Magazine*, April 1995
'The Kobe earthquake', *Wide World GCSE Geography Review*, September 1995
www.reliefweb.int
www.news.bbc.co.uk
www.city.kobe.jp
www.volcan.und.nodac.edu
www.cdera.gov
www.pubs.usgs.gov
www.neic.usgs.gov
www.earthquakes.usgs.gov.4kids/4grownups

Exam-style questions and sample answers for Chapter 5

Question 1 – F Tier

(a) What is an active volcano?

A volcano which is still erupting regularly **(1)**

(This is accepted as correct, though 'likely to erupt again' is a better answer.)

(b) Explain three reasons why people live close to active volcanoes. **(6)**

1. *Heat from hot springs can be used to heat up glasshouses, which will allow crops to be grown early.*
2. *The steep sides of a volcano could make ski runs for tourists bringing in money in the winter.*
3. *Volcanoes make good soils.*

(Two good responses here, but the last part does not explain that volcanic rocks can break down to become fertile soils for farming. This answer is worth 5 out of 6 marks.)

Using Figure 1, the diagram of a plate boundary:

(c) Name

1. the type of plate boundary shown

   _____*Destructive*_____ plate boundary.

2. the zone marked at A

   _____*Subduction*_____ zone. **(2)**

   Both are correct

(d) Explain why volcanoes and earthquakes both form at this location. **(4)**

Friction between the two plates as the ocean plate goes under the continent, causes earthquakes if the land slips suddenly. The great heat from the friction and from the mantle below forces the magma to rise. This escapes through faults and vents to form volcanoes. This happens off Peru and Chile in South America.

(A sound answer that explains why both earthquakes and volcanoes occur here. There is even an example given, so earning 4 marks out of 4.

This candidate is gaining high marks with 12 out of 13 overall.)

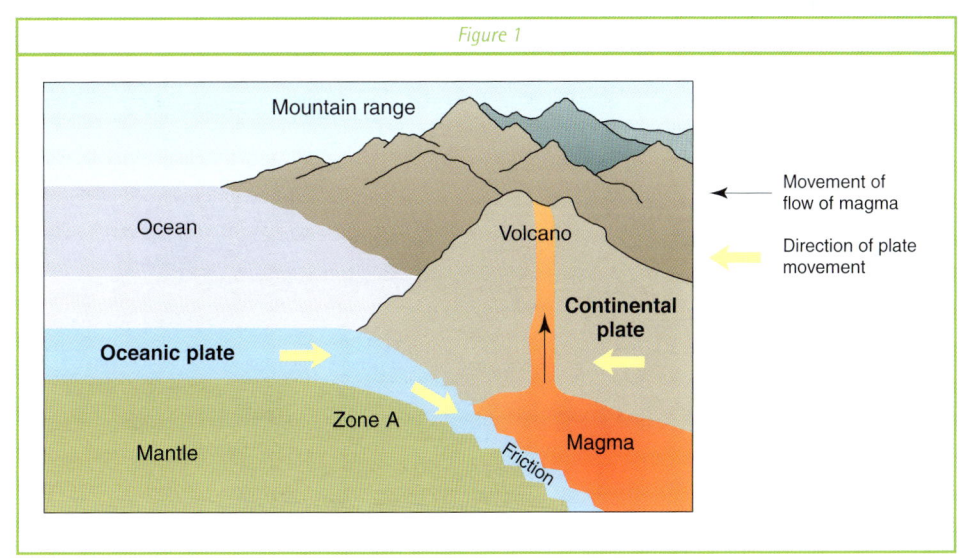

Figure 1

Chapter 5: Exam-style questions

Question 2 – H Tier

(a) Using the map below (Figure 2), outline some of the effects which this earthquake damage may have had on people's lives. **(4)**

Fires, buildings and roadways collapsed. The BART tunnel under San Francisco Bay began to leak and the baseball stadium had to be evacuated.

(This answer only got 2 of the 4 marks possible. The reason is that the question provided information about the damage and all that the candidate did was to 'lift this off' the map. To get 4 marks the effects on people needed to be explored – e.g. Damage to the transport system meant that people could not get to work and businesses would be disrupted. Residents may have needed re-housing and insurance claims dealt with.)

(b) Using examples you have studied, explain how people are trying to reduce the harmful effects of earthquake hazards. **(6)**

It is a good idea to be prepared. Have an emergency pack and a radio. Hide under tables and turn off gas and electricity. Many people now work in offices with earthquake-proof structures that can sway in earthquakes, not crumble. Emergency services need to be good.

(This response is worth 3 out of 6 marks. It has a number of correct ideas and some knowledge of how to prepare for an earthquake. The answer has little structure and there are no examples given. It also lacks details, for instance, about building structures like the 'X' frames and flexible foundations in Tokyo's tall buildings, or the cable fastenings and spiral cores in San Francisco's motorway support legs. Earthquake 'drills' and first aid training for school children in Colombia might also have been mentioned.

Overall this candidate gained a mark of only 5 out of 10, and missed opportunities to do better.)

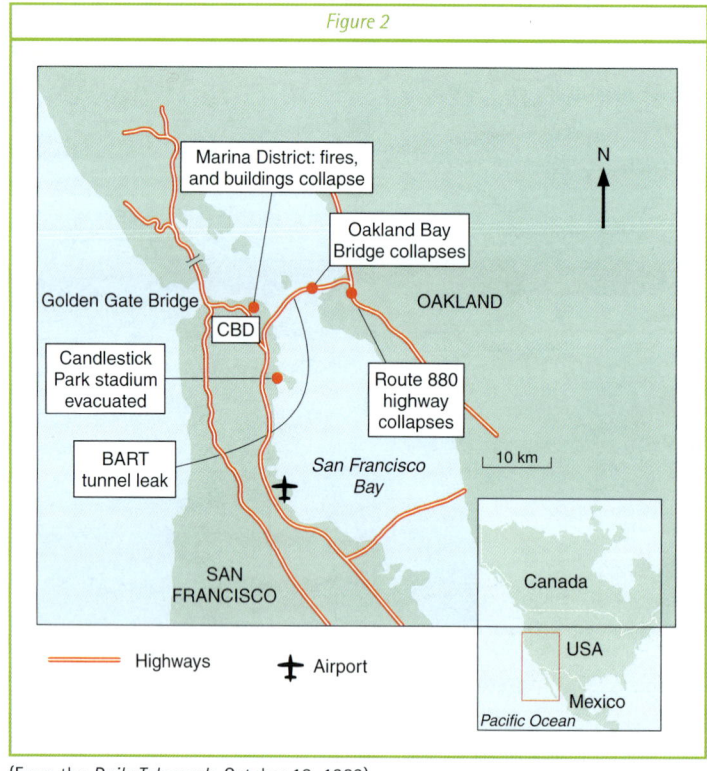

Figure 2

(From the *Daily Telegraph*, October 19, 1989)

Weather hazards   Chapter 6

Violent storms and floods are amongst the most terrifying weather hazards we face. Tropical storms, depressions and river floods threaten people in many parts of the world. How can people cope with these large-scale natural disasters, especially those who live in the less economically developed countries?

Issues surrounding weather hazards

Storms and floods

- The worst single flood event in the UK started on August 15, 1952, in the small Devon holiday resort of Lynmouth. A **depression** and its thunderstorms caused over 300 mm of rain to fall in 24 hours, on an already very wet Exmoor. When it was all over, 35 people had died and 1 000 others were left homeless.
- In October 1998, Hurricane Mitch became the fourth strongest **hurricane** of all time, when it battered much of Central America. Wind speeds of 180 mph and rainfall of over 600 mm in 36 hours brought widespread damage. In November, Honduras reported 7 000 people dead and 8 000 still missing.
- In April 1991 a **tropical cyclone** hit the coast of Bangladesh. The 140 mph winds and 7 m tidal wave swept across the delta. The death toll of 150 000 resulted from drownings and the disease and food shortages that followed.

Not so natural disasters

Environmental groups argue that many recent floods are due to human not natural causes. Here are some questions they ask:

- **Why do we cut down so many trees?**
  – this speeds up surface run-off, washes soil away and increases the chance of rivers flooding.
- **Why do we build more and more cities?**
  – so many concrete surfaces and storm drains bring rainwater into rivers too quickly.
- **Why do we live on river floodplains?**
  – building more houses there just increases the risk for people.

Figure 6.1 Flooding in Bangladesh

Figure 6.2 Hurricane Mitch, one of the strongest hurricanes in recorded history, left a trail of devastation across Central America

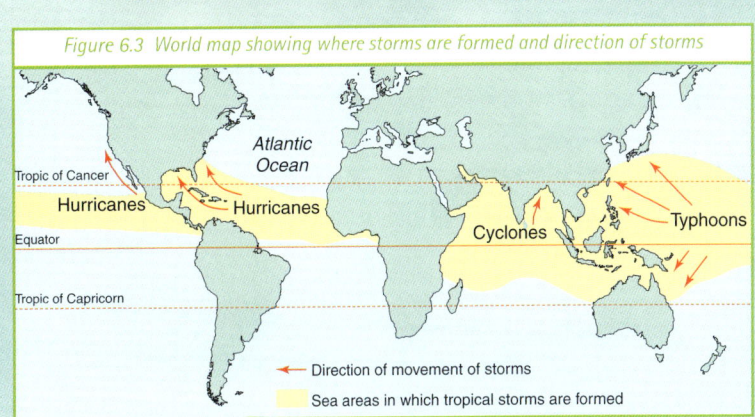

Figure 6.3 World map showing where storms are formed and direction of storms

Section 2: Managing the Physical Environment

Hurricanes

A **hurricane** is perhaps the most well known type of tropical storm. With continuous wind speeds of 120 kmph or greater, hurricanes are most common in the Atlantic Ocean and Caribbean Sea in late summer and autumn. They are named, over a six-year period, using an internationally agreed list of male and female names in alphabetical order. Disastrous names are not used again.

In other parts of the tropics such storms are called typhoons or cyclones (see Figure 6.3).

Hurricanes develop from areas of low atmospheric pressure called **depressions**. Winds blowing over the warm sea (over 27°C) push moist air upwards, developing the depressions into tropical storms, and eventually into hurricanes. Over the sea, wind speeds may increase to over 250 kmph (a category 5 hurricane), but over land hurricanes tend to weaken and decay. Hurricanes are unpredictable and can suddenly change their overall speed or direction (or 'track').

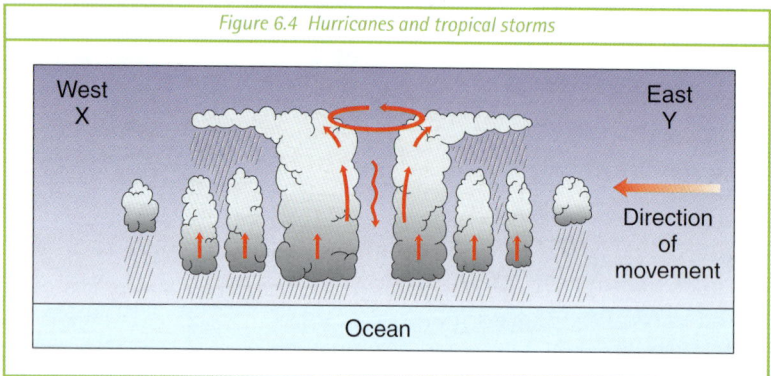

Figure 6.4 Hurricanes and tropical storms

What are hurricanes like?

Research using satellite photography (see Figure 6.5) and aircraft shows that all hurricanes are distinctive, but many do have typical features:

- a strong spiral of winds spinning anticlockwise (north of the equator)
- though varied in size, they are on average about 500 km across
- a calm and clear 'eye' at the centre, often between 30 and 60 km across
- a near vertical 'eye wall' of dense clouds and very strong winds
- a surrounding area of intense rainstorms and strong winds
- the front right area of an Atlantic hurricane often brings storm surges and tornadoes.

Figure 6.5 Typical Features of a hurricane

ACTIVITIES

1 Describe how a hurricane develops. (C 1.1)
2 Enlarge Figure 6.5 putting labels on it taken from the list of typical features given in the text. (C1.3)

Hurricane Mitch

Hurricane Mitch illustrates how a hurricane develops and the effects it has on countries and on people's lives. Obviously strong winds can damage and destroy homes and property, but wind strength is not always related to the amount of damage done (see Figure 6.7) or the number of deaths caused. Often the heavy rainfall, landslides and river floods cost most lives. Along the coast, **storm surges** bring massive tidal waves ashore (see Figure 6.5).

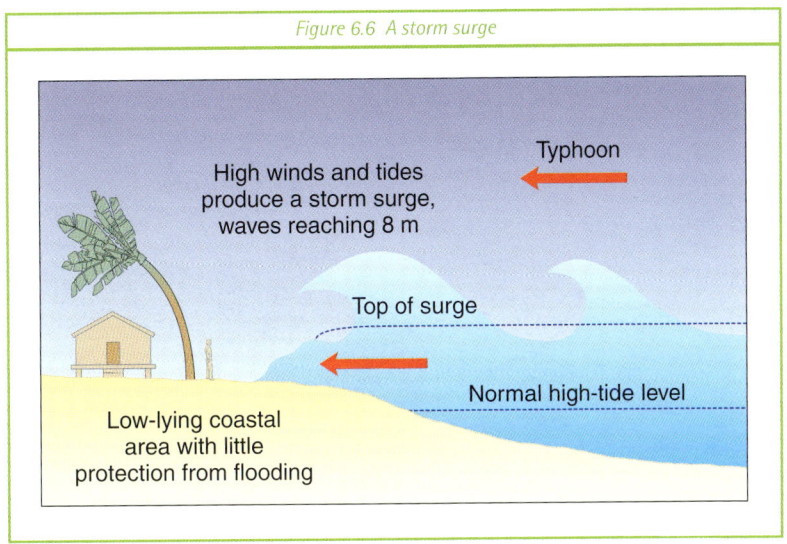

Figure 6.6 A storm surge

In LEDCs, shortages of food, clean water and shelter, medicines and emergency services add greatly to the problems. This was seen in September 1998 when Hurricane George hit the Caribbean Sea and the Gulf of Mexico with the same wind speeds. The difference between the effects in two countries was striking.

Figure 6.7 Comparing impacts in different places

country	deaths	other effects
Dominican Republic	500	• 100 000 people left homeless • 70% of the country's bridges destroyed • 90% of banana exports lost
USA	4	• 1 million people evacuated quickly and safely • most power supplies reconnected in a few hours • insurance and federal disaster funds made available

Figure 6.8 Damage along the Pan American Highway in Nicaragua

ACTIVITIES

1 Suggest why the effects of hurricanes vary from place to place. (We shall return to this idea later.) (C1.2)

Section 2: Managing the Physical Environment

Figure 6.9 The 'track' and effects of Hurricane Mitch

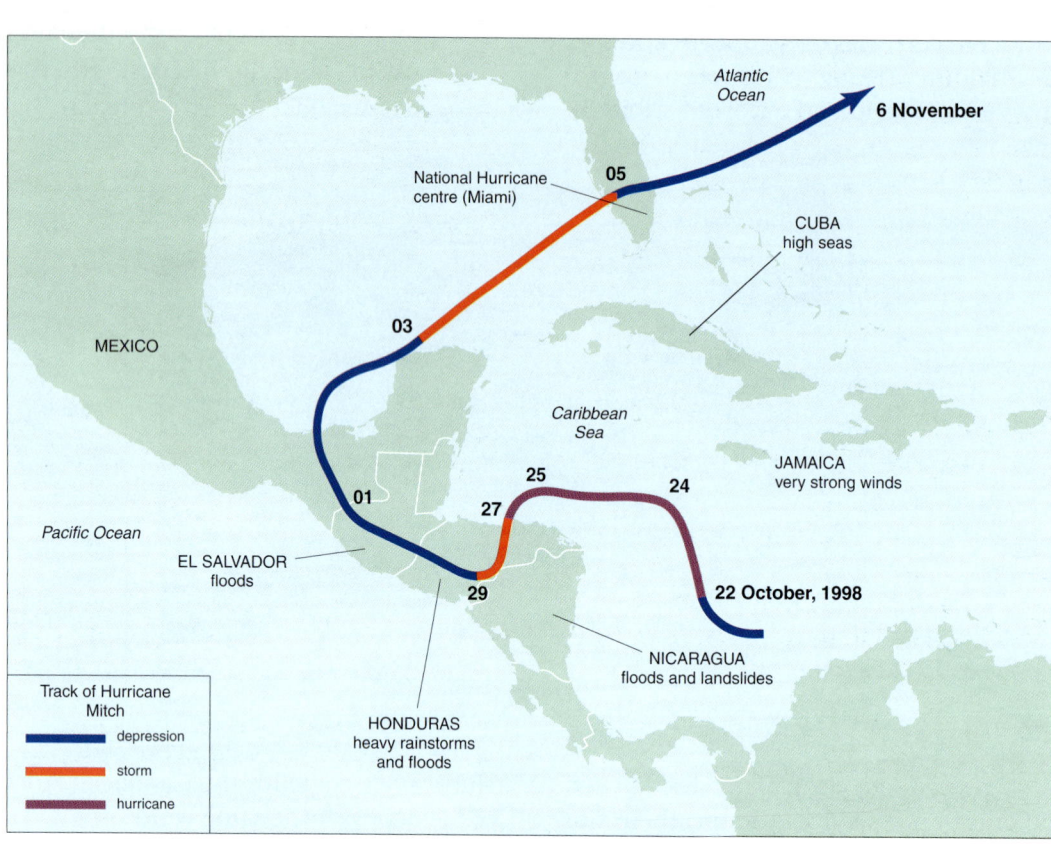

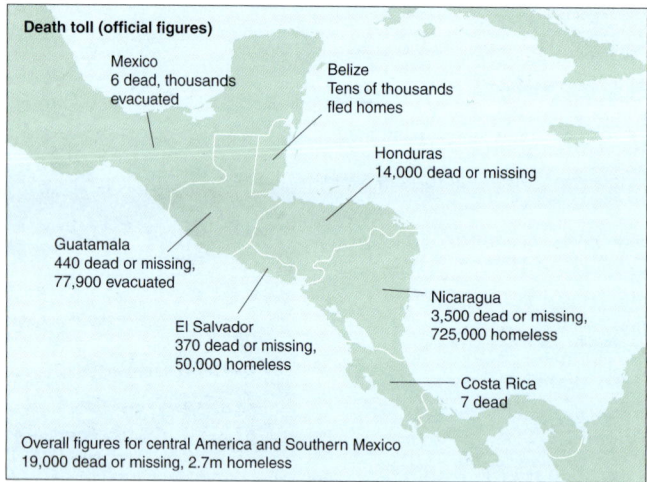

ACTIVITIES

1 Which country shown in Figure 6.9 do you think was worst affected? Explain why (C1.2)

2 Draw up a table that tells the story of Hurricane Mitch. The headings and first entry have been set out for you below. (C2.3)

dates	classification	track	weather effects	effects on people
October 22–24	Tropical storm	Northwards near Jamaica	Very strong winds	n/a

Depressions

Depressions are areas of low atmospheric pressure. As far as the UK and Europe are concerned these tend to arrive from the Atlantic, and involve the mixing of warm moist air from the tropics (tropical maritime air) and cooler air, from polar or arctic areas. The boundary between these different air masses is called a **front**.

Depressions are easy to recognise on weather maps shown in newspapers or TV forecasts. They appear as a series of rings – lines showing air pressure – called **isobars** (see Figure 6.10). It is important to check that the numbers fall towards the centre of the rings (LOW) and not confuse them with high pressure systems and their calmer and drier weather.

Though less violent than hurricanes or cyclones, depressions are similar, often bringing strong winds (as in England's Great Storm of October 1987) and heavy rainfall. This rainfall is caused by air being lifted up and cooled. The fall in temperature turns any water vapour back into water droplets (condensation) and leads to precipitation. In the winter of 2001, the continual arrival of these depressions caused very heavy rainfall and flooding throughout the UK.

Weather maps are quite specialised, but there is another simpler way to look at depressions, by using **satellite photographs**. These show clearly the pattern of clouds as they swirl around the centre of a UK depression in an anticlockwise direction (see Figure 6.11). Looking at a plan view of a depression, however, doesn't give you the whole picture. To understand the processes going on within a depression, you need to look at it in cross-section.

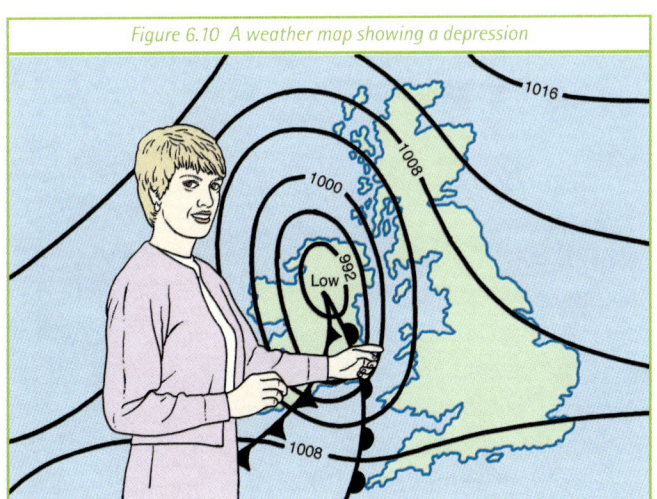

Figure 6.10 A weather map showing a depression

Figure 6.11 A satellite photograph of a depression

This allows you to see in more detail how the temperatures, winds, clouds and rainfall change.

The features of a depression

The cross-section of a depression (shown in Figure 6.12) shows two fronts, where different types of air meet. This depression is moving from left to right. The fronts are referred to as the **warm front** and the **cold front** as they signal the arrival of first the warmer tropical air, and then later the return of the cooler polar air from behind. Notice how the type of cloud, the temperatures, rainfall and winds change as the two fronts pass. As a depression weakens the faster-moving cold front may catch up with the warm front and lift all warm air clear of the ground. This situation produces a mixed or **occluded front**.

Synoptic charts are used by the Meteorological Office to show weather patterns and to help in forecasting what may happen next. Information from ground stations, aircraft, ships, and weather balloons is plotted on to these maps. A special series of symbols have traditionally been used to show this data (see Figure 6.13).

A much simpler, though less detailed, set of symbols is now being used in newspaper and television forecasts. Modern radar, satellite images and computer models are also able to give more accurate and even real time information.

Section 2: Managing the Physical Environment

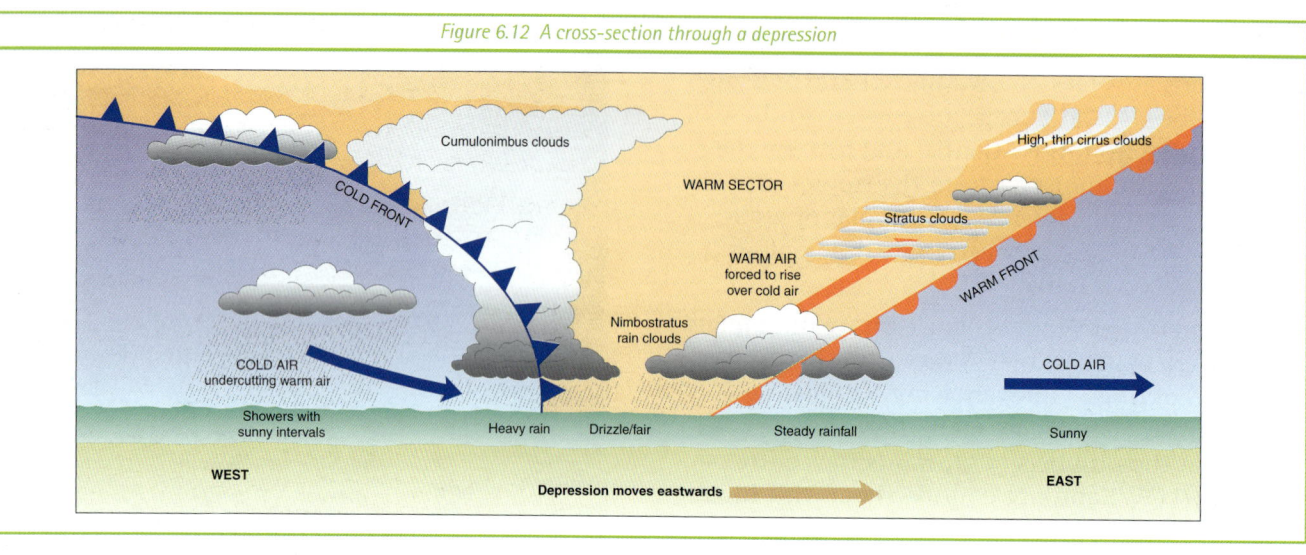

Figure 6.12 A cross-section through a depression

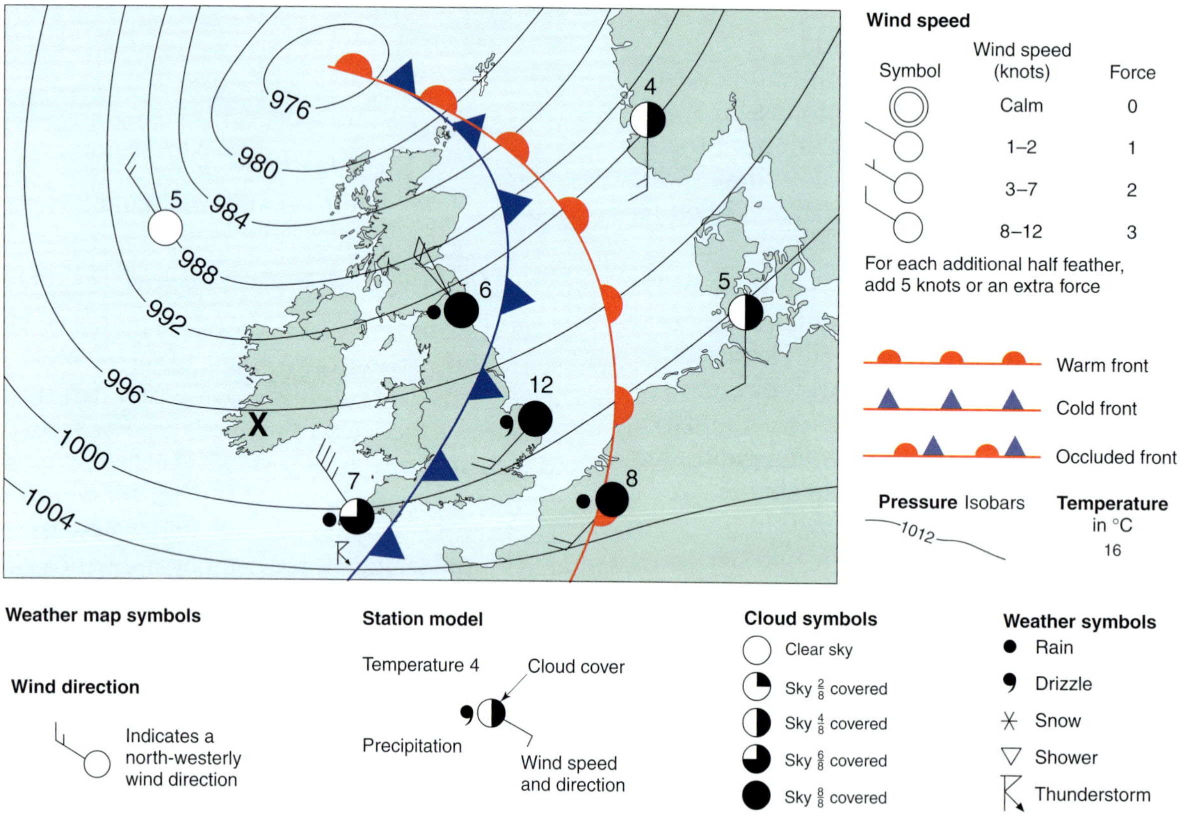

Figure 6.13 A synoptic chart and weather symbols

ACTIVITIES

1. List some similarities between depressions and hurricanes. (C1.3)

2. Describe how the weather changes as the depression shown in Figure 6.12 passes over (remember to work from right to left). (C1.2 and N1.1)

3. Use Figure 6.13 to carry out these tasks:
   - Describe the weather in France
   - Explain the weather in Cornwall.
   - Station X in SW Ireland is recording a temperature of 6° C, force 5 winds from the NW, rain and showers, and has 4/8ths cloud cover. Draw its map symbols. (C1.2 and N1.1)

Floods and their causes

Floods happen when the amount of water in a river increases, until it can no longer fit into the river channel, and spills out on to the floodplain.

It is important to understand the processes by which rainwater gets into rivers and what situations make floods likely (this forms part of the wider **hydrological cycle**, see Research page 62). The reasons why floods happen can be natural, but can also be caused or made worse by people.

Natural causes of flooding

- Heavy rain from storms is an obvious cause of flooding in rivers.
- A sudden thaw, especially in the spring, will cause snow to melt rapidly.
- Rain over a long period saturates the ground preventing the **infiltration** of further rain
- **Impermeable** rocks cause **surface run-off**.
- Little or no vegetation cover means there is little **interception** of rainwater.
- Tropical climates often have almost all of their rain in one season of the year.

Human causes of flooding

- **Deforestation** – the cutting down of trees for timber or to allow farming reduces interception, and water quickly runs off the land into rivers. This can also lead to soil erosion.
- Some types of agriculture can also increase the risk of flooding – overgrazing, growing only one crop, ploughing downhill, draining land or leaving the soil bare.
- **Urbanisation** – the continued growth of towns also creates more rapid run-off. Impermeable surfaces (roofs and streets) prevent infiltration, and storm drains carry water rapidly into the river.

Living or working close to rivers may not only cause more flooding, but it will also increase the risk of flooding for people and property. If rivers are not managed properly, schemes to prevent flooding upstream can cause floods downstream.

The impact of rainfall on rivers can be dramatic, increasing both the depth of water and the speed at which it flows. The amount of water, which passes down a river in a given time, is called the **discharge**. It is measured in cubic metres per second (cumecs).

Storm Hydrographs (see Figure 6.14) are used to show how river discharge changes when a storm occurs. A steeply rising graph and a high **peak flow**, show that water is arriving rapidly and suggest flooding may be likely. When this happens the delay between the highest rainfall and the highest discharge, called the **lag time**, is quite short. In wet weather the graph (and the river level) may take a long time to go down. Hydrographs are often drawn automatically using gauging stations and computers. They help the Environment Agency to monitor rivers, and give flood warnings to residents and businesses.

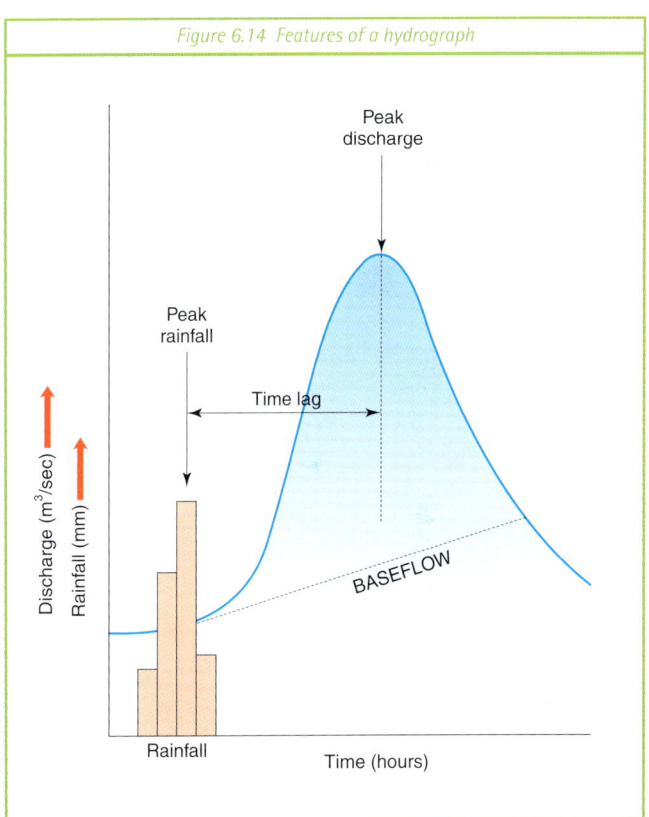

Figure 6.14 Features of a hydrograph

ACTIVITIES

1. List and explain all of the words on this page written in **bold type**. (N1.1) (C2.2)

Section 2: Managing the Physical Environment

Figure 6.15 Changing landscapes and flooding

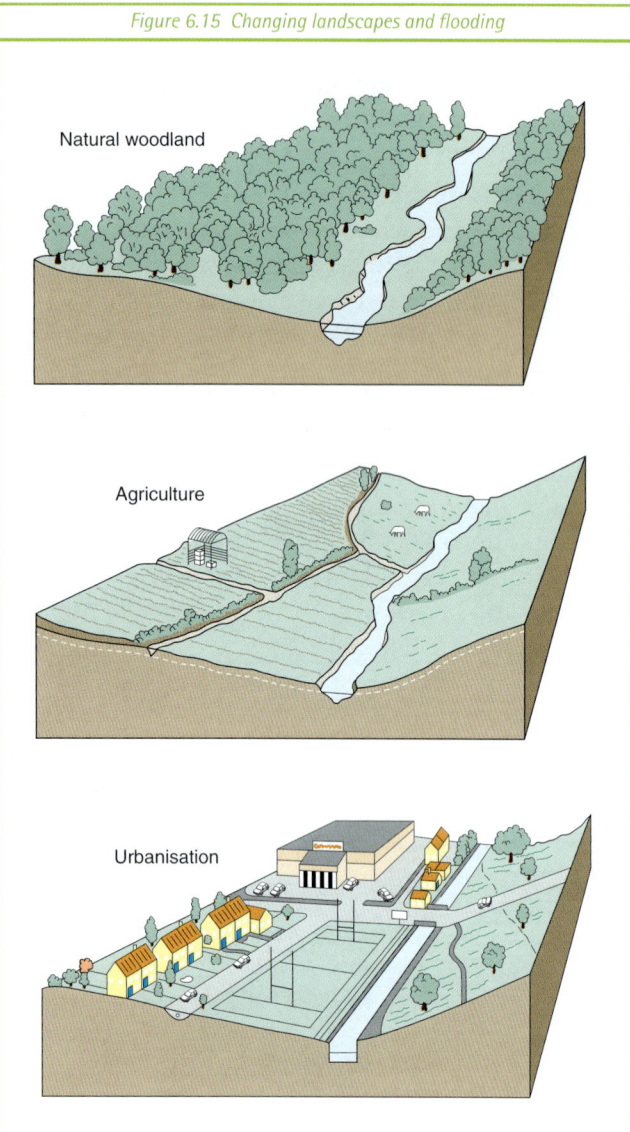

Figure 6.16 Changing hydrographs

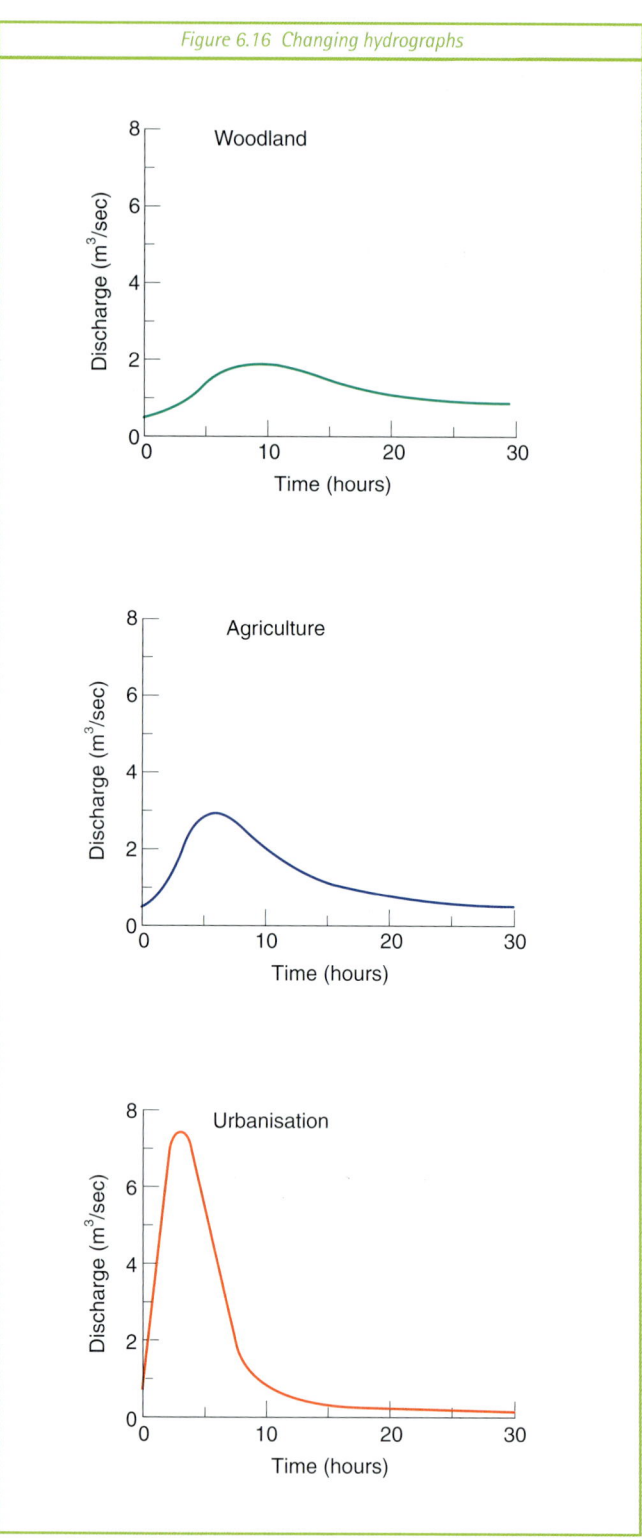

ACTIVITIES

1. The three hydrographs (Figure 6.16) match the sketches in Figure 6.15. Taking each one in turn, describe how and why they are different. (C2.2)

Flooding in Bangladesh

Bangladesh is a less economically developed country (LEDC) of over 100 million people. Most live on the floodplains and delta of three very large rivers – the Ganges, Brahmaputra and Meghna. Almost three quarters of the country is less than one metre above sea level. The monsoon climate of this region of South East Asia brings heavy seasonal rainfall and tropical cyclones. Figure 6.17 has been annotated to show some of the reasons why Bangladesh floods, as it did in 1988, 1992, 1998 and again in 1999.

ACTIVITIES

1. Using Figure 6.17, list the causes of flooding in Bangladesh, dividing them into (**a**) physical (natural) and (**b**) human (caused or made worse by people). (C1.1)
2. Give some examples of (**a**) immediate and (**b**) longer term effects, from Figure 6.18. (C1.1)
3. Why do you think so many people live in this region despite the risk of flooding? (C1.1)
4. Suggest reasons why Bangladesh feels that it not able to control its own future. (C1.1)

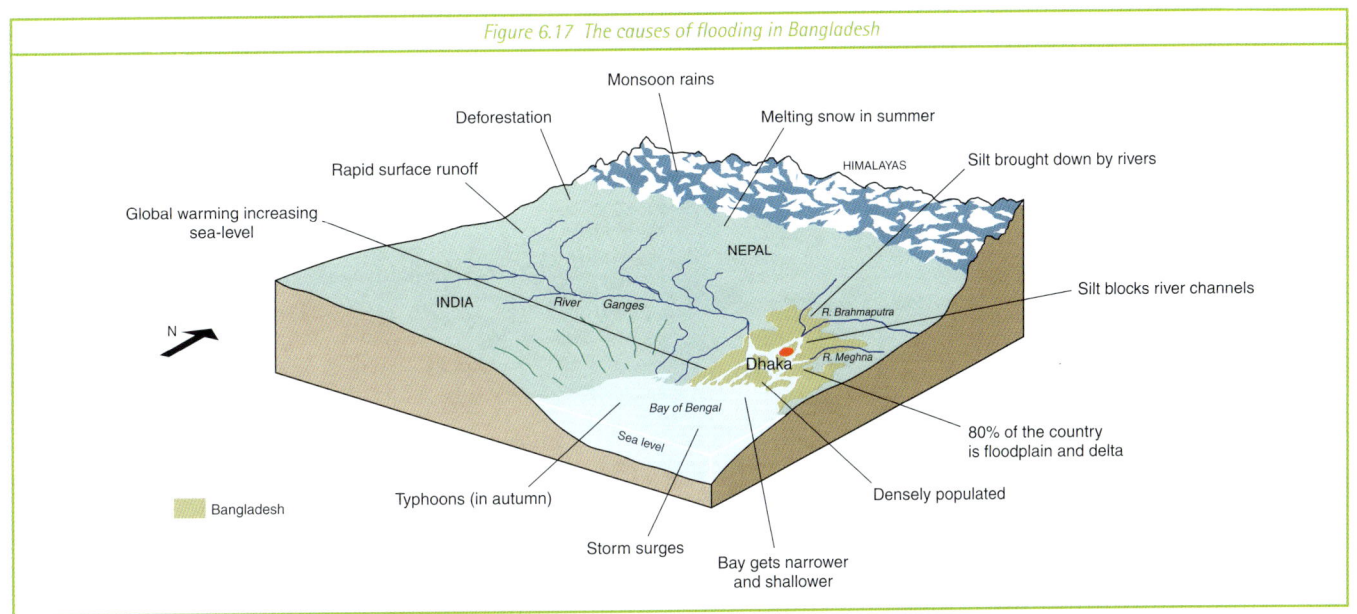

Figure 6.17 The causes of flooding in Bangladesh

We often refer to the primary and secondary effects of flooding:

- **Primary** effects happen quickly and include flood damage to houses, roads and crops, and people drowning.
- **Secondary** effects happen later and include homelessness, poverty, lack of food and illness from dirty water (cholera and typhoid).

Why do people take these risks?

Despite the disastrous effects of flooding, many people especially in LEDCs like Bangladesh still live or work in areas where floods occur every year. The reasons include:

- river silt creates fertile farmland
- the land is flat for building
- there is nowhere else to go

Figure 6.18 The effects of 65 days' flooding in 1998

- 1 million km² of land flooded
- over 30 million people affected
- nearly 1 million houses damaged
- over 1000 dead
- over 25 000 cattle lost
- over 2 million tonnes of rice lost
- over 15 000 km of roads destroyed
- nearly 5000 km of river embankments damaged
- over 5000 factories flooded
- nearly 15 000 schools flooded

Section 2: Managing the Physical Environment

Flooding along the Yorkshire Ouse

The river Ouse, and its many tributaries, has a history of winter flooding that affects businesses, homes and farmland, from York to the Humber Estuary. Records show that there were serious floods in York as long ago as 1263, though in more recent times, 1978, 1982, 1991 and 2000 were particularly damaging. Selby downstream of York, and Malton on the River Derwent are other nearby victims of flooding.

A variety of physical and human factors lie behind this flooding (see Figure 6.20).

Figure 6.19 Flooding in York, Winter 2000

Figure 6.20 Causes of flooding around York

Some of the causes of York's floods

Above York:
- Winter depressions bring rain
- Melting snow in Spring
- Steep hills upstream
- Flat open floodplains
- Trees cut down
- Land drained by farmers

Around York:
- More concrete and buildings
- Several rivers joining up
- Low bridges over the river
- Intensive farming of land

Below York:
- Land almost down to sea level
- Lots of river meanders
- High spring tides from estuary

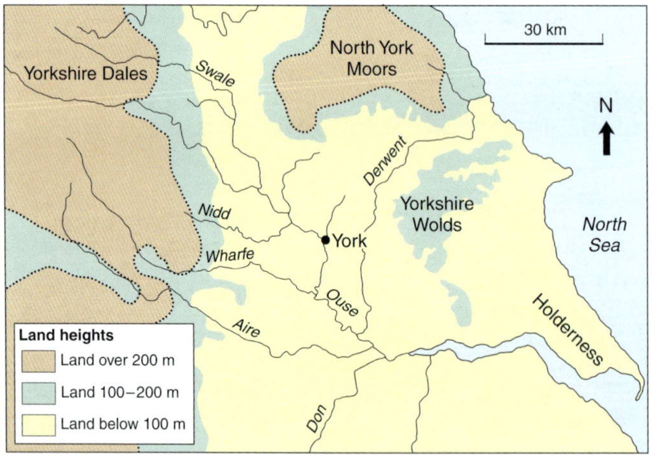

ACTIVITIES

1. Using Figure 6.20, re-write the list of causes of flooding around York again, dividing them into physical (natural) and human (caused or made worse by people). (C1.3)

2. Which causes seem to be different from those in Bangladesh? (C1.3)

Weather watch

It seems unlikely that we will ever be able to prevent storms or even floods happening, but accurate weather forecasting, modern building technology and proper planning do allow us to cope better with their effects.

Miami Beach in Florida is hit by a hurricane, on average, every four years. Strong winds cause widespread damage, and the heavy rainfall and massive waves flood coastal areas. Yet despite this threat, the warm climate and sandy beaches attract almost 40 million visitors each year and over half of the population has retired there.

Protecting Miami Beach is not easy because:

- most people don't take the threat seriously
- 5 narrow bridges link it to the mainland
- the protective beach has been eroded
- most land is less than 3.5m above sea level

Some measures however are in place:

- a National Hurricane Warning Centre
- beach nourishment (new supplies of sand)
- new buildings are built 6 m above sea level
- evacuation plans, escape routes and 50 hurricane-proof shelters

The **National Hurricane Warning Centre** based in Miami uses its NOAA (National Oceanic and Atmospheric Administration) satellites to track tropical storms in the Atlantic. From June to November the Hurricane Advisory Service monitors their growth and warns other countries around the Caribbean about the arrival of potential hurricanes.

The **Federal Emergency Managment Agency** (FEMA) advises people and local councils about how best to cope in an emergency. They produce leaflets and codes that help save lives in storms and floods:

- If you evacuate your home, pack food and water, cooking utensils, toiletries, medicines, blankets, important papers, valuables, a rope, flashlight and radio, etc.
- Only stay if your home has been properly protected from wind damage (see Figure 6.21) and flooding.

The **Meteorological Office** provides weather information and warnings for weather hazards in the UK. Severe risk of fog and frost, as well as gales, heavy rain and snow are seen in TV forecasts.

The **Environment Agency** carries out flood protection, monitors river levels and warns of potential floods – which are then classed as being red (very serious) or amber alerts. Information about an area's risk of flooding is shown on maps, which can be viewed on the agency's Internet site.

ACTIVITIES

1. Why do people in Miami risk living and working in Florida's hurricane zone? (C1.3)
2. Name four government agencies that watch out for storm and flood problems. What do they do? (C1.3)

Figure 6.21 A typical FEMA publication

Preparing your home for a hurricane

If you live along the Atlantic Ocean or Gulf of Mexico coasts you should be prepared for hurricanes before they threaten. With common materials, you can easily protect your home from hurricane-force winds.

1. If your home doesn't have hurricane shutters, cover your windows with ½-inch-thick marine plywood.
2. Drill holes for screws 18 inches apart.
3. Remove outdoor antennas.
4. Bring in lawn furniture, outdoor cooking equipment, toys and garden tools that could become missiles during hurricane-force winds.
5. Store drinking water indoors in clean bathtubs or in jugs and bottles.

Source: Federal Emergency Management Agency   Adapted from Bob Laird, USA TODAY

Section 2: Managing the Physical Environment

Flood Watch

Trying to manage flooding (see Figure 6.21) can involve two different approaches to the problem:

Hard engineering means building dams and cutting new channels to try to prevent flooding. Engineers on the Mississippi and Colorado rivers in the US have used this approach, though today, China's Three Gorges Dam is one of the few large-scale schemes under construction. Hydro Electric Power, cheaper water supplies and industry will help pay the huge costs involved. Reservoirs use up valuable land, damage ecosystems or trap silt behind the dams. New channels and embankments allow the river to carry more water, but there are risks here too as embankments can fail (see Research on page 62).

Soft engineering reduces flood risks by keeping many of the natural features of rivers. Planting large numbers of trees, for example, will intercept the rainfall and delay the arrival of water in the river. Where it is too expensive to prevent flooding, it is best not to use the land for building. River floodplains are better used as wetland, livestock farms or recreational areas. This is seen as being **sustainable**, meaning that, hopefully, it will not lead to problems in the future.

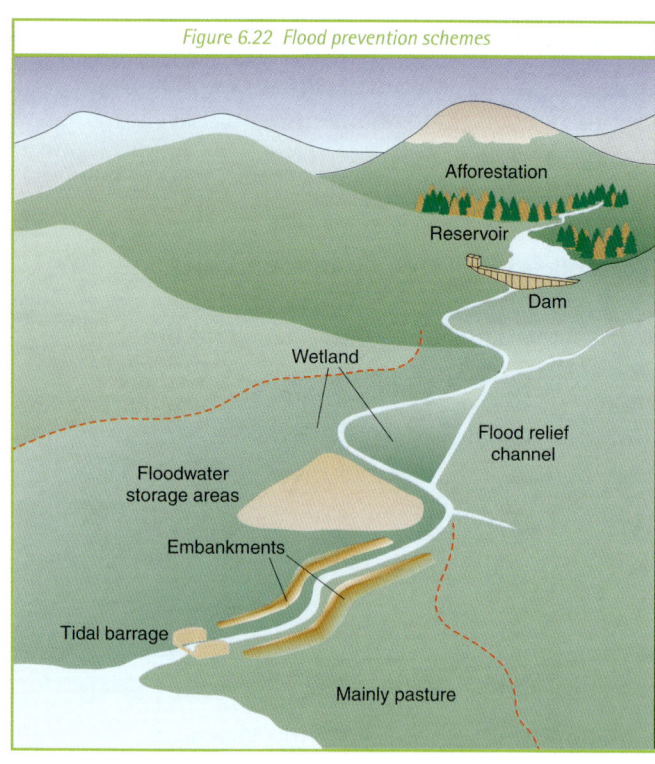

Figure 6.22 Flood prevention schemes

Flood Management around York

With the River Ouse flooding its valley more often, the Environment Agency has spent over £7 million to try to prevent and reduce flooding.

Upstream there has been reforestation in Swaledale, and farmers in Wharfedale are being encouraged to 'set aside' land rather than drain it. Dams and reservoirs have been constructed in Nidderdale.

In and around York there is a new flood warning system. Important historic buildings have been flood-proofed and insured, and there is strict control on new building near the River Ouse. A large area of pasture at Clifton Ings, north of the city, is used as washland and flood water is allowed to collect there. In the city centre the river channel has been widened and embankments built up and there is a new barrier to prevent water from the Ouse backing up into the smaller River Foss.

Downstream there are more flood embankments which protect farmland and villages between York and Selby. One embankment is over 5 km long as part of the river here is also tidal.

Figure 6.23 The river Ouse in the centre of York

Flood Management in Bangladesh

Bangladesh needs to spend its money very carefully because:

- it is located in a part of the world where natural disasters like flooding occur all too often (see Figure 6.17)
- as an LED country it has many demands upon its resources including poverty, food and health problems

If disaster does strike it will have to depend upon emergency help from other countries and agencies like Oxfam or Save the Children. Blankets, tents, food, water and medicines will all be needed (see chapter 7 for more details about international relief).

ACTIVITIES

1. Read the information about flood management around York, and describe the different methods used. (C1.3)
2. Using only Figure 6.23, describe how flooding is being prevented in the city of York. (C1.3)
3. Study the list of schemes in Bangladesh's FAP and find:
   - three hard engineering schemes
   - two schemes which allow flooding
   - one which looks sustainable
   - one which will cause conflicts
   - one which needs agreements. (C1.3)

The Flood Action Plan

Bangladesh has used funds provided by The World Bank to try to cope with river and coastal flooding. The most recent scheme, the The Flood Action Plan (FAP) began in 1990, and plans to spend another $ 500 million by 2005. This involves:

1. Building up its embankments to protect the main cities, roads and towns
2. Dredging channels to hold more floodwater
3. Allowing controlled flooding over the floodplain
4. Building storage compartments to hold floodwater
5. Use of ponds to rear fish
6. Irrigation in the dry season to reduce reliance of flood season crops
7. Disaster preparation, building escape centres in raised school buildings
8. Improving flood forecasting using weather radar and transmitting flood warnings
9. Constructing dams upstream
10. Redirecting river channels to prevent erosion of banks or bends
11. Planning water sharing with India
12. Planting mangroves along the coast.

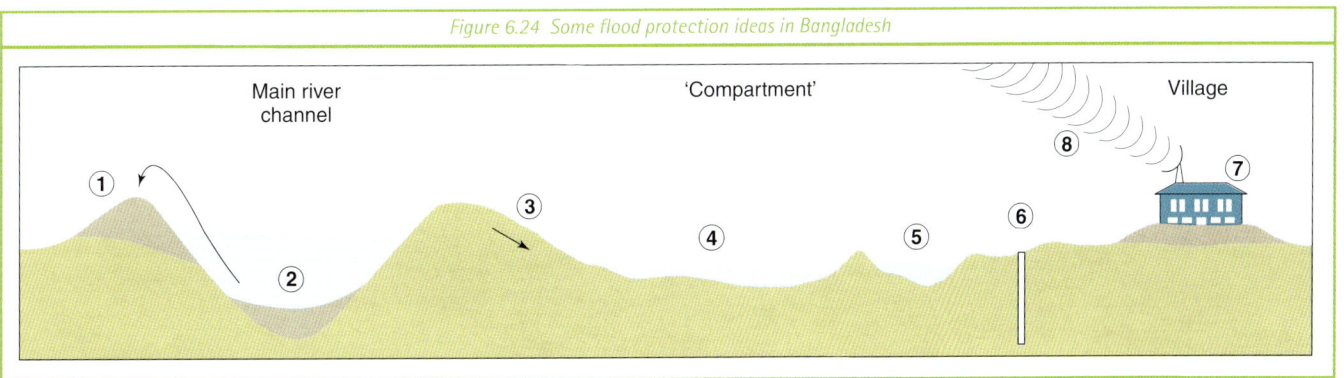

Figure 6.24 Some flood protection ideas in Bangladesh

Section 2: Managing the Physical Environment

Further tasks

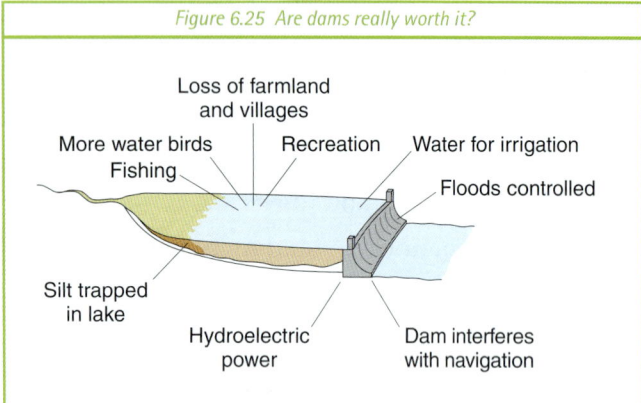

Figure 6.25 Are dams really worth it?

ACTIVITIES

Look at Figure 6.25. It shows some of the costs and benefits of building a large dam.

1. Research an example like the Aswan in Egypt or the Volta Dam in Ghana.
2. How many of these ideas fit the dam you have chosen?
3. Do the benefits outweigh the costs? (IT1.1, Wo)

Research

1. Research a case study of a weather hazard from books, the Internet or 'live' if one occurs opportunely. Hurricanes or UK floods in the autumn. (IT2.1, Ps, Wo)
2. Research and draw up a diagram of the **hydrological cycle**, labelling the processes that influence flooding. Learn them thoroughly. (It2.1, Ps, Wo)
3. Research Environment Agency flow and hydrograph data for a local river or a larger case study you want to use. (IT2.1, Ps, Wo)
4. Research the work of aid agencies like Oxfam via the www.oneworld.org/portal. (IT2.1, Ps, Wo)

Decision-making
The most useful topics could centre on flood management, deciding which types of flood scheme might suit a particular river or town. Another idea might be to hold an enquiry about whether or not to build a reservoir, perhaps linking this to your work on water supply, developments in a LEDC or to an area of attractive countryside. (Ps, Wo)

References
'The Shrewsbury floods' – *Geography Review*, September and October 1999.
www.nhc.noaa.gov
www.fema.gov
www.metof.gov.uk
www.environment-agency.gov.uk
www.nwl.ac.uk
www.irn.org
www.rivernet.org
www.shrewsburyfloods.co.uk
www.york.gov.uk
www.oxfam.org.uk/coolplanet

Exam-style questions and sample answers for Chapter 6

Question 1 – F Tier

Study the table below which shows information about two storms.

event	date	windspeed (km per hour)	death toll	damage ($US millions)	other information
Cyclone Gorky Bangladesh	May 1991	232	131 000	1 700	LEDC
Hurricane Andrew USA	August 1992	264	60	20 000	MEDC

(a) Comment on the differences between the strength of the winds and the impacts they had. **(3)**

Winds were stronger in Hurricane Andrew, and this explains the greater damage caused (it was over ten times worse). But the death toll was actually much lower.

(This compares the different figures correctly and is worth all three marks.)

(b) Explain these different effects. **(4)**

The high death toll in Bangladesh is because it is an LEDC and there are few storm or flood defences when a cyclone strikes. In the USA people are warned and can move away from the coastal areas. Damage however is very costly in places like Florida because people have more possessions and more buildings will have to be repaired afterwards. In Bangladesh they are poor and if you have only a little it is cheap to replace.

(This shows good understanding of the data given, earning all four marks.)

(c) Using an example, explain the short-term and long-term effects which people face if a hurricane strikes. **(6)**

Hurricane Mitch struck countries in Central America in October 1998. Strong winds brought heavy rain and floods to coastal areas. Inland in Nicaragua there were landslides and buildings were destroyed. In Honduras, 800 000 people died. By November there were more casualties as others became ill as food and clean water ran out. The effects of the hurricane were also felt in the following year because the farmers' food crops had been flooded. Worse still, so were the bananas, which were the country's main exports. Poor countries easily get into debt and will have to pay this back in the future.

(The candidate starts with a clearly named example. Countries are also named and some accurate figures used. Different effects like storm damage and flooding are referred to. What makes this an excellent answer is the way it has been structured, looking first at short and then longer term effects. This is well worth all of the six marks offered.)

This candidate has gained full marks on this question, 13 out of 13.

Chapter 6: Exam-style questions

Question 2 – H Tier

Study the sketch map below which shows the River Severn at Shrewsbury.

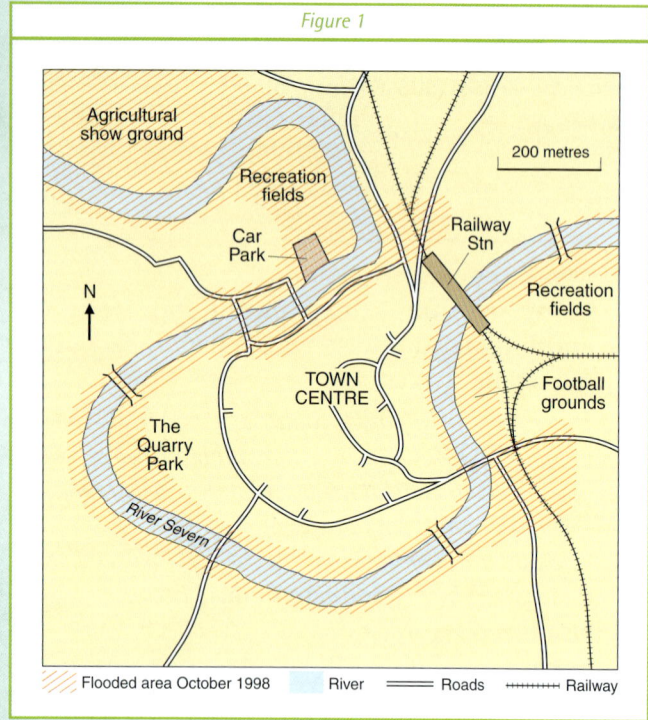

Figure 1

(a) Name two different ways in which the land is used along the river and suggest why. (3)

1. *A Park*   2. *Recreation fields*

This is because if they flood there is less damage and they can recover easily.

(This answer is correct, three marks awarded.)

(b) Use the map and your own ideas to describe how a flood would affect people in the town? (6)

Peoples's homes would be flooded and the shops and offices would be damaged. There would be no football matches and the showground would have to close.

(The candidate could have used the map more to point out problems with travel (there are many roads, bridges and a railway station). People not getting to work and the loss of business are other ideas. It is worth saying that the town centre would not have flooded at all. This answer is not complete and would only have gained two out of the six marks possible.)

(c) The Environment Agency plans to build a dam upstream. What will local people feel are the advantages and disadvantages of this proposal? (6)

The new dam will hold back the river and prevent Shrewsbury's houses being flooded. It may also give a good supply of water and allow boating on the lake. Local farmers might not want this as they would lose land and environmentalists say wildlife will be affected.

(There are both advantages and disadvantages here, and the answer does refer to different local people and what they might feel. The two disadvantages are a little vague not saying which wildlife might suffer or even how. This is worth four out of six marks.

[For further ideas refer to Figure 6.25]

The overall score of this candidate was 9 out of 15, and shows how important it is to read the question carefully.)

Water and food supply — Chapter 7

Water supply

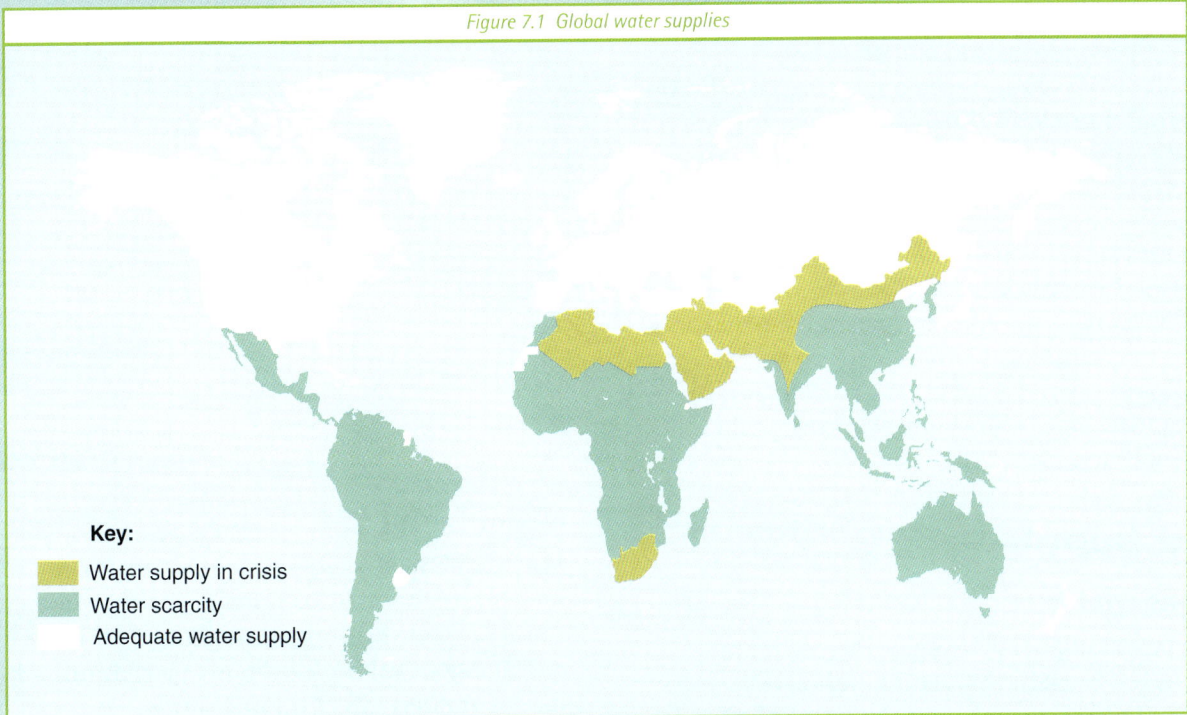

Figure 7.1 Global water supplies

Key:
- Water supply in crisis
- Water scarcity
- Adequate water supply

Water facts

- 70% of the earth's surface is covered in water.
- Less than 3% of the earth's water is fresh, i.e. not part of the seas or oceans.
- Of this 3%, most is stored in ice caps and glaciers.
- Only 0.008% of the earth's fresh water is found in lakes and rivers.

- 'The world has no more fresh water than it did 2000 years ago, when the population was less than 3% its present size.'

Alex Kirby, BBC Environmental correspondent

Water is a vital resource. We rely upon a supply of clean water for drinking. It is essential for human survival. We use water to **irrigate** the land in order to grow crops and rear animals for food. Water is used by our manufacturing industries and in our homes. Increasingly, water is used in our leisure time for recreation.

ACTIVITIES

1. **Quiz time!**
   How many litres of water does it take to grow/produce a serving of each of these items?
   - Chicken
   - French fries
   - Hamburger
   - Milk
   - Rice
   - 1 egg

   Choose your answers from this list: 24, 140, 260, 480, 1 600, 5 200 (all figures in litres)

   Answers: Chicken 1 600, French fries 24, Hamburger 5200, Milk 260, Rice 140, 1 egg 480.

ACTIVITIES

2. What percentage of the world's water is available for people to use? (N1.1)
3. List five ways in which people use water. (C1.2)
4. Give two reasons why the demand for water is increasing. (C1.2)
5. Look at Figure 7.1. With the help of an atlas, name the following:
   a. Three countries where water supply is in crisis.
   b. The **continents** where water supply is in crisis.
   c. Three countries that suffer from water shortage.
   d. Three countries that have plenty of fresh water.

Section 2: Managing the Physical Environment

All these uses are increasing the demand for water across the earth. In some parts of the world, this precious resource is becoming increasingly scarce.

Hundreds of millions of people, in both LEDCs and MEDCs, face water shortages. The problem is now spreading beyond the arid and semi-arid regions. The demand for food to feed a growing population means that more land is being irrigated. Irrigation is the artificial watering of the land. Irrigated farmland will increase 20% by the year 2025; by this date, one third of all the world's people could be facing regular and severe water shortages. The areas at greatest risk are Asia and Sub-Saharan Africa. However, parts of the Mediterranean countries, the Indian sub-continent, China, North and South America and Australia are also at risk. To add to the problem, this precious resource is being polluted by human activity.

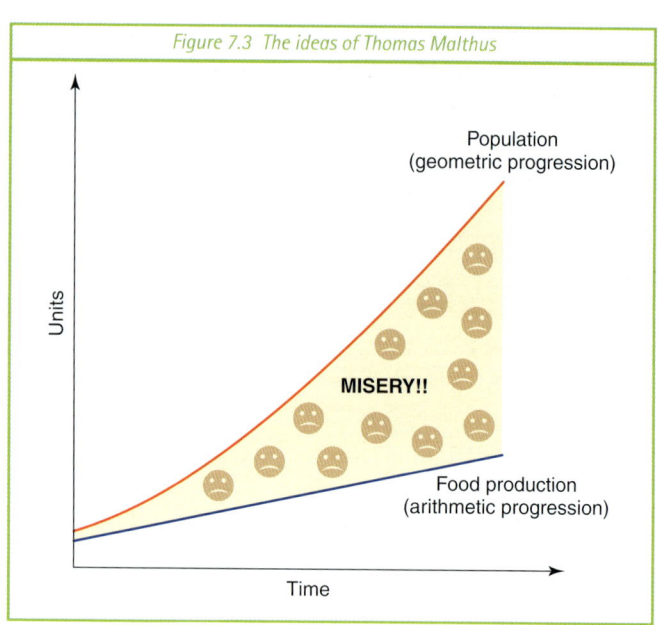

Figure 7.3 The ideas of Thomas Malthus

Food supply

Thomas Malthus

People have long been aware of the problems of trying to feed an ever-increasing population. Thomas Malthus wrote about the problem as long ago as 1798, when he first published his theory. He said that population growth will always be greater than the amount of food that could be produced. He argued that, if population growth was allowed to continue unchecked, the population would increase in a geometrical progression. However, food production would only increase in an arithmetic progression. The wider the gap between the two, the more people, he said, would exist in 'misery'. Malthus said that the checks to population growth would be 'famine', 'ill health' and 'war'. Are his ideas still relevant today?

Malnutrition

Malnutrition is where the diet lacks some foods which are essential for human health. The main cause of malnutrition is hunger, which leads to a deficiency of calories, proteins or vitamins. Hunger has its greatest effect upon children, as it can affect both their mental and physical growth. This may disadvantage them for the rest of their lives, as it may result in them having less opportunity for work or education.

Rickets is a bone softening disease, which affects children. It results from a lack of vitamin D in their diet. In parts of Africa, protein deficiencies develop because many people's diet consists mainly of one type of starchy food, such as cassava. Protein is important for growing children. A lack of protein may lead to diseases such as kwashiorkor. This is a common cause of infant deaths. In adults, the

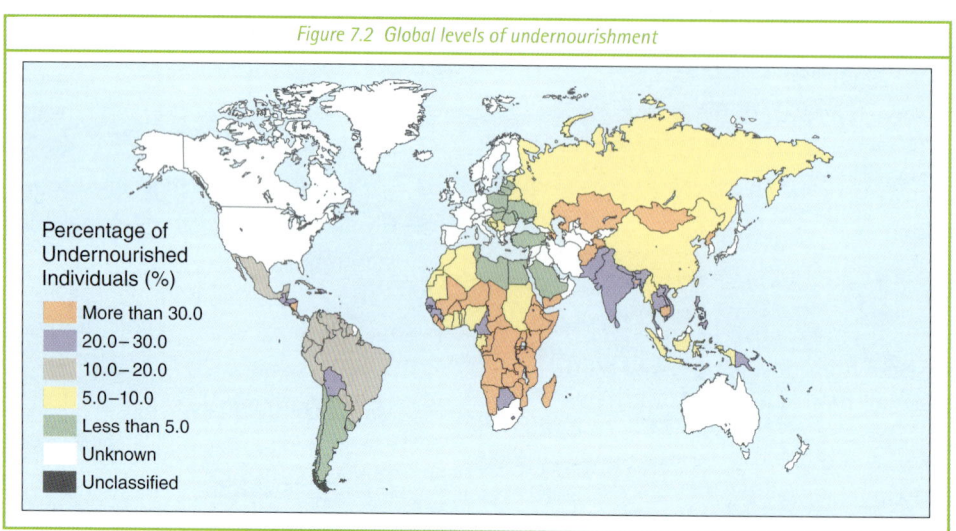

Figure 7.2 Global levels of undernourishment

symptoms of illness may take many years to show, either to the people themselves, or to health workers. Eventually, diseases develop. These may result in a person being unable to work, making their situation even more desperate. Beriberi is a disease of the nervous system which is caused by a lack of vitamin B1. In some parts of East Asia, white, polished rice is now produced. This has no husk. The husk contains vitamin B1 and as this is no longer a part of the diet, the disease is widespread. The results of malnutrition are a reduced ability to work due to disease and illness. This leads to a loss in production and also increases the costs of health care within the country. Both of these issues hinder the attempts of governments of LEDCs to improve the quality of life for the people of the country. They use up money and valuable resources, which are already in short supply.

ACTIVITIES

1. Look at Figure 7.2.
   a Describe the pattern of world undernourishment.
   b Describe the relationship between the pattern shown in Figure 7.2 and that shown in Figure 7.1 (page 65).
   c With the help of an atlas, give the percentage of population that are undernourished in each of these countries: Afghanistan, Brazil, Ethiopia, India, Tanzania, Turkey. (N1.1)
2. What is malnutrition? (C1.2)
3. Describe three effects of malnutrition upon people. (C1.2)

Water supply: effects on people

1. Dirty water can lead to disease

In most MEDCs, clean drinking water is readily available. Water is filtered to remove organic matter and harmful organisms. It is chlorinated to kill micro-organisms which can harm people. Where water is not treated, it can carry infection. It may contain many harmful bacteria and viruses.

Water-related diseases

Water-related diseases claim the lives of millions of people each year. In Sub-Saharan Africa, they claim more lives than do war and AIDS in the whole of Africa.

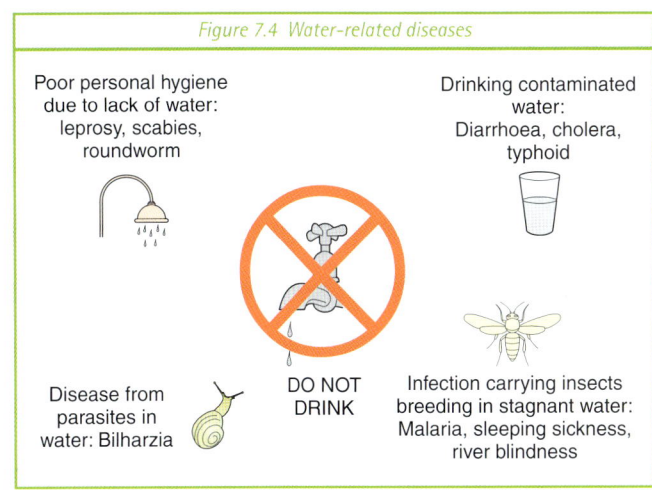

Figure 7.4 Water-related diseases

1. Water-borne diseases are dirty water diseases. They are caused by water supply becoming contaminated by untreated human sewage, animal waste and chemicals from fertilizers and pesticides.
2. Water-based diseases are caused by organisms, such as tapeworms and roundworms. These spend part of their life in water and infect people when swallowed or by boring through the skin of bathers. They breed very successfully in stagnant water behind dams.
3. Some water-related diseases are carried by insects or other animals. Of these, malaria is the most common and is carried by mosquitoes. It affects over one hundred LEDCs and costs them billions of pounds each year. Often, building dams and flooding areas creates ideal conditions for the mosquito to breed.
4. Water-scarce diseases thrive in areas where water is in very short supply. Here there is too little water available for washing hands, etc.

Figure 7.5 Dangerous water-related diseases

disease	type	areas affected	deaths per year
Diarrhoeal disease	water-borne	world wide	4 million
Malaria	carried by insects	Africa, SE. Asia, India, S. America	2 million
Typhoid	water-borne	Asia, Latin America, Africa	600 000
Ascarrasis	water-based	Africa, Asia, Latin America,	60 000
Denque	carried by insects	Asia, Africa, Latin America,	24 000
Bilharzia	water-based	Africa, Near East, SW. Asia	20 000
Cholera	water-borne	S. America, Asia, Africa	20 000

Section 2: Managing the Physical Environment

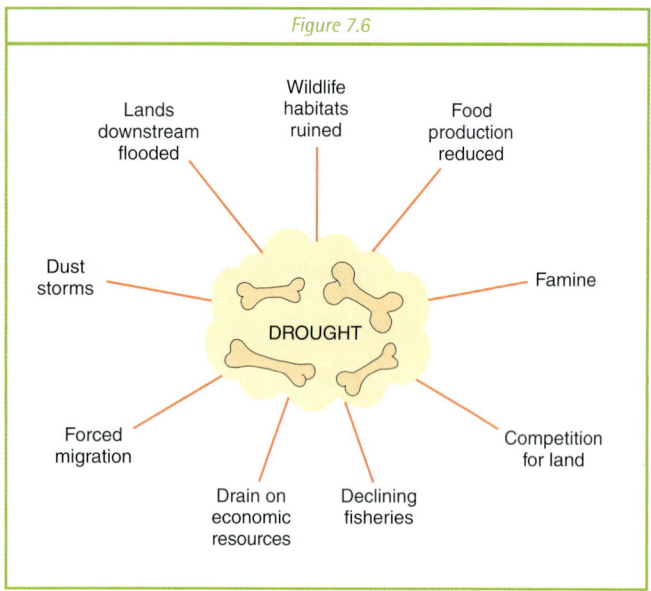

Figure 7.6

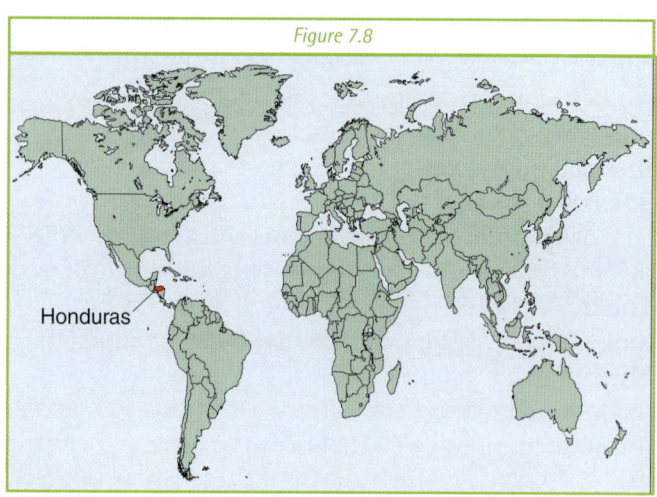

Figure 7.8

Many of these diseases do not always lead to death, but victims can become paralysed and unable to work. There are 270 000 cases of blindness caused by water-related diseases each year.

2. Drought can affect people

The effects of drought on people

Drought can lead to famine and on some occasions, starvation. A lack of water means that farmers find it difficult to grow crops and rear animals, and as rivers dry up, fisheries are lost. All these factors lead to a decline in food production.

Animals become very thin and lose their value. Herders are forced to move weak animals, in search of any remaining green pasture, which, in a recent drought in Kenya, was on the grass verges and roundabouts in Nairobi, the capital city. Forty percent of Kenya's cattle were lost in the drought. Children, especially, are at risk as there is no milk available for them. People's livelihoods are lost, as it may take years for them to rebuild herds or re-establish crops.

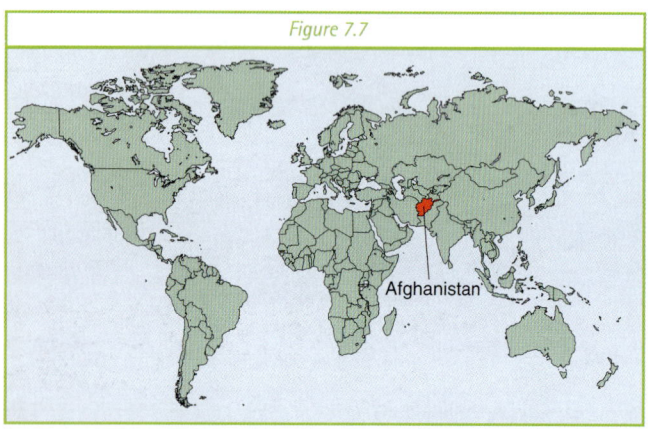

Figure 7.7

A famine can lead to civil unrest. Competition for any land that is available becomes very intense. This occurs particularly on the banks of rivers, where pastoralists and crop farmers compete for land. Areas may also become unstable, as many people have no choice but to migrate elsewhere to find work or food. The presence of large numbers of refugees creates

Drought in Afganistan — Case Study

By 2001, the North of Afghanistan had suffered from three years of drought, affecting five million people. In addition to this problem, the people of the region also had to deal with the effects of living in an isolated area, war and a devastating earthquake which hit the region in 1998. Many people were living in **absolute poverty** and depended solely on the land in order to subsist. Growing enough food, or rearing animals, was difficult enough in a year when there was rain, but with the continued drought, the United Nations declared the region a 'food crisis area'.

This is the story of Abdul, living in Abagarda village in Northern Afghanistan. Abdul's wife died in 1999 and he was left to bring up his eight children. The family depended on crops grown on their small field, milk from one cow and fruit from four mulberry trees. In 2000, Abdul had to sell his cow and two mulberry trees in order to buy food, as he was unable to produce enough to feed his family. In 2001, he was not able to sow his field as he had no seeds. None were left from the previous year. Nobody in the village could help the family as they were all in the same situation. Abdul's children had to collect wild grass in the mountains. This was eaten, mixed with flour and water. Roots and leaves were also eaten. Abdul's son became very ill after eating poisonous leaves.

Faced with having no food, Abdul had to leave his children and look for paid work elsewhere in the region. He managed to find eleven days' work and was able to buy 21 kilos of wheat, with which he had to feed his family for several months.

Source: www.oxfam.org.uk

Chapter 7: Water and food supply

Case Study: Drought in Honduras

CHOLUTECA, HONDURAS – When it finally started raining in southern Honduras in late August, Alejandro Fuentes walked through the stubble of his failed corn crop, poking the ground with a sharpened stick and dropping in his last seeds of corn and beans. His eight-year-old son followed behind, listless, his hair discolored by malnutrition. Fuentes said he prayed with each seed he dropped, asking God to let the rains continue. Asked what would happen if the rains stopped again, he fell silent and looked at the ground.

Source: Pacific News Service, September 4, 2001, Paul Jeffrey
www.pacificnews.org

costs for the host country or region. These include the economic costs of running the refugee camps and the environmental costs, as the land around the camps comes under severe pressure.

As soils dry out, problems of soil erosion occur (see pages 72–4). This can lead to silting and flooding in other areas. Dust storms become more common and as a result, eye infections increase. Animal habitats are lost and species move away, or die. In some cases, this represents a further loss in food supply for some groups of people. Water supply can also be further affected, as groundwater levels fall. Wells dry up, or quality of water decreases. As river flow is reduced, less **Hydro-Electric** Power can be generated, leading to problems with power supply.

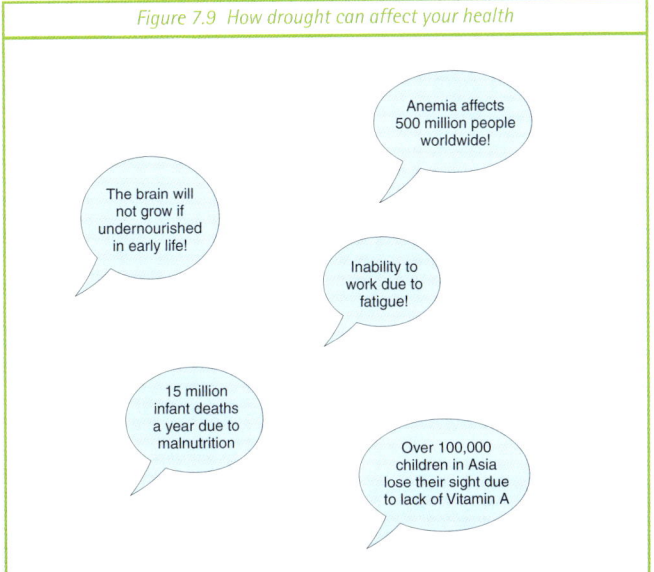

Figure 7.9 How drought can affect your health

Figure 7.10

ACTIVITIES

1. Copy and complete the diagram below by writing these labels in the correct boxes.
   Less food produced, drought, unable to work, famine

   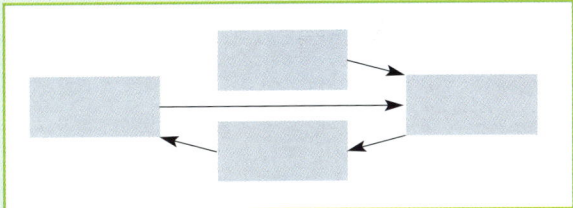

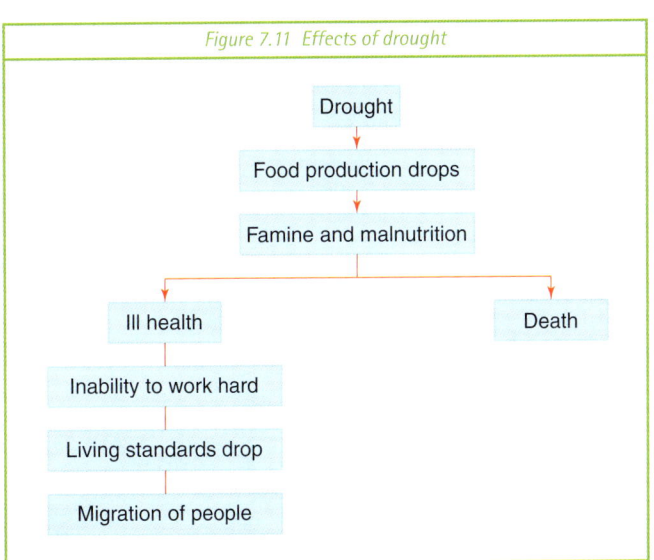

Figure 7.11 Effects of drought

2. Look at Figure 7.5 (page 67), which gives information on dangerous water-related diseases.
   a. Draw a graph to show the information in the table. (N1.1)
   b. Which *type* of disease causes the most deaths per year? (N1.1)
   c. Which regions of the world are most affected by water-related diseases? (N1.1)
3. Describe four effects of drought and famine on people. (C1.2)
4. What is meant by absolute poverty? (C1.2)

69

Section 2: Managing the Physical Environment

Why are some places hot deserts?

The world's hot desert regions are found mainly along the Tropic of Cancer in the Northern Hemisphere and the Tropic of Capricorn in the Southern Hemisphere. Some geographers say that deserts are areas that have less than 25 cm of rainfall in a year. Others prefer to say that they are places with much more evaporation than rainfall.

The two features of the hot desert climate are high temperatures in the daytime and lack of rainfall. These conditions persist throughout the year and there are a number of reasons why they occur at some places in the world. The Earth's surface is heated by the sun. The process by which the sun's rays heat the Earth's surface and the air above it, is called insolation. At the tropics, air which has risen into the atmosphere in equatorial regions is descending. Due to this, there is little condensation and, therefore, little cloud cover. In the tropics, therefore, the rates of insolation are high.

Temperatures are also affected by latitude. The location of the tropics on the Earth's surface, means that the sun is high in the sky and the sun's rays are highly concentrated in a small area, heating the ground very quickly. Also, in the lower latitudes, i.e. between the tropics, the sun's rays have to pass through a lesser amount of the Earth's atmosphere.

Britain is in the upper latitudes where the sun's rays strike the Earth at a much lower angle. These rays have to heat up a far greater surface area and, therefore, Britain is not as warm as places in the tropics. The sun's rays also have to pass through more of the atmosphere and, therefore, lose heat. Heat is absorbed by gases in the atmosphere as the sun's rays pass through it.

In hot desert regions, rainfall is scarce. When it does rain, evaporation rates are high. These conditions make it very difficult for plants to survive. Only those which are adapted to the harsh conditions are able to survive. There are three reasons why some areas in the tropics receive such low amounts of rainfall:

1. The descending air at the tropics creates high pressure conditions over the land, e.g. over the Sahara desert. These conditions last for most of the year. Therefore, the prevailing winds blow from the land, out towards low pressure areas over the oceans. Due to this, they are dry winds, picking up little moisture as they blow across the land.

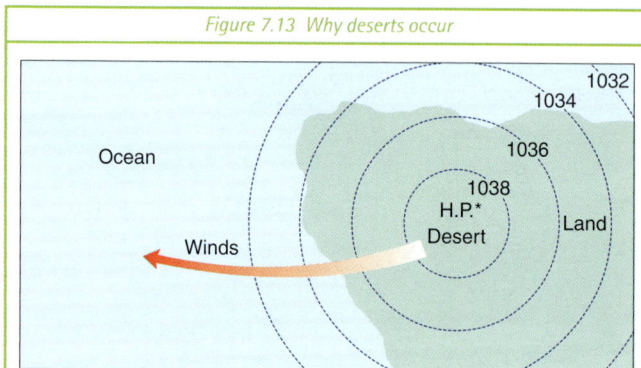

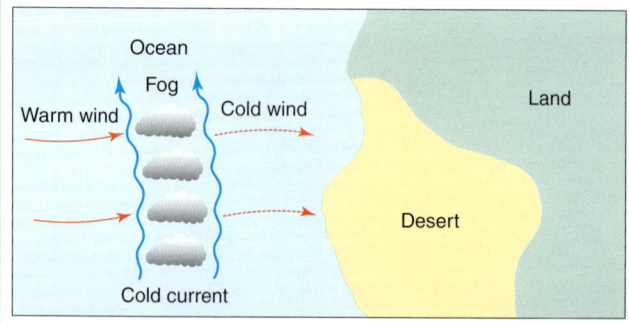

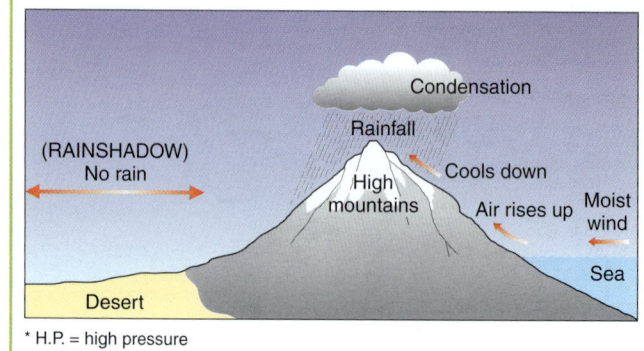

Figure 7.12 Why some places are hot deserts

Figure 7.13 Why deserts occur

* H.P. = high pressure

2 Transfers of heat take place in the world's oceans. Warm currents take warm water away from the tropics and cold currents carry cold water back towards the tropics. Warm, moist winds, blowing from the ocean towards the land, cool down quickly when they cross these cold ocean currents. The water vapour condenses, sea fogs form, and it rains. Moisture is returned to the sea and does not reach the coastline, giving rise to hot desert conditions on the western edges of continents. In this way the Peruvian current off South America has helped create the Atacama Desert.

3 In areas where winds blow across the ocean, if when they hit land they have to cross a mountain range, they are forced upwards. The warm, moist air cools down as it rises and condenses, clouds form and there is precipitation over the highland area. As the winds travel inland, they become increasingly drier and a rainshadow area develops. This area receives little rainfall. Such areas sometimes occur outside the tropics, as in the Patagonian desert of Argentina.

Often a combination of some, or all of these three factors leads to a place receiving low, unreliable rainfall.

ACTIVITIES

1 What is a desert?
2 Look at Figure 7.14. With the help of an atlas:
   a Name the hot desert regions, shown as numbers 1 to 6.
   b Name the lines of latitude, shown as A, B and C.
3 Describe the location of the world's hot desert regions.
4 Explain how each of the desert regions shown in Figure 7.14 as 2, 3 and 6, have been formed. (C1.3)
5 Explain the effect of latitude on climate. (C1.3)
6 Look at Figure 7.15, a **climate graph** for a desert region.
   a What is the annual range of temperature? (N1.1)
   b What is the annual rainfall total? (N1.1)
   c Copy and complete the passage below. Use four of the words/terms in the wordbank.

WORDBANK
extreme   longitude   latitude   59 mm
temperate   10°C   0 mm   31°C

Hot desert regions have a more _____ climate than other areas in the same _____ . The highest monthly temperature is _____ and the lowest monthly rainfall total is _____ .

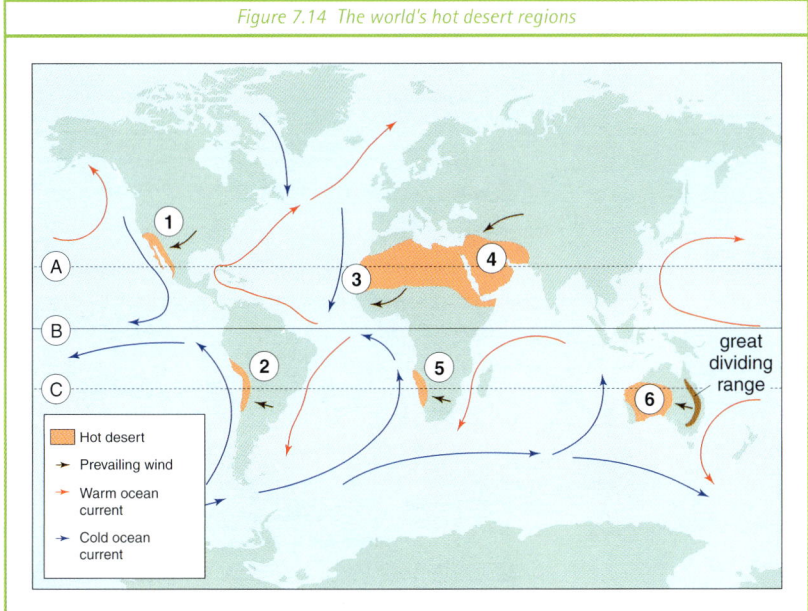

Figure 7.14 The world's hot desert regions

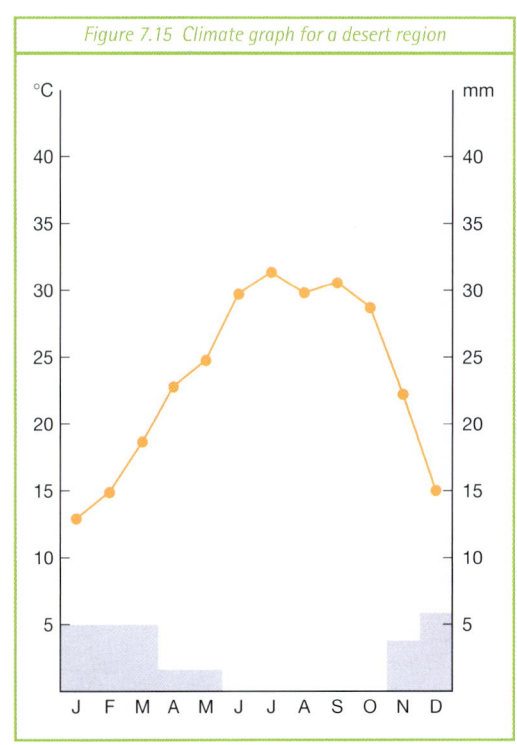

Figure 7.15 Climate graph for a desert region

Section 2: Managing the Physical Environment

Desertification

As Figure 7.16 shows, many areas of the world are at risk from **desertification**. This is when the land becomes unproductive desert, due to lack of water and soil. The risk is greatest at the fringes of the world's hot desert regions, such as the Sahara desert in Africa. The fringes of hot deserts are dryland areas; they are part of the zone of transition from savanna regions to arid (desert) areas. In Africa, the area is known as the Sahel. Drought started in the Sahel in 1968, when virtually all the crops were lost and up to 70% of the cattle killed. In 1973 there was practically no rain and 100 000 people died. There were severe droughts between 1983–5 and in the 1990s.

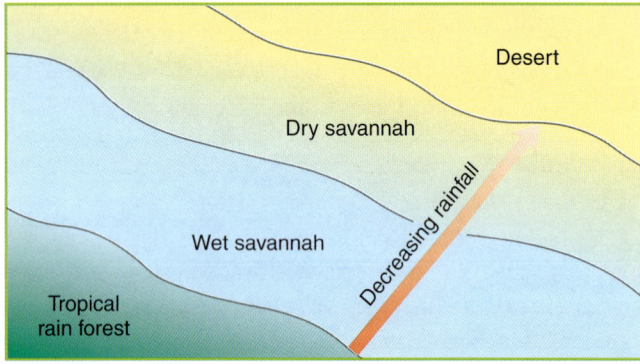

In the dryland regions, the ecosystem is very delicately balanced. Plants and animals have adapted to be able to withstand the high temperatures and dry conditions. Drought in these regions is not uncommon. However, plants are able to survive and re-grow when the drought ends and the rains return.

People have also adapted to these conditions. For thousands of years, people have farmed these lands successfully using shifting cultivation and nomadic herding. In both of these practices, farmers move on to farm another area, leaving the land time to re-grow or refresh itself.

In the Sahel and on the fringes of desert regions throughout the world, these drylands are becoming degraded. It is estimated that 70% of the world's drylands are now degraded, in that they suffer from **soil erosion** and desertification. It becomes increasingly difficult for plants, animals and people to survive. In the last 50 years, the Sahara has spread south, to cover an extra 65 million hectares. It is not clear whether this land will be able to recover, or has become permanent desert.

Causes of desertification

There are two main reasons why desertification takes place in drylands:

1. **Climate change**

    Scientists feel that natural climate variations may increase drought in some areas. There is strong evidence that rising levels of emissions of greenhouse gases may lead to global warming. This may lead to some areas becoming wetter, but others becoming drier, leading to an increased risk of desertification. Much research has been done on the ways in which the ocean currents may affect climate. Dramatic changes, such as El Niño events (changes in ocean

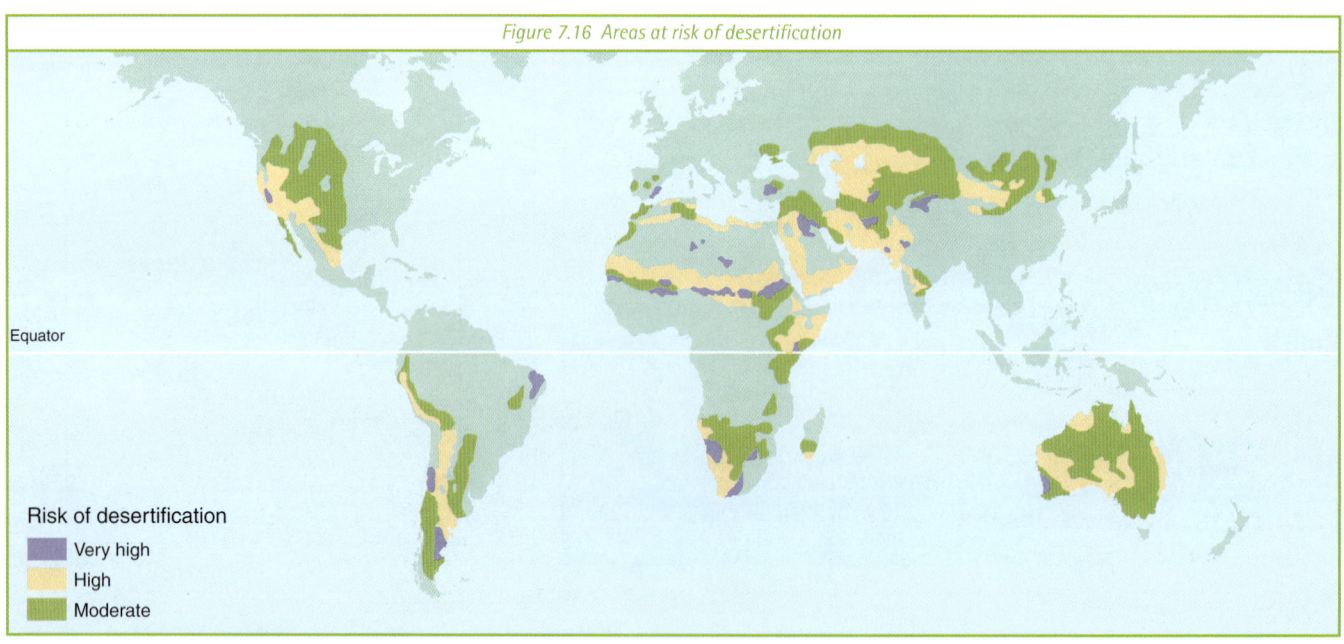

Figure 7.16 Areas at risk of desertification

currents, which can affect climate patterns) have been widely studied, helping us to predict periods of drought to enable the countries affected to be better prepared for them.

2 Unsuitable human activities

The population in the Sahel has increased by more than 30% in the last 50 years. All these extra people need to be fed and housed. This increased population pressure has led to changes in human activity, which have increased desertification.

Permanent settlement

New technology has made it possible to dig deep wells to tap groundwater. This has encouraged nomadic herders to settle near the wells, where a water supply was always available.

Deforestation

In the permanent settlements, there is an increased demand for fuel, in the form of wood. Traditional energy sources, such as fuelwood, are those upon which most people in rural areas of LEDCs depend. High levels of poverty mean that people have little alternative but to use the natural resources available to them. On the land around the settlement, trees are removed on a large scale. Often people have to travel long distances on a daily basis in order to collect wood.

Overgrazing

Availability of water from wells has enabled farmers to keep more animals. **Overgrazing** has resulted as large herds of goats, or other livestock, search for increasingly scarce plants to eat. Plants are torn up by their roots and the ground becomes compacted from being trampled by hooves.

Overcultivation

With the increased demand for food, farmers have reduced the amount of time for which land was left fallow. In the past, land was left fallow for up to twenty years, to give it time to recover its fertility. This has been reduced to as little as one year when farming practices were changed to meet the demands of the increased population. Often the best land has been used to grow cash crops, such as cotton and peanuts for export. Farmers have been forced on to poorer quality, marginal land. When overused, the soil quickly loses its fertility. The soil

Figure 7.17 Overgrazing

eventually becomes degraded, as it loses its structure and its ability to hold moisture to allow plants to grow.

As a result of these activities, there is a loss of vegetation. Plants provide most of the organic material from which new soil is made. The plant roots help bind soil together and keep moisture in the soil. Once the protection of plants has been lost, it can easily dry out and be blown away by the wind, or washed away by heavy rain in flash floods.

Deep gullies can be formed as **surface run-off** is increased. It is difficult for water to **infiltrate** into soil which has been compacted by animals, or baked hard by the sun because of lack of plant cover and moisture. Therefore, large amounts of water flow over the surface, carrying the soil with it, causing soil erosion. This problem is made worse when farmers plough up and down the hill on sloping land. The furrows act as channels for rainwater to run down.

Figure 7.18 Gulleying

Section 2: Managing the Physical Environment

ACTIVITIES

1. What is meant by the term desertification? (C1.2)
2. Copy and complete the flow diagram (Figure 7.19) by writing these labels in the correct spaces:
   soil erosion; overgrazing; vegetation dies; long term; overcultivation; more fuelwood needed; soil exposed; vegetation cleared. (C1.2)

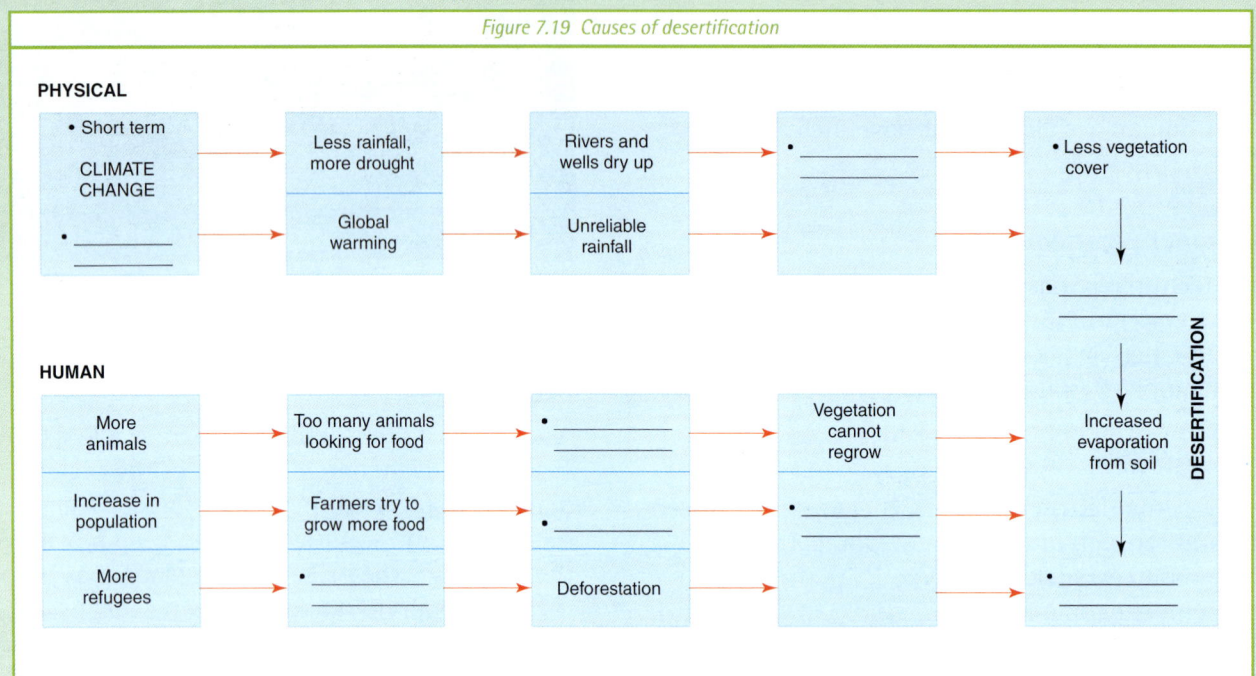

Figure 7.19 Causes of desertification

3. Explain how overgrazing and deforestation can lead to soil erosion. (C2.3)
4. Explain how soil can become exhausted. (C2.3)
5. Look at the climate graph of a semi-desert region (Figure 7.20.)
   a. What is the total annual rainfall? (N1.2)
   b. What is the annual range of temperature? (N1.2)
   c. Describe the pattern of rainfall and temperature throughout the year. (N1.3)
6. Look at Figures 7.21 and 7.22, precipitation and vegetation maps of Africa.
   a. With the help of an atlas, name six countries in the Sahel region.
   b. You are going on a journey from Lagos, northwards to Algiers.
      i. Describe the changes in precipitation levels you would experience.
      ii. List the types of vegetation regions you would pass through.
      iii. Find out which countries you would pass through on your journey.
      iv. How many of these countries are at risk from desertification?

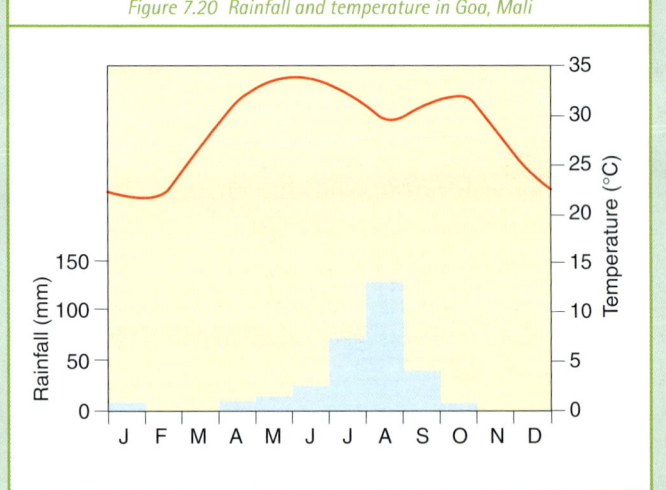

Figure 7.20 Rainfall and temperature in Goa, Mali

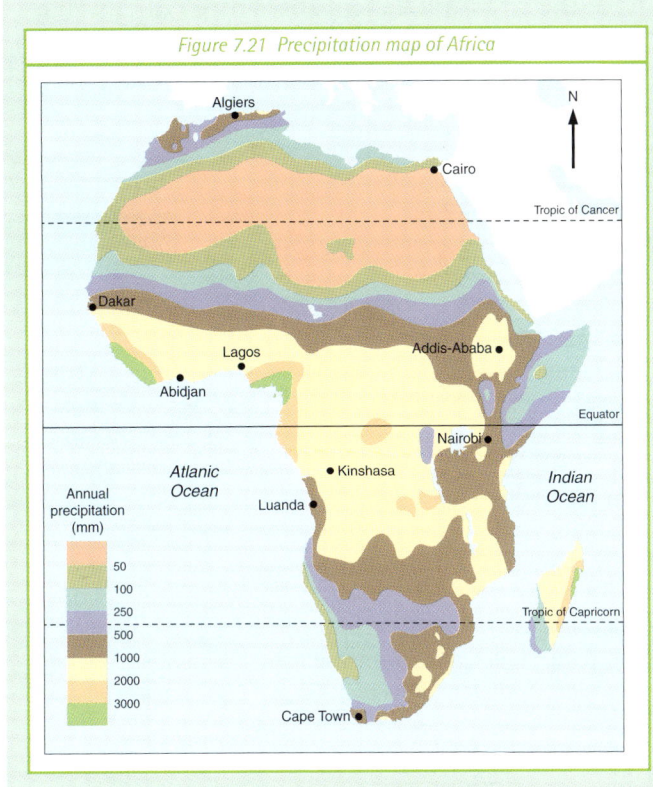

Figure 7.21 Precipitation map of Africa

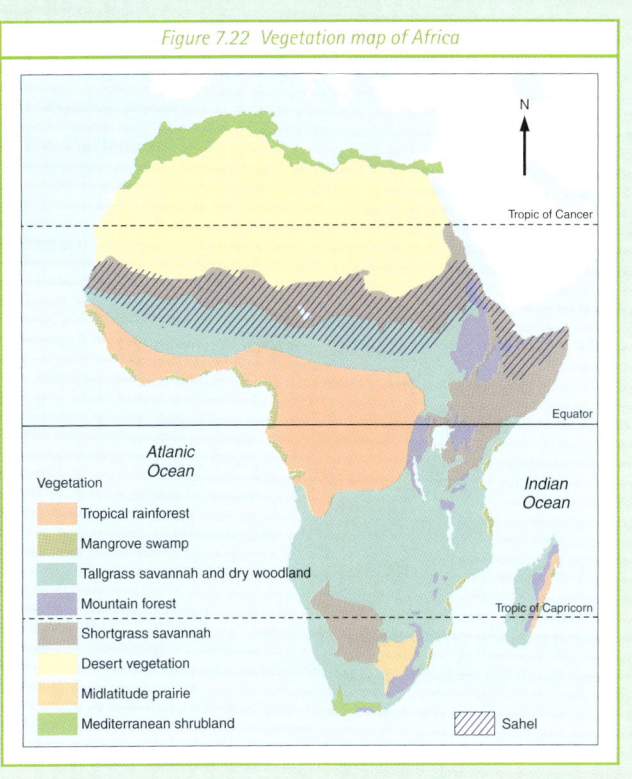

Figure 7.22 Vegetation map of Africa

Water management and soil conservation

There are many methods by which governments and communities can attempt to manage water supply and conserve soil. All of these schemes bring their own benefits, but some may bring disadvantages or **costs**. The success of any scheme should be measured against its **sustainability**. If a scheme is to be sustainable, it should be easily kept going for many years into the future by the people it is designed to benefit, i.e. local people are able to run and maintain the scheme without help from overseas experts. The benefits of the scheme should be far greater than the costs to local people and their environment.

Figure 7.23 A sakia

Irrigation

Irrigation is the artificial watering of the land. For thousands of years, people have lifted water from rivers or pumped up water from the groundwater supply. This water then flows down irrigation canals, which have been dug out and lead to the farmers' fields. The Egyptians have long had methods, such as the *shadoof* or the *sakia*, for extracting water from the River Nile. The shadoof consists of a long pole, mounted like a see-saw. On one end is a bucket hanging from a rope and on the other is a weight. A person pulls on the rope to lower the bucket into the river. When released, the weight lifts the bucket up, which is then swung round, to drop the water into the irrigation canal. The sakia has buckets attached to a large wheel, which is turned by oxen or horses. As it turns, water is lifted from the river and drops into the irrigation canal.

The benefits of these irrigation techniques were that water could be lifted from the Nile, even when water levels were low. Therefore, land could be watered at all times, not just when the seasonal floods occurred. With the land being irrigated, two or three crops a year could be grown.

Section 2: Managing the Physical Environment

The costs of irrigation come from overuse. The area of land under irrigation is rising yearly. As a result, rivers are drying up and water tables are falling on every continent. In parts of Spain the groundwater levels have fallen by up to 5 metres. The water from the La Mancha aquifer (groundwater store) is being used up to three times as fast as rainwater can replace it naturally. As rivers shrink, fishing grounds are lost and grazing lands next to the rivers dry up, forcing cattle herders to look for pastures elsewhere. Whole communities are no longer able to make a living on the flood plain and are forced to migrate.

Little of the water flowing down the irrigation canals actually reaches the crops in the fields. Most of the water is lost to evaporation due to the heat, or escapes as canals are often not lined with concrete.

The problem of salinisation occurs as the natural salts contained in the irrigation water builds up as it seeps into the soil. This makes the soil too salty for crops to be grown and the land has to be abandoned. Many large scale irrigation projects involve the building of high technology dams such as the Aswan High Dam on the River Nile. Water is stored in a reservoir behind the dam and can be used for irrigation throughout the year (see chapter 6 for the costs and benefits of dams). Other water storage and management schemes can offer more benefits and are more sustainable than the high cost, prestigious dam building schemes.

Mossi farmers — Case Study

Mossi farmers of Burkino Faso, West Africa

In the 1970s and 1980s, after a series of droughts, a scheme was set up by Oxfam to encourage farmers to put up lines of stones (bunds) on their land. These were not new; they were first used in the early 20th century, but were abandoned because of war and the political situation in the country. The bunds were revived, and other systems introduced by 'experts' were no longer used. The stone bunds were built during the dry season, and over the years, have come to reach up to one metre in height. They allow time for the scarce, heavy showers of rain to seep into the land, reducing surface run-off and, therefore, soil erosion. In the droughts of 1983 and 1984, crops still grew on land with bunds, whilst nothing grew in other fields. With international funding, the idea was publicised throughout the country and now 150 villages have stone lines. In these areas, yields have risen by 40%.

India — Case Study

The Watershed development took place in three villages in India's Maharashta state. Simple, sustainable, but effective farming techniques were employed in an attempt to conserve water and soil and to increase food production.

Key factors in the success of the development:
1. Use of local technology, which was easy to adopt.
2. Outside input into the village is only in the form of a small cost for seeds etc.
3. There is the participation of villagers at each stage of the project.
4. It generates local employment.

Several techniques were employed on land near the villages. These were: grassland development, gully plugs, tree plantation, land levelling and earth dams.

Grassland development

This took place on hilly land, where farmers were getting very low yields from growing crops. Traditional crop farming was stopped and grass planted. The grassland helped to conserve soil and water, as much of the seasonal, heavy rainfall is intercepted and can then slowly infiltrate into the soil, instead of running over the

Figure 7.24

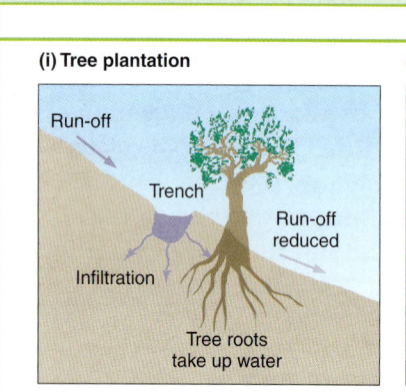
(i) Tree plantation

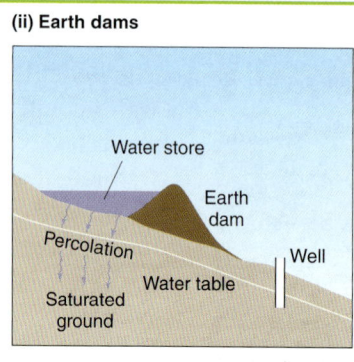
(ii) Earth dams

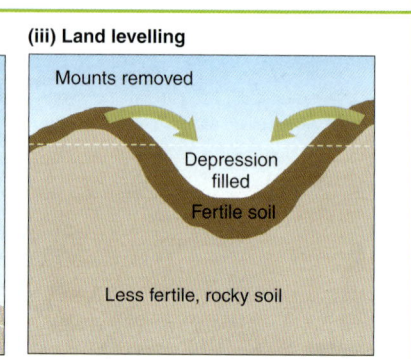
(iii) Land levelling

Chapter 7: Water and food supply

India — Case Study

surface. Cattle are not grazed on the grass, as this could lead to erosion. They are fed in stalls, on grass collected from the fields. The returns for the farmer, from dairying, are far greater than from crop farming. As there is less run-off and evaporation of water, well levels rise, improving water supply. The cattle also provide a source of organic manure for the soil, increasing its fertility.

The project involved no high-technology methods, but costs came from buying seeds, animals and feeding stalls. Fences were made from locally produced materials, further increasing the involvement of local people.

Gully plugs

Figure 7.25 Building a gulley plug

These are barriers, made from local materials, which are built at intervals down the slope. The dry stone wall has sufficient spaces in it to let some water through, but slows up the movement of water sufficiently to allow some to infiltrate the soil. Soil, carried by the water, collects behind the wall, slowly filling up the gulley. This fertile silt is excellent for growing young, crop plants, which later can be replanted in the fields.

Tree plantation

This scheme took place on the steepest slopes, where the building of walls or bunds is very difficult. A trench was dug to reduce the speed of surface run-off and hold the water long enough for it to infiltrate the soil. Trees were planted downhill of the trench, the water stored in the trench helping them to grow. The falling leaves also add organic material to the soil, increasing its fertility. The benefits of the scheme are not as immediate as others, in that villagers have to wait many years before gaining any harvest from the trees. A programme was used to explain clearly the long term benefits of tree planting.

Land levelling

Figure 7.26 Land levelling

This scheme took place in some fields where the land was particularly uneven. The tops of the mounts were stripped off to fill in the depressions. The scheme does reduce surface run-off, as water has more time to infiltrate on a level surface. It also provides a better growing surface for crops. However, trained engineers are required and the scheme involves the use of some heavy machinery. Drains have also got to be put in place to remove excess water in times of heavy rain. This is an additional cost. A further disadvantage of this scheme is that the best soil at the surface becomes mixed in with rocky and less fertile soil from below.

Earth dams

These are built to store water where rainfall is irregular. As water is held in storage, the rate of percolation through the underlying rock is much greater. This raises the level of the water table and raises levels in wells. Silt that builds up behind the dam can be put on the fields to improve their fertility. However, the scheme demands high technology, with costly topological surveying and dam designing to be done. It also requires heavy machinery.

Figure 7.27 An earth dam

ACTIVITIES

1. Explain how any one of the schemes given in the case study helps conserve water *and* soil. Use a labelled diagram. (C2.2)
2. Which scheme do you think is the most sustainable? Give reasons for your answer. (C2.2)
3. Which scheme would be the most expensive to carry out? Give reasons for your answer. (C2.2)

Section 2: Managing the Physical Environment

Managing food supply

In 1996, a World Food Summit was held in Rome. The meeting considered how countries could secure their food supplies, i.e. 'when all people, at all times, have physical and economic access to sufficient, safe and nutritious food.' The aim was to reduce the number of malnourished people in the world to half the 1996 level by 2015.

How can governments and other organisations work towards this aim?

> **ONE VIEW**
>
> Although world population has increased by 20% over the last 30 years, progress has been made in increasing the quantity of global food supply and improving the nutrition of populations. In LEDCs, where population has nearly doubled in the last 30 years, the proportion living in a chronic state of undernourishment has been reduced by half, from 36% to 18%.
>
> **ANOTHER VIEW**
>
> 790 million people, 1 in 5 people in LEDCs, still do not have enough food to meet daily nutritional needs. Development has not benefited everybody. Some countries have made great progress; hunger has increased in others. Even in countries where food supplies are adequate at the national level, there are shortages in some regions.
>
> **ARE WE MAKING PROGRESS?**

The Green Revolution

The '**green revolution**' was a planned international effort to increase food supplies, funded by the USA and the governments of LEDCs. The revolution began in Mexico and spread to countries around the world. It was particularly successful in India.

The worst ever food disaster was the 'Bengal famine' in India in 1943. Four million people died of hunger in that year alone. The cause was that the population was growing at a much faster rate than food supply. Efforts began to improve the food supply and success came in 1967, with a number of farming changes, now known as the 'green revolution'. There were three main areas of change:

1. **Expansion of farming area**
   The area of land under cultivation was increased by bringing unused land into production.

2. **Double cropping**
   A decision was made to have two crop seasons each year. To do this, there would have to be two wet seasons each year. One would be the natural monsoon, the other an 'artificial monsoon.' This was achieved by building large scale dams and irrigation facilities.

3. **New varieties of seeds**
   High yielding varieties (HYVs) of seeds were developed in research laboratories. These were mainly of wheat, rice and corn. The new varieties yieded far more, but needed much more water, chemical fertilizer and pesticide in order to survive.

Here are three reasons why some people would argue that the green revolution is not an example of sustainable development.

1. Food cannot always be distributed to isolated areas of LEDCs.
2. The poorest farmers cannot afford to buy HYV seeds, fertilizers and pesticides.
3. The poorest people have no money to buy the extra food.

By 1990, 75% of Asia's rice areas and 50% of the wheat areas in Africa, Asia and Latin America were using the HYVs. 70% of the world's corn is also from HYV seed. The methods used in the green revolution have been successful, in that they have increased food production. However, the rapid increases in yields have slowed down, yet population is still rising. People are still dying from hunger. The world still needs more food. Perhaps the solution lies at a local level, where communities can take control of their own food production.

Chapter 7: Water and food supply

> The only alternative is to create productive small farms. This is the only way to end rural poverty, feed everyone and protect the land for future generations.

Zambia's drought — Case Study

New seeds and traditional methods beat Zambia's drought

A drought-resistant legume crop, called cowpea, is helping to feed people in Zambia. Thirty farms are taking part in the scheme, where crops are rotated. Maize (corn) will grow better in a field previously used for legumes, as these fix nutrients in the soil.

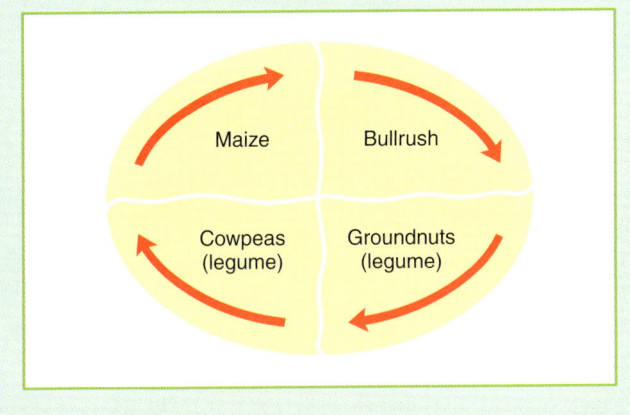

Figure 7.28 What LEDCs need

Figure 7.29 Opinions about technology and farming

Beyond the green revolution

Many farmers in LEDCs are given little say over the use of GM crops. In March 2000, at a meeting in India, farmers were given evidence from interested groups, for and against the use of GM crops. After one and a half days of deliberation, the verdict was 9 to 4 to reject the use of GM crops on their land until their safety could be guaranteed. They suggested **organic farming** and crop diversification as a more environmentally friendly option for India. This view was published widely in Indian newspapers.

GM crops: will they benefit LEDCs?

Some geographers argue that the world produces enough food. They state that people are still hungry and malnourished on account of poverty and the food not being in the right place at the right time. They argue that the best way to solve problems of food supply (sometimes called food security), are to use sustainable schemes, run by local people. These put people in charge of their own food supply. Some scientists argue that it is morally wrong to tamper with nature and that we are unsure of the environmental effects of introducing genetically modified crops. Will we be able to control the spread of pollen from the crops? They also argue that the companies who produce GM seeds will control most of the world's food production.

However, others argue that GM crops should be a part of a sustainable method for meeting the world's food needs. Crops can be 'designed' to grow well in a particularly dry region. They can be 'designed' so that they are more nutritious, solving problems of malnutrition.

In Kenya, many luxury crops are grown to sell to other countries, e.g. vegetables and flowers. These make a lot of money for the producers. However, this means that fewer crops, such as maize and potatoes, are grown for local people. This leads to food shortages and an increase in food prices. Some luxury crops, such as tomatoes, also use up a lot of water, which is already in short supply. In 1998, tomato growers in Kenya were given access to the GM 'FlavrSavr' tomato. This grows and sells well and more farms are now switching to growing crops such as this to sell abroad.

Food technology

Many experts believe in changing the type of crops, away from those traditionally grown, towards plants which are easier to cultivate and are more nutritious. Soya and hemp (also used in textile production) have been introduced into some areas. Both plants are very nutritious and tasty and can be processed to make a range of foods that contain nutritional balance.

The use of simple, sustainable technology to process locally grown food is felt by many to be a long-term solution to the problem of malnutrition. The production of soups, spreads and purees give food 'added value'. It puts local people in control of their food supply and helps feed people with cheaper, cleaner food than the raw product. It also creates local employment and can assist in educating people about nutrition.

ACTIVITIES

1. What evidence is there of progress made in feeding the world's population? (C1.2)
2. What was the green revolution? (C1.2)
3. Describe three methods of increasing food production used in the green revolution. (C1.2)
4. Explain how crop rotation may help farmers produce more food. (C1.2)
5. Could the use of GM crops help improve food supply in LEDCs? Copy and complete the table below, to give arguments for and against their use. (C1.2)

For GM crops	Against GM crops

6. Explain how changes in food technology could reduce levels of malnutrition.

Food and water aid

Overseas aid

Many organisations give aid to LEDCs. The World Bank has funded many large-scale development projects, such as the building of dams. Governments of MEDCs set money aside for overseas development. Many people feel that MEDCs have a responsibility to give aid to LEDCs. First, in that we are all global citizens and as such, are bound to help each other. Secondly, to stop areas of the world becoming unstable. Thirdly, as the legacy of colonialism; many feel that LEDC countries were exploited by MEDCs when they were colonies and that this has kept these countries poor. It is time to repay this debt.

Much of the aid given to improve water and food supplies is from Non-Governmental Organisations (NGOs) such as Oxfam, WaterAid, ActionAid and Comic Relief.

British overseas aid

The British government has a Department for International Development (DFID). This department is allocated billions of pounds each year, to implement 'national strategies for sustainable development in all countries …. so as to ensure that current trends in the loss of environmental resources are reversed … by 2015'.

Britain is a major aid donor, particularly in Commonwealth countries. Here, money is provided as grants for schemes to improve the growing of crops and rearing of animals. Money is also available for schemes which reduce soil erosion and for scientific research, education etc. The department also contributes to other international bodies with an interest in improving food and water supplies. These include the European Union's aid programme, the United Nations and non-governmental organisations, such as Save The Children. Figure 7.30 shows some examples of schemes funded by British aid.

Non-Governmental Organisations (NGOs)

These are charitable organisations. They receive donations from the public and raise money through campaigns, collections and charity shops. This money is then used to fund schemes, many in LEDCs, and mainly concerned with improving food and water supplies in these areas. The NGOs provide short-term food aid in drought stricken areas and long-term help to improve food and water supplies in a sustainable fashion.

Chapter 7: Water and food supply

Figure 7.30

Rice paddies

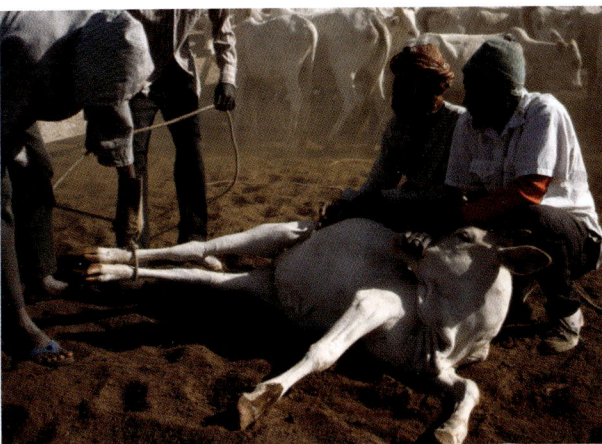
Branding animals

Growing clove trees

A village well

Other international bodies such as the World Trade Organisation (WTO) can also play a role in world food supply. There is now a global economy. Most countries now trade outside their national boundaries. This international trade can affect food supplies in a country. The WTO has developed international trading rules, which 142 countries have signed up to. Many MEDCs have decreased or scrapped tariffs on imports from LEDCs. A declaration signed at Doha, in Qatar, in 2001 states that LEDCs should be given special rights, to increase their trading opportunities.

ActionAid   Case Study

Helping poor people get access to food is a priority for ActionAid. It provides farmers with seeds, basic tools, and loans for livestock and fertilisers (where essential), and helps them learn how to use the most appropriate methods.

This approach is vital when many of those whom ActionAid works with have inherited only a part-share of an already small area of land.

In countries such as Ghana, Kenya and Ethiopia, ActionAid monitors local livestock and grain prices, school attendance, and rainfall for signs of food shortages, so it can take action in time to avert a crisis.

ActionAid works with local groups and community organisations to ensure that change is appropriate and long-lasting, including:

- encouraging farmers to cultivate crops which are most suitable for their local conditions and which help the land regain its fertility
- help farmers take practical steps to reclaim and safeguard new land for agriculture

cont.

Section 2: Managing the Physical Environment

ActionAid — Case Study

cont.
- training farmers to build terraces, plant hedges, compost, and use suitable seeds for local conditions, that give good yields
- encouraging the use of demonstration plots to teach others how to use new production techniques
- providing advice on day-to-day animal care, and training veterinary assistants
- helping to vaccinate animals against common but preventable diseases

Figure 7.31

Global Water — Case Study

Organisations such as Global Water manage projects to supply clean water to isolated areas. The Rural Outreach Water Supply Programme consists of drilling wells and installing pumping systems. These may be windmill, solar, diesel or animal/human powered. Water treatment equipment, such as filters, purifiers and disinfectant are provided, storage tanks and containers installed and local people trained to use and maintain the equipment.

ACTIVITIES

1. Why do some people think that MEDCs should give aid to LEDCs? (C1.2)
2. Describe ways in which the British government helps to improve food and water supplies in an LEDC. (C1.2)
3. What is an NGO? (C1.2)
4. Describe ways in which an NGO can help to improve the food or water supplies in an LEDC. (C1.2)

Comic Relief — Case Study

Comic Relief's website is www.comicrelief.com, where you can find out more about the cause. Their mission statement indicate their aims: 'Comic Relief is seriously committed to helping end poverty and social injustice in the UK and Africa.

The Central Dryland Development Project

Background

The Republic of Dryland is a small country in the Southern Hemisphere. The Great Desert covers approximately half the country to the North and East. The Central district of the country is in an area of unreliable rainfall. Very little rain falls in the area for much of the year and, often, the Southern River is reduced to a trickle. In the short rainy season, flash floods occur and, at Capitalville, the river channel is filling up with silt. Soil erosion is a serious problem in the Central district. The area has suffered from a two-year-long drought and food is in short supply. The situation is made worse as there are large numbers of refugees escaping from the civil war in neighbouring Wetland.

The people of Central Dryland have appealed to the government for help to develop their region. Dryland is a Less Economically Developed Country (LEDC) and the government has limited money and resources available with which to help its people and develop the country. Any decisions about spending must be very carefully considered.

Proposed new developments

The Regional Development Council is meeting in the city of Capitalville. There are four proposed new developments for them to consider. They have to decide which development they will allow to go ahead in the Central district of the country.

1. It is proposed to ask a foreign government and the World Bank to help fund the building of a 100-metre-high dam across the Southern River, at Homestead village. This would create a 75-kilometre-long lake behind the dam.

2. It is proposed to build a system of earth and stone bunds on 50 farms around Homestead village. The materials will be from local sources. The Regional Development Council will provide money for technical help and any equipment needed.

3. It is proposed to create a system of irrigation canals on two large farms, near Homestead village. Concrete lined canals will carry water from the Southern River to the fields. Electric pumps will also draw water from the groundwater store. This will then be pumped into the irrigation canals. It will also feed a new sprinkler system.

4. It is proposed to drill a new well at Homestead village. An American aid agency has offered to supply well drilling and water purification equipment. The agency will also install two large, polythene water storage tanks, just outside the village. The foreign government will fund and build a new community and education centre, next to the new well. They will also fund a water distribution project. The aid agency has offered to train local people to manage the project and to build an irrigation system, using local materials.

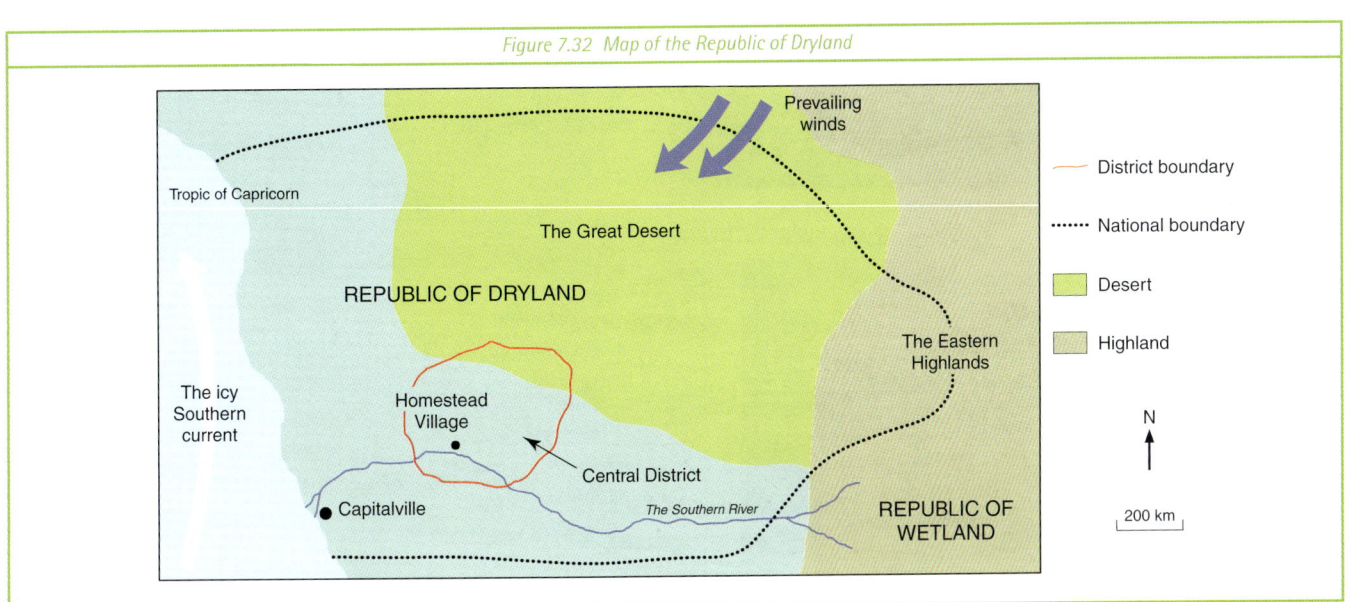

Figure 7.32 Map of the Republic of Dryland

Section 2: Managing the Physical Environment

Figure 7.32

High dam

Earth bund

Irrigation canals

New well

People interested in the proposed developments

Business manager

I want my business to make a profit. I need plenty of fresh water to put on my fields, in order to grow green beans and genetically modified tomatoes. These are for export abroad.

Local councillor

I want a scheme which will bring fresh water and food to the whole community. Our people need training opportunities. They should be able to help themselves.

Local farmer

I just want to be able to feed myself and my family.

Government minister

I want to put Dryland on the map. I want people from other countries to visit our country. We need a development, which will give us some prestige.

Factfile — Dryland

- GNP (per capita) US dollars — 320
- Adult literacy (per cent) — 54
- Population employed in agriculture (per cent) — 71
- Birth rate (per thousand) — 56
- Death rate (per thousand) — 15
- Infant mortality rate (per thousand) — 64

Chapter 7: Water and food supply

Tasks

You are an adviser to the Dryland Regional Development Council.

1. Draw a series of sketches to explain to the people of Central Dryland why their area receives such little rainfall.
2. Write a report for the Dryland government describing, 'The state of the nation'. Include information about the country's problems and living conditions. (C2.3, IT1.2, PS1.2)
3. Look at the needs of the four people interested in the development plans. For each, suggest which of the four developments they would favour. Give reasons for your choices. (C2.3)
4. Write a report to the Drylands Regional Development Council. In your report, state which of the four development plans you think should go ahead. Give the advantages and disadvantages the scheme will bring for the people and the environment. (C2.3, PS1.2, IT1.2)

Research

The 'Central Drylands Development Plan' is an example of a decision-making exercise. It will also provide opportunities for the use of ICT in the presentation of work. Further use of ICT could be in the research of issues surrounding food and water supplies. This can be carried out by drawing on information found on Internet websites; there are many hundreds dealing with the topic. Run a search on key words such as these:

famine, drought, desertification, desert, soil erosion, malnutrition, green revolution, food supply (security), GM crops, sustainable development, overseas aid.

Here are a few examples of web sites that may be useful in research:

www.comicrelief.org.uk
www.oxfam.org.uk
www.oneworld.org
www.learn.co.uk
www.unccd.int
www.padrigu.gu.se
www.actionaid.org
www.foodfirst.org
www.fao.org
www.greenpeace.org
www.cafod.org.uk
www.jhuccp.org
www.malnutrition.org
www.kerrcenter.com

Some areas for further research

1. Why is the location of the Sahara desert different to that of other deserts?
2. Sudden heavy rainstorms occur in deserts. What type of rainfall is this and why does it occur?
3. Deserts have a large diurnal range of temperatures (day/night). Why is this so?
4. What is de-salinisation? How does it help in increasing fresh water supply?
5. In what ways is water, a scarce resource, polluted? What can be done to reduce levels of water pollution?
6. Investigate a case study of a large-scale dam.
7. Compare a scheme to improve water and food supplies in an MEDC with one from an LEDC.
8. Investigate the work of an aid agency or international aid organisation. Describe how the scheme improves food and/or water supplies.
9. Do you agree with the large-scale use of GM crops in order to improve nutrition in LEDCs?
10. Investigate ways in which food technology techniques could improve food supply.

Exam-style questions for Chapter 7

Question 1 – F Tier

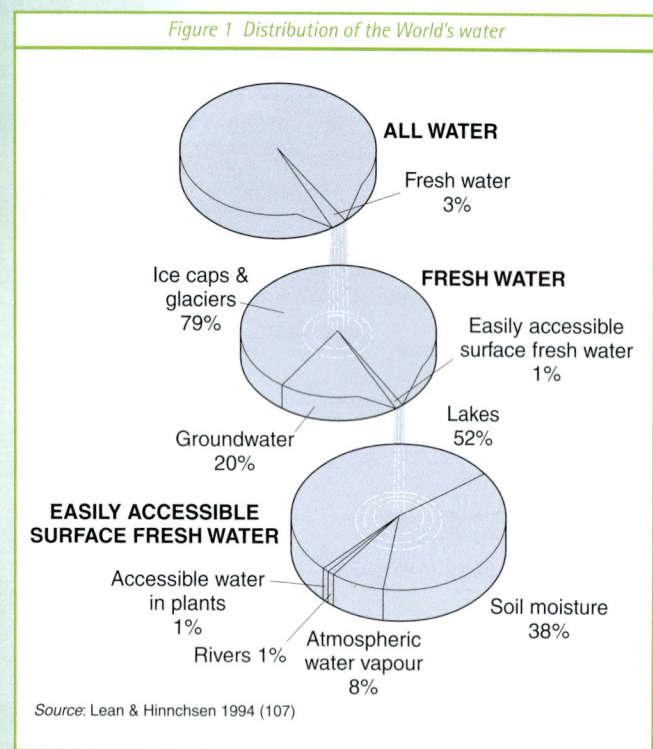

Figure 1 Distribution of the World's water

Source: Lean & Hinnchsen 1994 (107)

(a) Study Figure 1.

  (i) What percentage of all the world's water is fresh water? **(1)**

  (ii) What percentage of the world's fresh water is not easily accessible? **(1)**

  (iii) What percentage of the world's surface fresh water is found in rivers? **(1)**

(b) Study Figure 2.

Figure 2 Population size and expected water use in some water-short countries

Country	Population 1995 (millions)	Cumecs of water per capita 1995	Population 2025 (millions)	Cumecs of water per capita 2025
Algeria	28	527	47	313
Egypt	62	936	96	607
Israel	5	389	8	270
Libya	6	111	13	47
Yemen	15	346	40	131

  (i) Which country will have the least amount of water per capita in 2025? **(1)**

  (ii) Which country's population will increase the most between 1995 and 2025? **(1)**

  (iii) Suggest **two** reasons why less water per capita will be available to many countries by 2025. **(4)**

(c) Study Figure 3.

Suggest reasons why some places get little rainfall. Use Figure 3 and your own knowledge. **(6)**

(d) What is desertification? **(1)**

(16 marks)

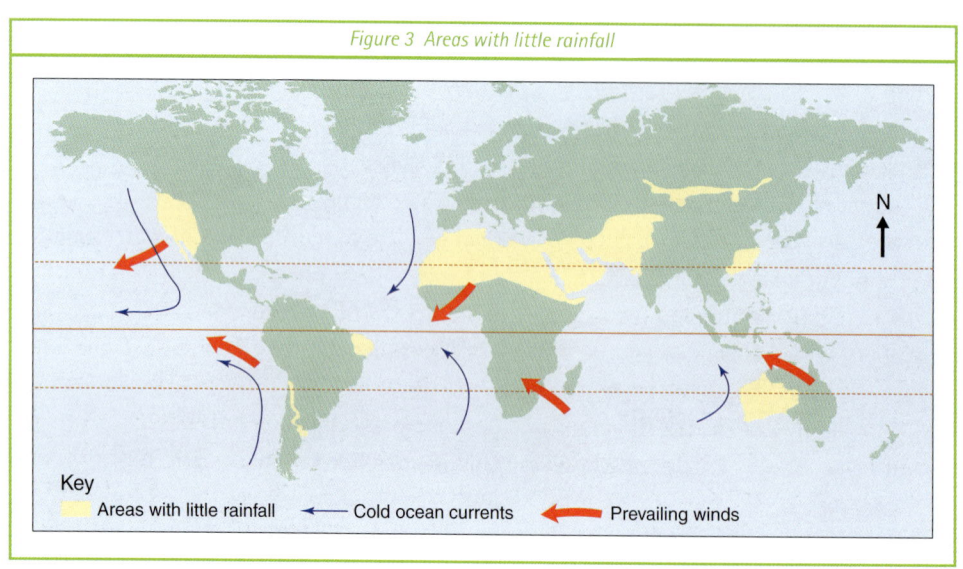

Figure 3 Areas with little rainfall

Key: Areas with little rainfall — Cold ocean currents — Prevailing winds

Chapter 7: Exam-style questions

Question 2 – H tier

(a) Farmers may mis-use farmland when they try to grow more food. This may lead to soil erosion and desertification.

Study Figure 4 below.

(i) Why might an area, such as the one shown in Figure 4, suffer from soil erosion? Use the information in Figure 4 and your own knowledge. **(6)**

(ii) Suggest **two** ways in which water and soil may be conserved, in an area such as that shown in Figure 4. **(2)**

(b) Many agencies are involved in the management of food and water supply. These include the following:

- Charity organisations
- National governments
- International organisations

Explain how one or more of these organisations may improve food and water supplies in LEDCs. Use examples from one or more LEDCs that you have studied. **(6)**

(30 marks)

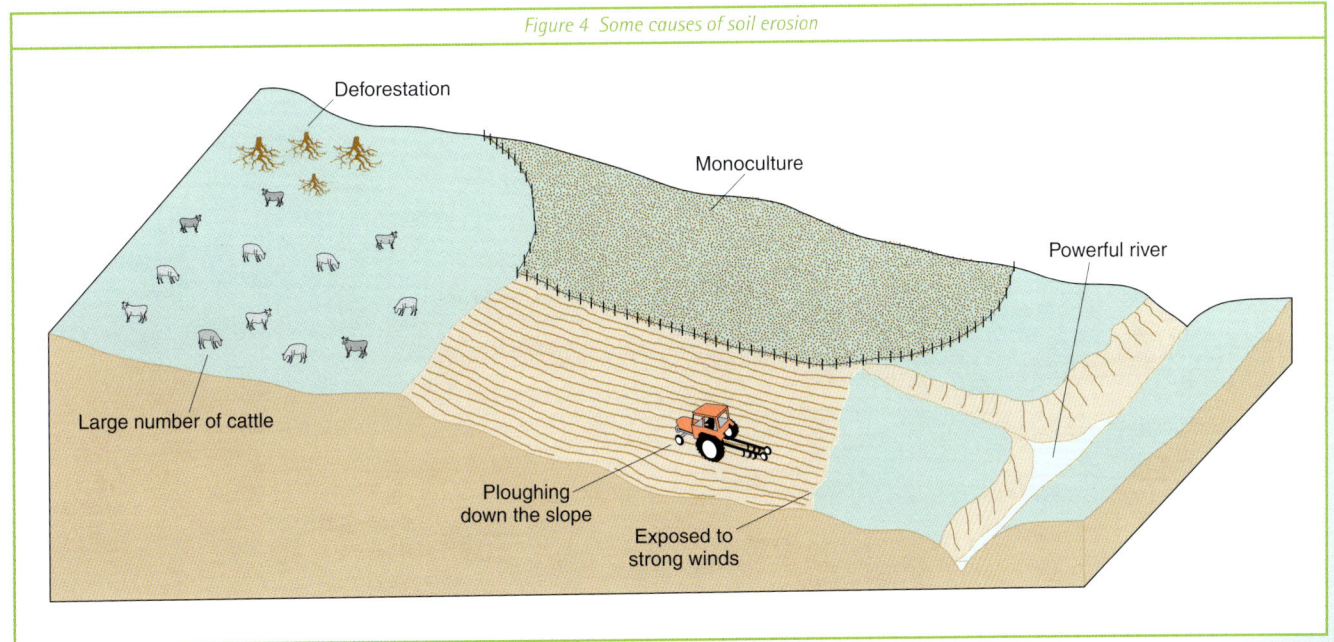

Figure 4 Some causes of soil erosion

Commentary on exam-style questions for Chapter 7

Question 1 (a) is testing the ability to interpret a pie chart and is, therefore, a skills question. There is one mark per question: (i) __3__ % (ii) __99__ % (iii) __1__ % At the H tier, a question testing higher order skills and carrying more than one mark may be set, e.g: Summarise the information shown in Figure 1. **(3)**

Parts (b)(i) and (ii) are also skills questions, testing the ability to extract information from a table. They carry one mark each: (i) Libya (ii) Egypt. The latter part is more demanding, as a calculation has to be done. Part (b)(iii) asks for two reasons, but carries 4 marks. The response will therefore have to be developed in each case, e.g. 'The populaton of the country is increasing rapidly (1) whereas the amount of water available remains the same (1)' Or, 'Climate change in some areas means that there is less rainfall (1) and groundwater stores are reduced (1)'. At the H tier, there may be no separate skills questions. A candidate may be asked to describe and explain the changes to water per capita for 6 marks. The mark scheme would be levels of response, looking at the quality of the overall answer, rather than individual point marks being awarded.

Question part (c) would be common to both the H and F tiers. It would be marked according to levels of response. Low level answers (1 or 2 marks) may be simple references to the latitude of a particular place, rains failing or, to rain falling only at 'certain

87

Chapter 7: Exam-style questions

times of year'. Middle level answer (3 or 4 marks) would be more specific and may refer to explanations of dry winds, rainshadow areas, descending air at the tropics, onshore winds crossing cold ocean currents, or they may discuss the non-effectiveness of rainfall due to high evaporation rates. Actual case study examples of these factors would take the answer to the higher level, as would further development of the point.

in part (d) a definition is required and 1 mark would be given for a reference to places being turned into deserts.

Question 2 (a)(i) would be a common question and would be marked using levels of response. Low level answers would merely state a problem. e.g 'Farmers overgraze areas of poor grassland'. Development of the point would be a middle level answer, e.g, 'Large numbers of cattle eat all the grass and roots, plants will not grow back and the soil is exposed to the wind or rain'. Continued development, i.e. references to the processes by which wind or rain would be able to erode soil, would access the higher level marks, as would actual case study examples. In part (ii), any two named schemes would access 1 mark each, e.g. 'contour ploughing' or 'animal husbandry'.

In part (b) levels of response marking would again be used. Low level answers may refer to short term emergency aid, whereas middle level answers may refer to funding for sustainable, locally run schemes or provision of appropriate technology. Actual case study examples would also take the answer to the higher level.

Pressures on the physical environment

Chapter 8

Tourism has become the world's fastest growing industry.

Attractive scenery encourages visitors

A recent survey of visitors to the Lake District National Park listed these reasons for their visit:

1. Scenery and landscape — 62%
2. Enjoyed previous visit — 34%
3. Fresh air — 29%
4. Peace and quiet — 26%
5. Outdoor activities — 22%

Honeypot pressures

A survey also discovered why people did not enjoy their visit to places like Windermere:

too commercialised, overcrowded, just like a seaside resort, too much traffic, and parking problems.

Conflicts in the countryside

How can the National Parks manage the conflicts between: farmers, visitors, local residents, the Ministry of Defence, quarry companies, the Forestry Agency and many other different groups?

Can conservation and recreation and development live side by side?

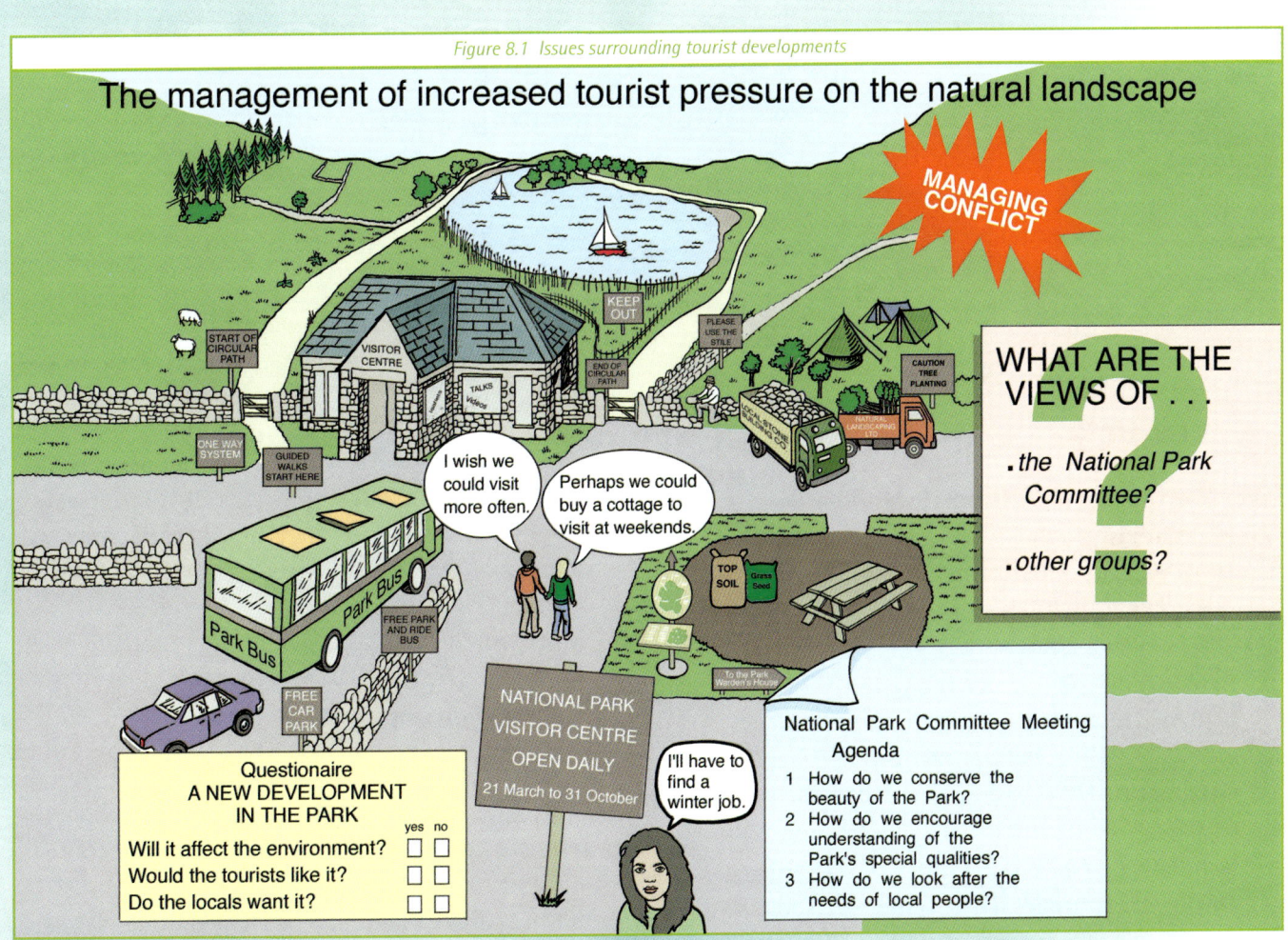

Figure 8.1 Issues surrounding tourist developments

89

Section 2: Managing the Physical Environment

Attractive physical environments 1 – Valleys and hills

In hilly areas, the land is high because the rocks have only been partly worn down or eroded. This may be because the rocks are hard like granite or some forms of sandstone. Rainwater falling on the hills runs overland into streams, which then cut down into the rocks creating **'v-shaped' channels**. Streams also move from side to side leaving **interlocking spurs** of land, which obstruct views along the narrow valleys. Upland streams appear to be fast flowing, but this is misleading. It is actually only turbulence using up most of the stream's energy.

In some places, harder layers in the rocks may resist erosion and form **waterfalls**. Water plunges over these falls often carving out a **plunge pool** in the softer rocks below (see Figure 8.2). In time the overhanging rock weakens and collapses, and the waterfall retreats slowly upstream. This may lead to a steep-sided gorge being formed downstream. Material in the stream is at this stage relatively large in size and angular in shape.

ACTIVITIES

1. Draw and label a simple diagram (side-view) of a waterfall like the one seen in Figure 8.2. (C1.3)
2. Add notes to explain the processes involved. (C1.3)

Figure 8.2 The spectacular valley and waterfall features at Thornton Force near Ingleton

Interlocking spurs

Narrow valley and v-shaped channel

Layer of hard limestone

Footpath erosion

Large rocks in streams

Processes in river valleys include:

- **Weathering** – this is caused by frost and rainwater, breaking down the rocks on the valley sides. These then fall or slide down (by gravity) into the valley below.
- **Erosion** by the force of the water is called **hydraulic action**. Rocks in the stream gradually wear away the riverbed and banks by **corrasion**. This erosion also makes the particles being carried by the stream smaller and more rounded. This process is called **attrition**.
- **Deposition** occurs when water levels fall, and the larger rocks are left behind as islands in the stream. This is process is called **braiding**.
- **Transportation** is when streams pick up material and carry it downstream. The current pulls this load along: a process called **traction**. When river levels are high, however, particles are literally 'bounced' along the riverbed (**saltation**). When a river floods, the water becomes discoloured and this is due to finer material being carried downstream **in suspension**.

Downstream **meanders** develop as the river's gradient becomes gentler and the current swings from side to side. Stronger currents undercut the outer bank to form **river cliffs** while **point bars** of gravel or sand are deposited near the inner bank where the current is slower. (See Figure 8.24, page 102)

Other hills and valleys may be shaped differently. Some are made up of permeable rocks like limestone, which produce distinctive scenery. Rocks stand out on the valley sides as vertical **scars** or cliffs. Beneath these are **screes** of loose rock, which has fallen after having been broken off by frost. Elsewhere there are horizontal **limestone pavements**.

Cracks and layers in the limestone let water pass through and allow streams to sink and flow underground, leaving **dry valleys** on the surface.

Beneath the ground, running water carves out **potholes** and **caves**. These underground streams later reappear from caves or as springs. Rainwater also dissolves the limestone (calcium carbonate): a process called **solution**. This produces 'hard' water. In drier conditions dripping water can reverse this chemical process and build stalagmites and stalactites. Heavy rainfall can soon fill up underground passages lifting the **water table** or water level in the rocks (see Figure 8.3).

In the western part of the Yorkshire Dales, different types of rocks have created interesting scenery. Ingleborough Hill (723 m), with its network of footpaths (the Three Peaks), cave systems (Gaping Gill) and limestone scenery, overlooks wooded dales (see Figure 8.4).

ACTIVITIES

1 Explain why cave systems form in limestone. (C1.1)

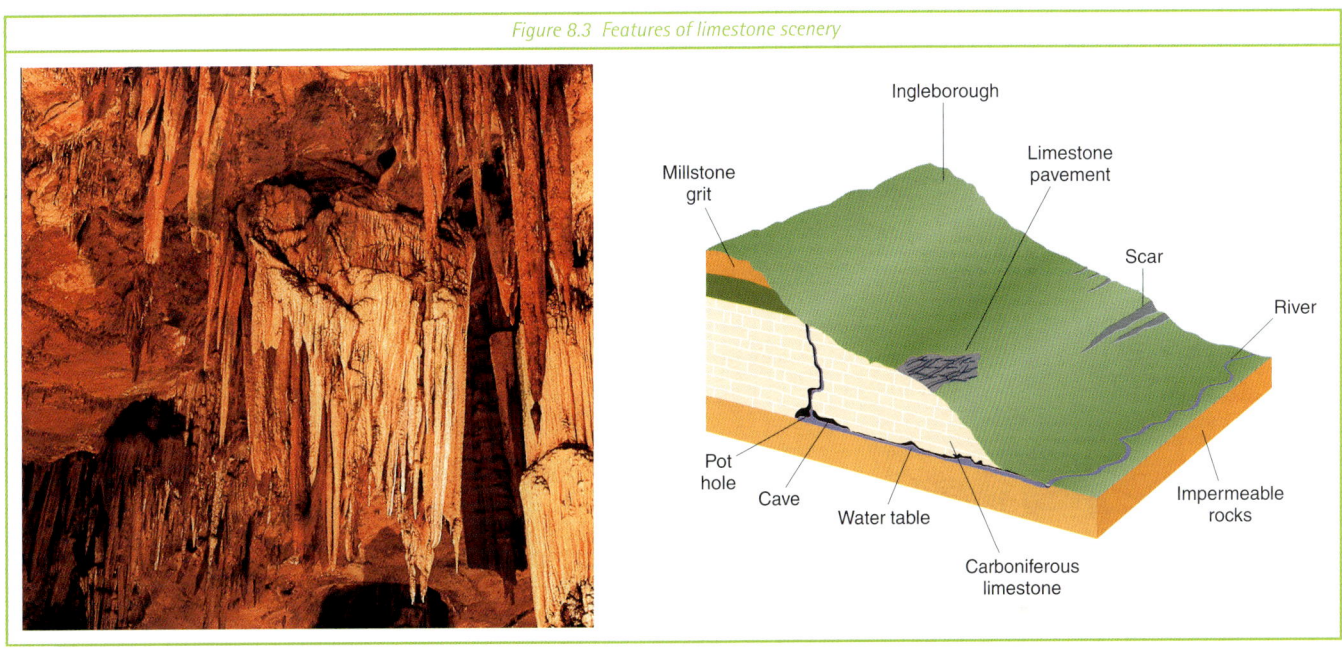

Figure 8.3 Features of limestone scenery

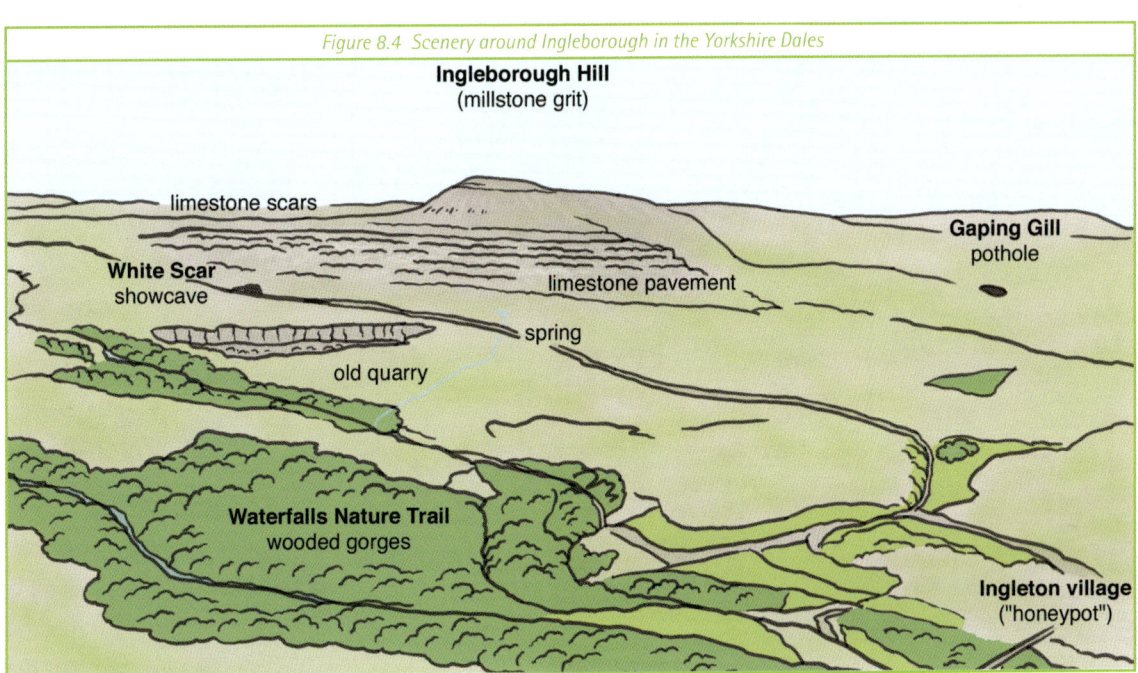

Figure 8.4 Scenery around Ingleborough in the Yorkshire Dales

Section 2: Managing the Physical Environment

This scenery attracts very large numbers of visitors in the summer months. In addition to the numerous walkers, climbers and potholers who come to the area, less adventurous visitors can enjoy the landscape by following the wooded Waterfalls Walk, a 4-mile nature trail, or by entering the show cave at White Scar. Both are near to Ingleton.

The village of Ingleton is a **honeypot**. Many of the local businesses here are increasingly geared to recreation and tourism – there are numerous cafes, souvenir and craft shops. There are also specialist outdoor activity businesses, and various campsites, caravan parks and other sorts of accommodation in this area. Jobs in these businesses are mainly seasonal or part-time.

ACTIVITIES

1. Using this map extract (Figure 8.5) give the grid references for Thornton Force and White Scar cave, (to 4 Figures) and Ingleborough Hill and Gaping Gill (to 6 Figures). (N1.1)
2. Explain what happens to Fell Beck as it flows south from Ingleborough, to Clapham village. (C1.2)
3. Suggest why this area attracts so many different sorts of visitors. (C1.2)
4. Identify three tourist facilities in or near to the 'honeypot' of Ingleton. (C1.2)
5. Make a list of the effects of tourism on (a) people and (b) the environment. (C1.3)

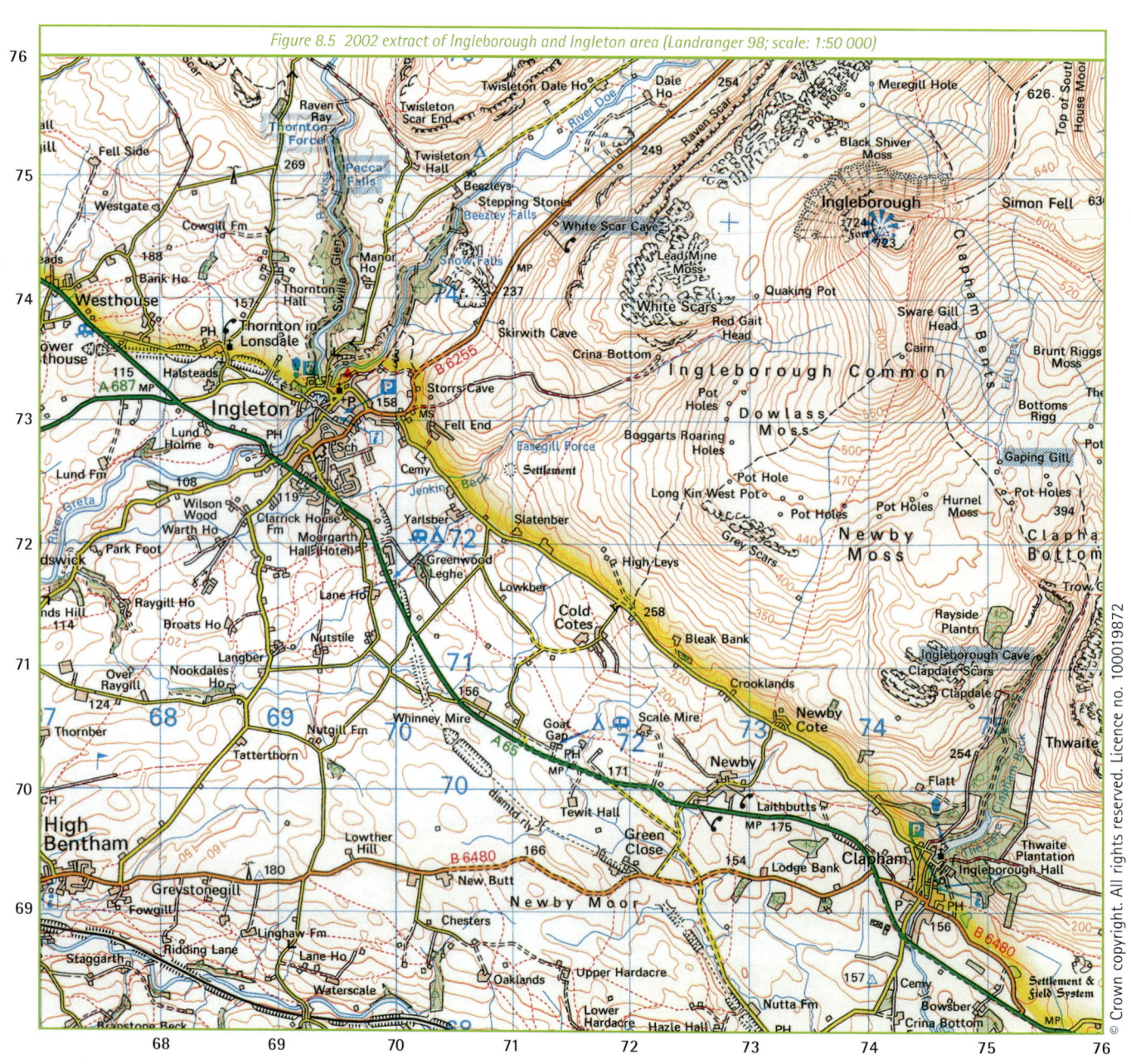

Figure 8.5 2002 extract of Ingleborough and Ingleton area (Landranger 98; scale: 1:50 000)

Attractive physical environments 2 – cliffs and beaches

Coastlines are a common feature of an island like Britain. They vary depending on the types of rock and the height of the land. The traditional picture we have of the coast is one of cliffs, beaches and holiday resorts. In eastern and southern Britain, Chalk rocks resist erosion and stand out as **headlands** along the coast. Clay on the other hand is much weaker and is easily worn away to forms **bays**.

However, over time even chalk cliffs will be eroded by the sea. One way in which this happens is when waves trap air in cracks in the cliffs. The pressure from this **hydraulic action** can open out caves or cut a **notch** at the foot of the cliff. Rocks will then fall onto the beach or the **wave cut platform** below (see Figure 8.6A). Even headlands can be eroded and this takes place in a series of stages, creating caves, **arches** and **stacks** (see Figure 8.6B).

Clay can be quickly washed away by waves, but much of the damage begins on land. There rainwater saturates the cliffs and **landslides** take material down on to the beach below. This may threaten houses or hotels built along the shore. The weight of these buildings will in fact make matters worse causing **slumping** (see Figure 8.7A).

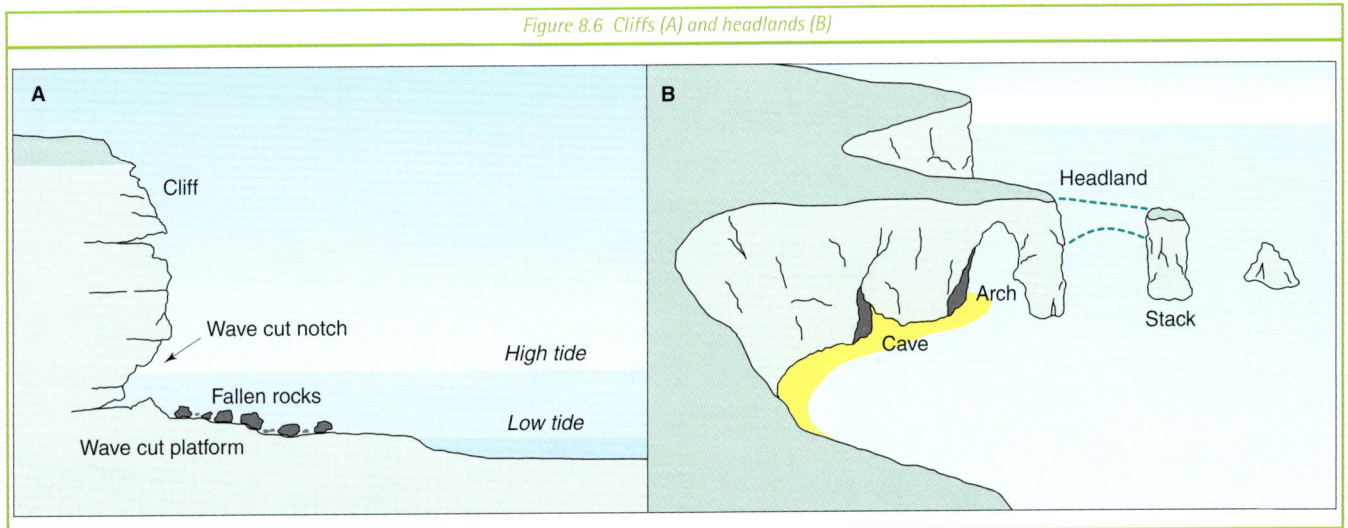

Figure 8.6 Cliffs (A) and headlands (B)

On the beach below, waves can either bring new sand or take it away, depending on the weather and other factors. Where they strike the coast at an acute angle, **longshore drift** occurs. The incoming **swash** brings sand up the beach at an angle, but the **backwash** removes it at right angles. This moves sand along the beach in a sort of zigzag pattern (see Figure 8.7B).

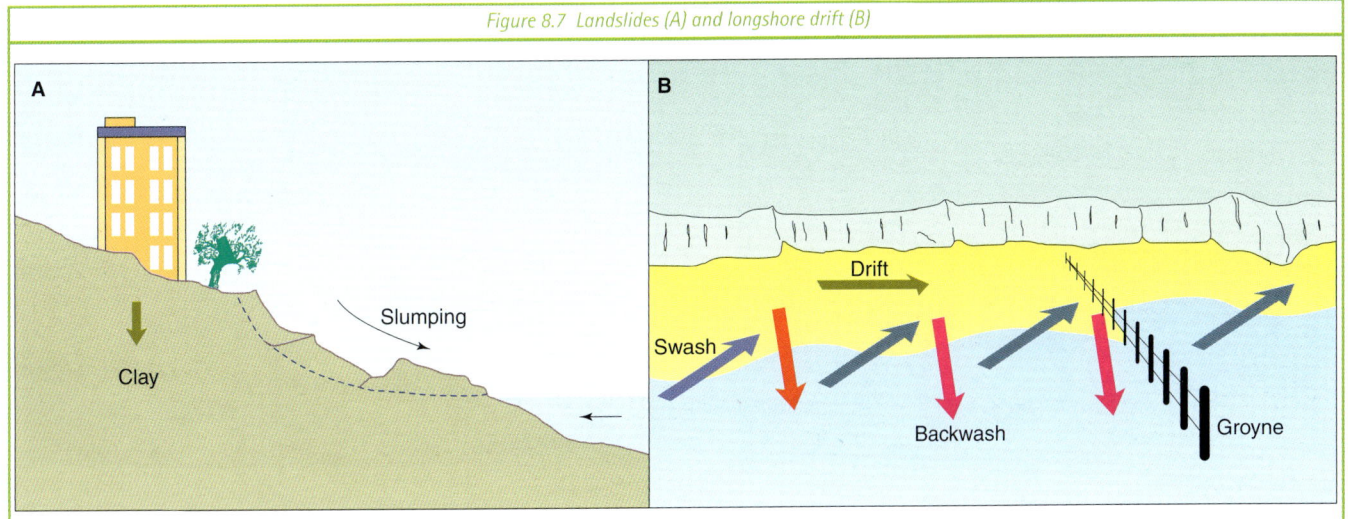

Figure 8.7 Landslides (A) and longshore drift (B)

Section 2: Managing the Physical Environment

Where waves and currents are less strong, and the water is calmer, sand and shingle are deposited. If there is a change in the direction of the coast or a river estuary, the longshore drift may build a **spit** (see Figure 8.8C). Silt from the river also collects here as the tide comes in. **Ecosystems** like sand dunes, salt marshes and mudflats form, providing places where wading birds and other wildlife can breed or rest whilst migrating.

The east coast of Yorkshire has many attractive coastal features along its 60 km Holderness coastline:

- In the north the impressive chalk cliffs at Flamborough Head are a popular beauty spot, and there are sheltered sandy beaches at nearby Bridlington (see Figure 8.8A).
- Hornsea and other small holiday resorts have built **sea defences** to prevent their beaches and clay cliffs being removed by waves and longshore drift (see Figure 8.8B). The sea has destroyed some thirty villages along this coast since Roman times.
- Spurn Head is a spit, which has grown out from where the coast meets the Humber estuary. It has a bird sanctuary and houses a small community of coastguards and lifeboat workers and their families. Winter storms threaten to break through the narrow neck of the spit, turning it into an island (see Figure 8.8C).

ACTIVITIES

1. Using Figures 8.6 and 8.7, describe how cliffs or headlands change. (C2.1)
2. What effects do different types of rocks have? (C2.1)
3. Suggest why a coastline such as that in Figure 8.8 might attract different sorts of visitors. (C2.1)

Figure 8.8 Attractive coastal features along the Yorkshire coast

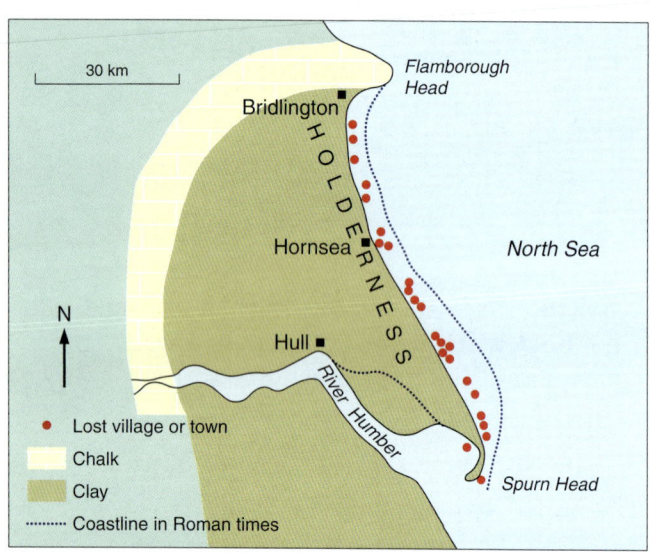

B Hornsea

A Flamborough Head

C Spurn Head

Chapter 8: Pressures on the physical environment

ACTIVITIES

1. Using Figure 8.8A, label the features numbered here on Figure 8.9. (C1.1)

2. From Figure 8.10, make a list of the effects of tourism on (a) people and (b) the environment. (C1.1)

Figure 8.9 Scenery and cliff features at Flamborough Head

Figure 8.10 The impacts of tourist development on people and the environment along the coast

95

Section 2: Managing the Physical Environment

The growth in recreation and tourism

More and more people have the time and money to go on holiday. Seven out of ten people in the UK take one annual holiday and many take two.

Figure 8.11 Number of holidays taken by UK residents per year

number	1971(%)	1991(%)
0	42	32
1	43	37
2	12	22
3 and more	3	9

Often this consists of an annual summer holiday abroad, and another 'short break' either in the UK, or on a skiing or winter sun trip. Many people make single day-visits to city attractions like the British Museum, coastal resorts like Blackpool, or rural Theme Parks like Alton Towers.

However, visiting the countryside remains the most popular outdoor activity in Britain (see Figure 8.12). Memberships of organisations like the Camping and Caravan Club have also risen significantly. The rural tourist industry in the UK is estimated to be worth around £6 billion per year.

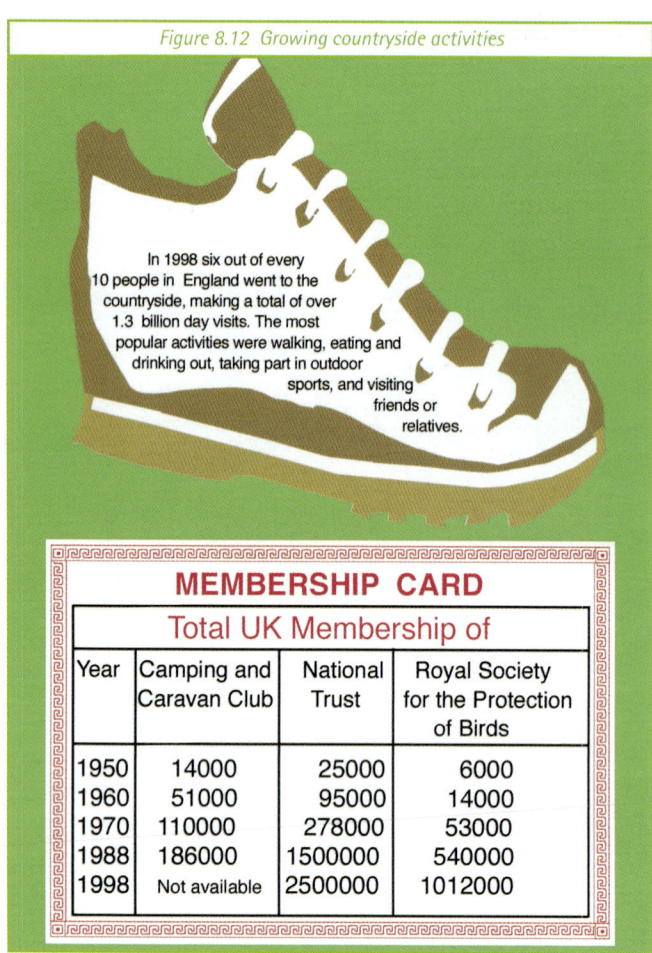

Figure 8.12 Growing countryside activities

In 1998 six out of every 10 people in England went to the countryside, making a total of over 1.3 billion day visits. The most popular activities were walking, eating and drinking out, taking part in outdoor sports, and visiting friends or relatives.

MEMBERSHIP CARD
Total UK Membership of

Year	Camping and Caravan Club	National Trust	Royal Society for the Protection of Birds
1950	14000	25000	6000
1960	51000	95000	14000
1970	110000	278000	53000
1988	186000	1500000	540000
1998	Not available	2500000	1012000

Why has there been this growth in recent years?

There are many reasons why more people go on holiday more often, but it is largely because we are better off and have more opportunities:

- more disposable income (to spend)
- more holidays with pay
- shorter working hours
- more interest in sport and leisure
- more car ownership
- more travel facilities
- more media coverage and advertising

ACTIVITIES

1. Use Figure 8.12 to describe how recreation and tourism are growing. (C1.1)
2. Explain why this is, using data from Figure 8.13. (C1.1)

Figure 8.13 Growth in income and opportunities

UK Manual workers' holidays with pay

Basic entitlement	1951 (%)	1990 (%)
Less than 2 weeks	31	0
2 to 3 weeks	68	36
3 weeks or more	1	64

Disposable household income

Year	£ per week
1993/4	287
1994/5	298
1995/6	307
1996/7	325

UK car ownership

Year	Number
1951	2.2 million
1960	5.5 million
1970	11.6 million
1995	20.0 million
2025	30.0 million (projected)

Chapter 8: Pressures on the physical environment

Accessibility – or how far you have to travel, and how easy it is to get there – is another important factor affecting people's visits to the countryside. The growth of the motorway network in England and Wales, for example, has certainly influenced the visitor patterns in the **National Parks** of England and Wales (see Figure 8.14)

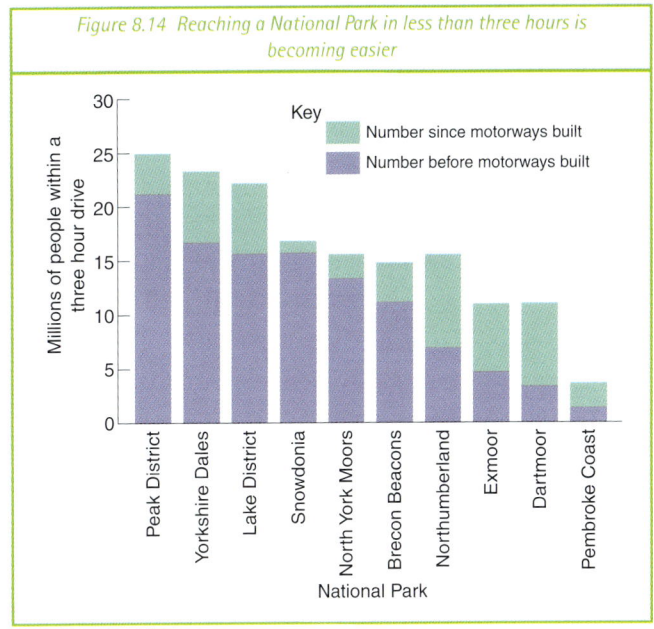

Figure 8.14 Reaching a National Park in less than three hours is becoming easier

ACTIVITIES

1. Using Figure 8.14, find which National Parks had the highest and lowest visitor numbers in the past? Which have seen the largest and smallest changes? (C2.1)
2. Suggest reasons for these answers, using Figure 8.15. (C2.1)

Figure 8.15 National Parks, cities and motorway developments in England and Wales

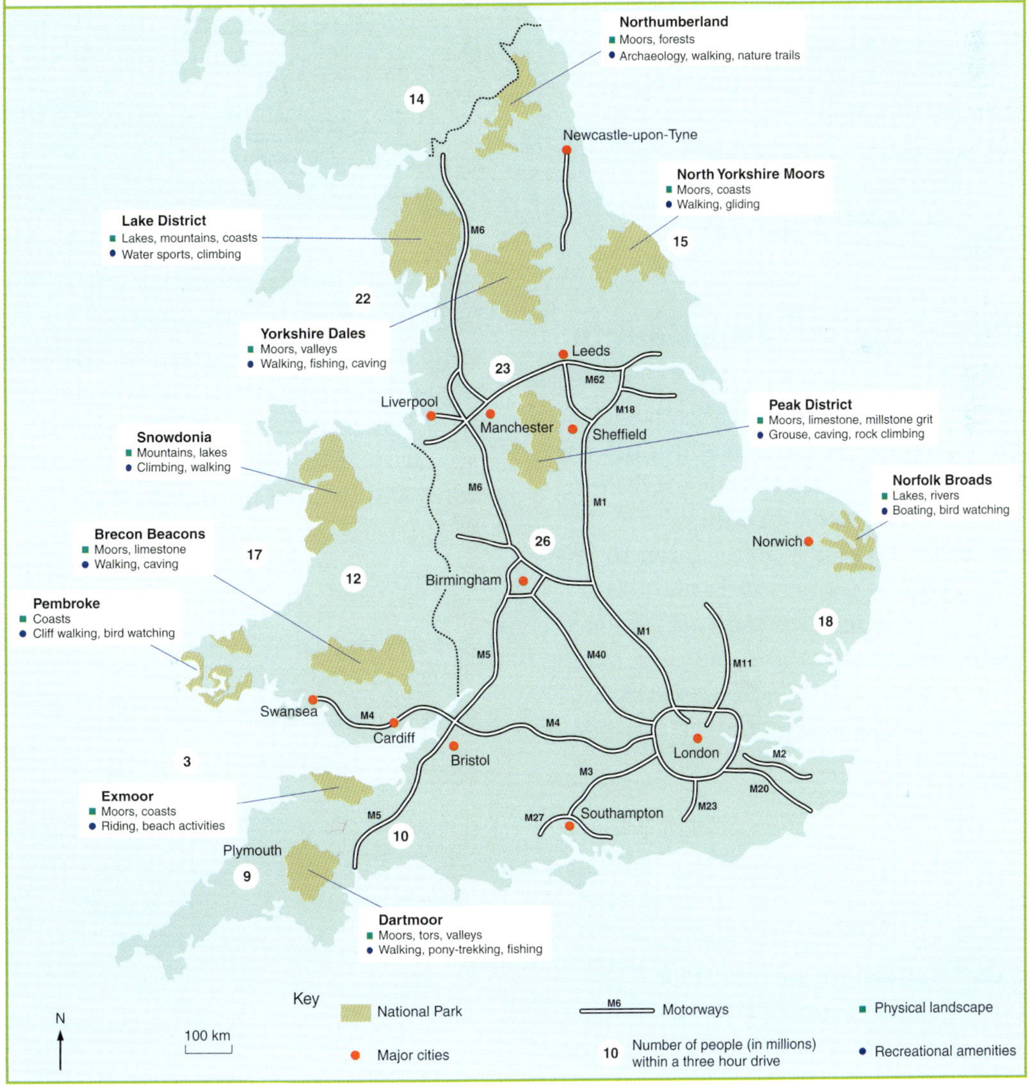

97

Section 2: Managing the Physical Environment

The effects of visitors

Visitors to the countryside make quite an impact. These impacts can be both good and bad:

- the economic benefits include the growth of many new businesses linked to tourism. These create a multiplier effect, which brings new jobs and services. Drawbacks include a change to more seasonal and part-time work and a decline in shops catering for the needs of local people.
- the environmental problems are increasing traffic congestion, greater demand for parking and increased air pollution. Footpath damage is another concern despite some successful repair projects.
- the social effects of a large number of visitors often causes conflict with local residents and farmers. Demand for second homes also pushes up local house prices.

Some people see this situation as being a battle between tourist development and conservation: 'economic gain versus environmental loss'. Other industries like quarrying and agri-businesses, have previously had this reputation. Organisations like the National Parks and the Countryside Agency are trying to help tourism to avoid becoming too commercialised and to develop in a more **sustainable** way.

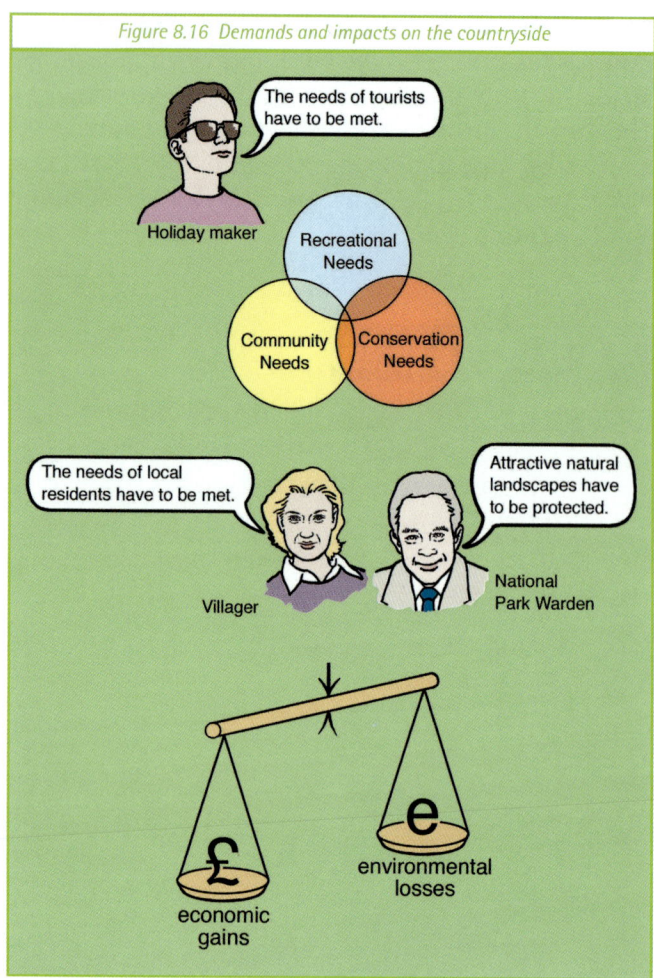

Figure 8.16 Demands and impacts on the countryside

Tourist businesses are very important to rural economies

Around 50% of the jobs in National Parks like the Lake District are in tourism compared with only 6% nationally. This is partly because traditional farming activities are declining. In 1999, farming's contribution to the UK economy actually fell to less than 1%. By 2000, 56% of farmers in England had begun non-farming activities like tourism. This has included providing accommodation, catering, and crafts and souvenir sales. In the National Parks tourism is the major earner and employer:

jobs. Many others, though, are part-time, seasonal, and poorly paid relative to urban areas.

Some businesses close down for the winter and lay off staff. This pattern encourages younger people, in particular, to move away. The focus on visitor services (see Figure 8.18) also threatens the survival of village stores and increases the costs of groceries, etc.

Figure 8.17 Visitor increases			
	Yorkshire Dales	Lake District	Peak District
visitor days	8.3 million	22 million	30 million
spending £	50 million	108 million	137 million

It is estimated that tourism in the Yorkshire Dales National Park has created around 1 000 full-time

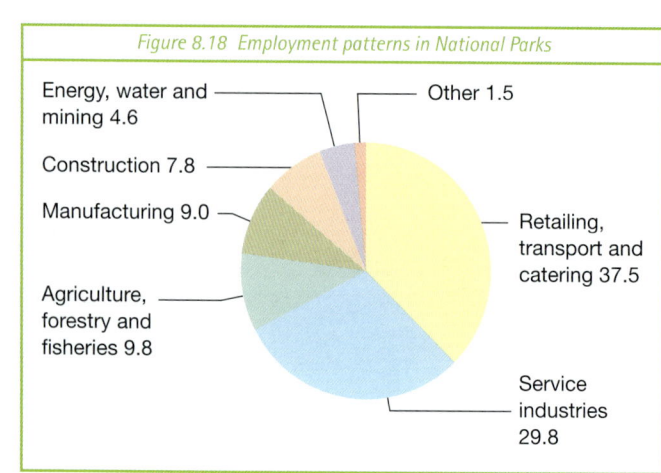

Figure 8.18 Employment patterns in National Parks

98

Chapter 8: Pressures on the physical environment

The environmental impact of visitors is linked to increased road traffic.

- This is because 90% of visitors to National Parks arrive by car. Congestion on narrow lanes leading into the National Parks is a growing problem, both at weekends and in school holidays.
- Peak flows at weekends are often around 11.00 am and 17.00 pm, with a marked lull around 14.00 pm (see Figure 8.19).
- Parking is a major concern. Once official car parks are full, the amount of roadside parking (grass verges and in field gateways) increases.
- Noise pollution (from motor cycles) and air pollution (from car exhausts) are other environmental impacts.
- There is little public transport for visitors or local people to use instead, and coaches do not easily negotiate country roads.

Pedestrian traffic also has impacts.

- Many of the footpaths in our National Parks are literally wearing out (see Figure 8.2). This is because the number of walkers is greater than the paths' **carrying capacity**.
- Other routes are being damaged by new forms of traffic (e.g. bikes and off-road vehicles); uses for which they were not designed.
- Once the damage gets too bad then the paths will not recover without repair work (Figure 8.19)

ACTIVITIES

1. Using this and the previous page, make two lists of (**a**) economic gains and (**b**) environmental losses. (C1.1)
2. Do you agree with the 'scales' cartoon in Figure 8.16? Explain your view. (C1.1)
3. Choose one environmental problem shown in Figure 8.19: Outline its causes and effects. (C2.1)

Figure 8.19 Environmental impacts of visitors to the Yorkshire Dales

Increased visitor traffic has led to a one-way system in Ingleton

Footpath repairs have been undertaken in the 'Three Peaks' area

Section 2: Managing the Physical Environment

The social impact of large numbers of visitors is not always easy to establish. Some locals may support development, whilst others oppose any changes. These different views depend upon people's age and occupation (see Figure 8.20). Younger people may want to see change whereas the old and retired will value 'peace and quiet'. Tourism is only one of many factors bringing change to rural villages.

ACTIVITIES

This could be a group activity where you, the villagers, decide what you want done.

1. Using Figure 8.20, suggest why each person in the list thinks the way they do. (Wo, Ps)
2. Decide who seems happy and who is not. (Wo, Ps)
3. If you were a parish councillor what would you do to try and help some of these people? (Wo, Ps)

Figure 8.20 Social impacts on a rural village

Village News

Farmer 1
It's hard to make a living rearing livestock. The land is not as good as in lowland areas. Visitors are also a nuisance, with their litter, dogs and careless behaviour. Subsidies used to help, but recent problems with BSE and Foot and Mouth have finished me off. I'm selling up.

Farmer 2
Farming is OK but you have to do other things. We are now an organic farm selling milk and our own cheese. I've diversified and we have a campsite for visitors, we do bed and breakfast and two old cottages are let out to families in the summer.

Retired couple
We used to come on holiday and last year we retired here. It's really nice and the sale of our old house outside London allowed us to buy this cottage and a new car, and live on our pension and savings.

Older local woman
I've lived here all my life. Money is very tight and I can't afford to run a car. There aren't many buses either. I am worried about other services like the doctor and whether the village shop will stay open.

Young married couple
We can't afford to live in the village as buyers from outside have pushed up prices. We will either have to live with my parents or move away. Wages in rural areas aren't good and jobs dry up in the winter.

Family group
We bought this cottage as a 'second home' to get away from the stress of city life at weekends. In fact we only seem to get here in the school holidays like at Easter. After all you want to be at home for Christmas and you go abroad in the summer.

Primary school teacher
It's really nice to teach here, you get to know everyone from children to grandparents. But smaller families and lots of retired people means there are fewer children. If classes get too small the school might close, as they have done in some villages.

Village store owners
We sell groceries, which saves people having to travel to town. We can't compete with prices in the town supermarkets so we have to look after our customers really well. We buy from the 'cash and carry' and sell newspaper and have 'off-licence' sales. The large number of old age pensioners helps the post office business. Like they say, 'use it or lose it!'

Craft shop owner
I took over an old shop in the village centre two years ago and I make and recondition furniture. My prices aren't cheap but it's good quality, and much of it is hand-made. I have a small showroom that brings trade from visitors. I employ two people already and may take on more staff during the summer.

Chapter 8: Pressures on the physical environment

National Parks in England and Wales

National Parks and Areas of Outstanding Natural Beauty have been set up to protect and manage areas of valuable scenery. The first ten National Parks in England and Wales were set up during the 1950s, beginning with the Peak District. In 1999 the Norfolk Broads was added to this list, and in the future the New Forest and the South Downs will join them. The scenery and attractions of each of them is summarised in Figure 8.15.

The National Parks have three duties:

1 to protect the environment
2 to promote enjoyment and understanding
3 to look after the interests of residents

This a tough task, as these are conflicting demands (see Figure 8.16). All National Park Authorities must try to find solutions to the problems this brings. Examples of these are shown in Figure 8.21.

Figure 8.21 Some National Park problems caused by visitors, and some possible solutions

Problems	Solutions
footpath erosion	resurfacing or re-routing
traffic congestion and full car parks	'park and ride' or one-way traffic schemes
'honeypot' sites under pressure	direct motorists or walkers to alternative sites
unsuitable developments	refuse planning permission
trespassing and litter	sign-posting and education
damage to fragile ecosystems (woodland)	protect in sites of special scientific interest (SSSI)
too many 'second homes' and holiday lets	provide low cost housing or only sell/let to locals

ACTIVITIES

1 Which are the biggest problems in Figure 8.21? Explain why? (C2.1)
2 Which solutions are most likely to work? (C2.1)
3 Choose *two* of the management strategies shown in Figure 8.22 and explain what they are and how they might help. (C2.1)

Figure 8.22 Three management strategies: planning controls, 'park and ride' and landscape conservation

Landscape conservation – the future

Money would be available to enable farmers to continue to farm livestock, whilst also maintaining traditional landscape features. Farmers would be helped by grants from conservation agencies. Farmers could also supplement their income from farm-based tourism. There would be more heather moorland and flowery meadows than today. Broad-leaved woodlands, walls and field barns would be well looked after. Footpaths would be monitored and repaired (see Figure 8.19).

FREE PARK AND RIDE BUS

Planning controls

- Use traditional building materials
- Designs are to match nearby buildings
- Barn conversions only if next to road
- Large buildings screened by tree planting
- Permission to park large vehicles
- B and B needs planning permission
- Developments must provide jobs
- Affordable housing a priority

Section 2: Managing the Physical Environment

Further tasks

1. Look at Figure 8.23. It shows the range of people who want to use land in the National Parks. Complete the conflict matrix by putting a cross where these groups' views and actions may lead to conflict, and a tick where they might get on. (C1.3)

2. Choose one type of land use. Suggest why it might conflict with the National Parks' duties to provide conservation, recreation and help for locals people and businesses. Choose another land use and suggest how it fits in. (C1.1)

3. Suggest how National Parks can preserve attractive landscapes and also meet people's needs for recreation. (C1.1)

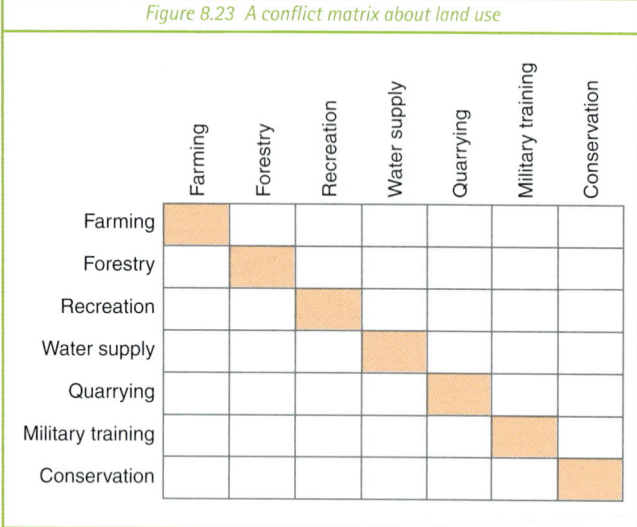

Figure 8.23 A conflict matrix about land use

Figure 8.24 A river meander

Research

1. Investigate what happens to rivers downstream. Use your research to help you label the features of the meander and floodplain shown in Figure 8.24. Describe how the river creates these features. (IT2.1)
2. Research different examples of beaches or cliffs and the coastal defences used to protect them. (IT2.1)
3. Research a different National Park, perhaps one from abroad like Yosemite in the United States. What scenery is found there? Are the problems the same or are they different? (IT2.1, Ps)
4. Research 'honeypot' problems or National Park management strategies. You could use the resources in Figure 8.1 (page 89) to explore issues surrounding tourist development. (Po, Wo)

References
Limestone Scenery in the Yorkshire Dales GeoActive, series 11, unit 227 – summer 2000
The Yorkshire Region, chapter 4, ISBN 0340705108
www.daelnet.co.uk
www.yorkshiredales.org.uk
www.lake-district.org.uk
www.countryside.org.uk
www.nps.gov/yose/guide
Contact the Education Officer, YDNPA, Colvend, Grassington, North Yorkshire, BD23 5LB

Exam-style questions and sample answers for Chapter 8

Question 1 – F Tier

Study the photograph below, which shows a small stream in a hilly area.

Figure 1

(a) Describe the features of the stream and its valley. **(4)**

The stream is flowing down hill steeply. It swings from side to side (meanders?) between steep sides called interlocking spurs. There is a small waterfall in the foreground, and the rocks in the stream are large. It is mostly grass with a few trees higher up, and rock outcrops at the top.

(This answer describes the features using some technical terms, earning all four marks.)

(b) Use the photograph to explain the processes at work. There are some headings provided. **(6)**

(i) **Weathering:** *On the left there are lots of loose rocks which have fallen into the stream.*

(ii) **Erosion:** *The stream is wearing away its channel making it V-shaped. The rocks cut into the bed of the stream*

(iii) **Transport:** *Boulders in the stream are not carried far. They are angular and still quite large. They may move in wet weather when the stream flows faster.*

(This shows good understanding of most processes. However the answer to weathering should suggest how this happens, e.g. by frost or tree roots (perhaps chemically). It therefore earns five out of the six marks.)

(c) Name a landform which you have studied: any river, ice, coastal or other landscape feature:

Name of landform: *Limestone Caves*

Describe and explain its main characteristics. (You may draw a diagram.) **(6)**

Carboniferous limestone is a hard, white rock. It has joints in it that let streams sink down into the ground. The rock is made of calcium carbonate, which dissolves in rainwater. This means that the streams can open up caves underground like Gaping Gill in Yorkshire. They later rise as springs, sometimes in another cave like at Clapham Cave. Stalactites and stalagmites are made when the lime water drips from cave roofs and makes rock again. Water levels in the caves rise and fall with the rainfall outside.

(The candidate has written a good summary of a limestone cave and how it forms. There is also a named example included. This deserves all six marks.

Overall this is excellent, earning 15 out of 16 marks.)

Chapter 8: Exam-style questions

Question 2 – H Tier

(a) Use lines or arrows to match these National Parks with their descriptions. (3)

National Park	Description
Lake District	An area of limestone and gritstone scenery with caves, dales and quarries. The most visited National Park in England and Wales.
Pembroke	An area famous for its glacial scenery, where activities include rock climbing and sailing. It attracts visitors despite its high rainfall.
Peak District	A Welsh National Park that has a coastal footpath and is a place where ornithologists enjoy watching large numbers of sea birds.

(The candidate has matched these correctly – three marks.)

(b) State two duties National Parks have to perform. (2)

1. *To conserve the countryside.*
2. *To provide facilities for visitors.*

(Both answers are correct.)

(c) When tourists visit National Parks, coastal areas and 'honeypot' sites, they put pressure on local people and on the environment.

Describe any two of these pressures using examples you have studied. (8)

1. **Traffic congestion.** *12 million visitors go to the Lake District every year. Many gather in centres like Bowness on Lake Windermere. Most people come by car and so there are problems with traffic. Parking is another problem especially on summer weekends. There are few local buses to help. Noise and air pollution is also caused and Ambleside is another bottleneck. The narrow, steep and winding roads cannot cope with the large volumes of cars and tour buses that arrive.*

2. **Second homes.** *Many properties in the Lake District are either a second home or a holiday cottage. This means that local people can't afford to buy a home in the village where they grew up because they have been bought up by others.*

(The candidate has a good knowledge of traffic problems around Windermere and describes this example well. The second part is less detailed and the problem is not really developed. (for example the problem affects one in six properties, and it is the low local wages that mean this problem affects mainly young people). This keeps the mark total down to five out of eight.

Overall this answer is very good earning 10 out of 13 marks.)

Section 3: Managing Economic Development

Chapter 9: Contrasting levels of development

Measuring differences in development

Understanding varies from person to person

To everyone development means change – change for the better. The trouble is that 'better' can mean different things to different people. We all have different ideas about which changes are improvements and progress.

It has been estimated that a fighter jet or a heavily-armed helicopter would cost an LEDC (Less Economically Developed Country) roughly the same as computer learning for one million people. Whether both, one or neither of these changes amount to development depends on the attitudes and understanding of development that we hold. Some people in Africa, Asia and South America argue that European communities, in which some families abandon their elderly members and in which old ladies are mugged in their homes, lack development.

Below are a number of statements from different people about their understanding of development:

- A UK factory owner and director: "Plenty of employment in industry and trade."
- A 15-year-old Thai schoolgirl: "Feeling secure that I can walk at night without being raped."
- A foreign aid worker in Africa: "A democratic political system so that people can decide about their own future."
- An Australian hairdresser: "All poverty eliminated."
- A UK trade union official: "Everyone having access to education and health care."
- A South Korean manufacturing export manager: "An infrastructure of roads, railways, ports and power supplies."
- A Ghanaian shoe mender: "Security! Having enough for the children to eat is a very important part of security."
- A retired Iraqi army general: "Powerful armed forces so that you can influence other countries."

105

Section 3: Managing Economic Development

ACTIVITIES

1. Which do you think are the most important and least important of these eight statements? Give reasons for your decisions. (C1.1, C2.1a, C2.1b)
2. Are there any other important understandings of development not covered by these statements? If so, what? It may help you to think about your own human needs and those of the community that you live in. (C1.1, C2.1a, C2.1b)

Measuring Development

Measuring differences in development is not easy but a range of indicators are frequently used in an attempt to show that countries and regions within countries do have different levels of development. The most frequently used is **GNP** (Gross National Product) or **GDP** (Gross Domestic Product) per person. These are money totals for the value of all goods and services produced in a year by either all the country's assets home and abroad (GNP) or purely at home (GDP). For an international trading country like the UK, its GNP is usually the larger total. In 1998 the UK's GNP per person was US$ 19 320 yet its GDP per person US$ 18 849. Using only the single indicator of GNP or GDP per person does not give a full and accurate indicator of development in a country or region. Alone it can be misleading. There are many other indicators that can be used alongside GNP or GDP per person to give a more realistic view of a country or region's level of development. Some of the more useful ones are:

1. The percentage of adults who are literate. This is the **adult literacy rate** and indicates what percentage of adults are able to both read with understanding and write a short simple sentence on everyday life.
   Japan has an adult literacy rate of 99% compared to Kenya's rate of 48%.

2. The number of people who die each year for every 1 000 of the population. This is the death rate. Kenya has a death rate of 10 compared to Japan's rate of 7.

3. The number of babies who die each year before their first birthday for every 1 000 live births. This is the **infant mortality rate**.
   Kenya has an infant mortality rate of 66 compared to Japan's rate of 4.

4. The average number of years a person can expect to live from birth. The term used to describe this is **life expectancy**. Japan has a life expectancy at birth of 79 compared to 59 in Kenya.

5. The percentage of the population who live in **urban areas** (towns and cities) rather than **rural areas** (countryside).
   77% of the Japanese population is urbanised compared to 25% of the Kenyan population.

6. The number of babies born each year to every 1 000 people. This is the birth rate.
   Kenya has a birth rate of 44 compared to a rate of 10 in Japan.

7. The percentage of the population aged under 15. This shows the **population age structure**.
   49% of Kenya's population is under-15 compared to 17% of the population of Japan.

8. The percentage of the working population employed in **primary sector** jobs (e.g. farming, fishing, mining).
   83% of Kenyan workers are employed in primary jobs compared to 11% of Japanese workers.

There are a number of other useful indicators but individually, GNP or GDP per person are the most important. It can be related to other indicators as a cause and/or effect. For example, high GNP or GDP per person provides the wealth to pay for better health care and education systems which in turn helps to raise GNP/GDP by making people more productive. Japan has a GDP per person roughly 65 times greater than that of Kenya, and uses around 40 times as much energy (fuel and power) per person as Kenya. The graph below shows the relationship between GNP per person and the percentage of working people employed in farming.

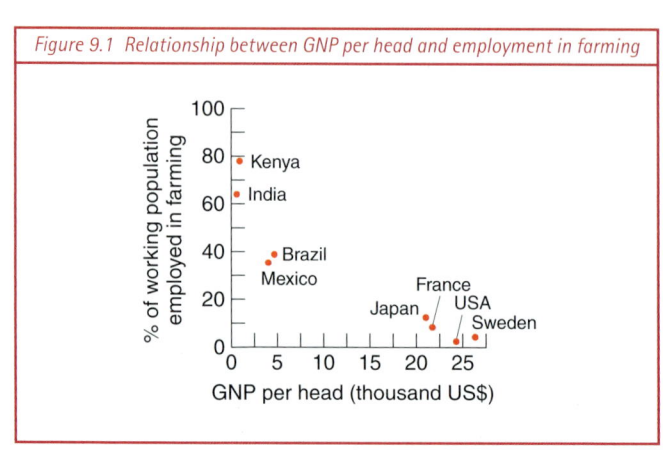

Figure 9.1 Relationship between GNP per head and employment in farming

Chapter 9: Contrasting levels of development

ACTIVITIES

1. Suggest why GNP or GDP per person alone do not give a realistic indication of development in a country or region.
2. Try to list a few more development indicators not given above. (WO1.1, WO1.2, WO2.1, WO2.2)
3. State the relationship shown by the graph in Figure 9.1 between GNP per person and the percentage of working people employed in farming. Can you explain this relationship? If so, do so. (N1.1, N1.3, N2.1, N2.3)
4. Put together a fact file comparing development in Kenya compared to Japan. Use the information on page 106 and then try to give reasons for the differences in development which your fact file shows. (IT1.1 – IT2.3)

How human welfare and economic development differ

Many people draw the distinction between development which is human and social, and development which is economic. Economic development focuses on wealth and how well off a country or region is in terms of money and material possessions. There is, however, more to life and people's well-being than monetary matters. Human happiness depends on a range of both monetary and non-monetary aspects of life. Aspects such as safety, security, cleanliness, health and friends come under the heading of **human welfare**. The difference between these two types of development can be seen in the statements of eight people about what development means to them which opened this chapter.

The term **quality of life** is concerned with human welfare and such development indicators as:

- life expectancy
- infant mortality rates
- basic health care programmes
- access to clean water supplies
- adequate food supplies

These quality of life indicators are likely to be linked also to economic development. High levels of economic development can provide the resources to pay for improvements in quality of life, and in MEDCS are generally passed on in higher human welfare.

It has become common practice to combine a number of these quality of life indicators into an index of quality of life. A recent index used by the Department of Environment to measure changes in the quality of life in the UK combines thirteen indicators:

- population of wild birds
- emissions of greenhouse gases
- days of air pollution
- road traffic
- rivers of good or fair water quality
- average life expectancy
- educational qualifications at age 19
- new homes built on previously developed land
- waste and waste disposal
- homes judged unfit to live in
- GDP growth
- social investment (e.g. roads, hospitals, schools …)
- employment

This index has become known as the **Skylark Index**. Indexes give a broader, more rounded picture of development than single indicators alone or economic indicators, that is GNP or GDP alone. The Skylark Index also stresses the importance of **sustainable development**. Development needs to create wealth without destroying the environment. Cleaner development creates jobs and prosperity but not waste, pollution and fewer skylarks!

Two other quality of life indexes are:

- the **PQLI (Physical Quality of Life Index)** which combines life expectancy, the infant mortality rate and the literacy rate to give a country a score out of 100. The UK scored 96; Ethiopia 16.
- the **HDI (Human Development Index)** combines life expectancy with education levels and the buying power of the local currency to give a quality of life level – High, Medium or Low – for each country.

Quality of Life indexes help to show how ordinary people's basic needs of life (e.g. housing, health, safety, literacy) are being met. Without them people experience **absolute poverty** and multiple **deprivation**. Relative poverty is less serious and occurs when people are only badly off or deprived in comparison to other people around them. Most poverty in the modern UK is relative. Absolute poverty can be found in LEDCs. For example, in Africa south of the Sahara, 44% of people are without access to any health services, 57% have no access to safe water and 50% are illiterate. About a quarter of the world's population live in absolute poverty.

Some people gain more benefits from general development in a country than do others. A country's political priorities help to decide who is advantaged or disadvantaged by economic development. Countries such as Iraq before the fall of Saddam Hussein and Somalia with high military spending give their people generally a lower quality of life than they could afford. Resources go to the armed forces rather than on health and education spending. On the other hand, economic development in Costa Rica has been limited but it has advantaged people generally. Its high quality of life for an LEDC has been because of its spending priorities; health and education have been give priority over military spending.

It is important to view the terms 'development' and '**living standards**' as widely as possible. Geographers and others are increasingly trying to broaden the idea about what they are about by trying to include welfare and quality of life ideas such as health and literacy as well as taking a longer term view of development. Any acceptable definition of development today must refer to the future, environmental change and what is sustainable.

ACTIVITIES

1. Consider which of the following indicators of development are human and social, and which are economic:
   GNP per person; Energy consumption per person; Life expectancy; Number of patients per doctor; Adult literacy rate; Infant mortality rate; Number of kilocalories of food eaten per person per day; Proportion of working people employed in primary occupations; Value of exports.

2. Suggest why the Skylark Index points to a higher quality of life in the North of England, especially North Yorkshire and the North East than in the South of England. Wealth is greater in the South of England. Give reasons for the high quality of life in some parts of Northern England.

MEDCs and LEDCs

The countries of the world can be grouped according to their level of development. Those with higher GNP/GDP per person, industrialisation especially tertiarisation, and generally higher human welfare tend to be found in Europe, North America and Australasia and are known as MEDCs, more economically developed countries. LEDCs, less economically developed countries, generally speaking have opposite characteristics to those of MEDCs.

Not all countries readily fit into this simple classification. **Newly Industrialising Countries** (NICs) such as the Tiger economies of South Korea, Hong Kong, Singapore and Taiwan have rising GNP/GDPs per person as a result of industrialisation. The **OPEC** (oil exporting) countries such as Saudi Arabia and Venezuela may have quite high GNP/GDPs per person but few other features of an MEDC. The former Communist countries in eastern Europe may be developing but cannot be described as either more economically or less economically developed at present.

How do the main indicators of development vary between MEDCs an LEDCs?

Uneven development is a feature of the world. Neither wealth nor human welfare are shared out equally. There is a development gap between MEDCs and LEDCs. Their levels of development contrast. The development indicators below point out these contrasts:

Figure 9.2 Development indicators		
Indicator	MEDC	LEDC
Birth rate	Low	High
Life expectancy	High	Low
Adult literacy rate	High	Low
Energy consumption per person	High	Low
Death rate	Low	High
Infant mortality rate	Low	High
Urban: rural population ratio	High	Low
Car ownership rate	High	Low
Primary sector employment	Low	High
Density of road network	High	Low
GDP per person	High	Low

The earlier section, Measuring Development, gives some idea for Kenya, an LEDC, and Japan, an MEDC, of what high and low mean as a number. For example, MEDCs might have birth rates of 12, death rates of 10 and adult literacy rates of 98. Birth rates of 40, death rates of 15 and adult literacy rates of 25 can be found in LEDCs.

The different industrial structures of MEDCs and LEDCS is an important distinguishing feature. LEDCs tend to have a higher proportion of their working population employed in the primary sector of economic activity (e.g. farming, mining, fishing). As Figure 9.3 below shows, as a country 'develops' from an LEDC to an MEDC the proportion of the working population employed in the **tertiary/quaternary sectors** (service industries) grows as that in primary activities shrinks.

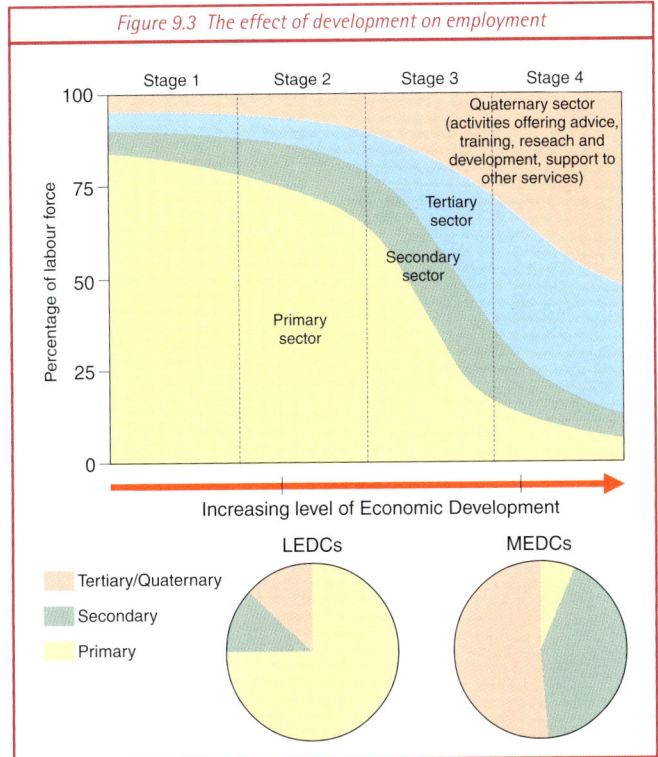

Figure 9.3 The effect of development on employment

The global distribution of different levels of development

There is a clear global pattern for development. The pattern is changing slightly, especially as the Newly Industrialising Countries continue to develop, but it is basically one of a 'Rich North' and 'Poor South.' The world map (Figure 9.4) shows the broad pattern of a divided world with wealth and high quality of life concentrated in the industrialised, mechanised and urbanised countries of the Northern Hemisphere. Much of the Southern Hemisphere is made up of economically poorer, rural, agricultural countries.

This gap between the economically rich countries and the economically poor ones is growing and has become so great that the USA and Japan have 42.5% of the world's wealth compared to the 1.6% in the whole of Africa.

'World's richest 358 people own as much as poorest half of world's population.'

Recent United Nations' report

Half of the world's population amounts to roughly 3 billion people. The poorest half are concentrated in these five areas of the world:

- south Asia, e.g. India, Bangladesh
- sub-Saharan Africa, e.g. Sierra Leone, Mali

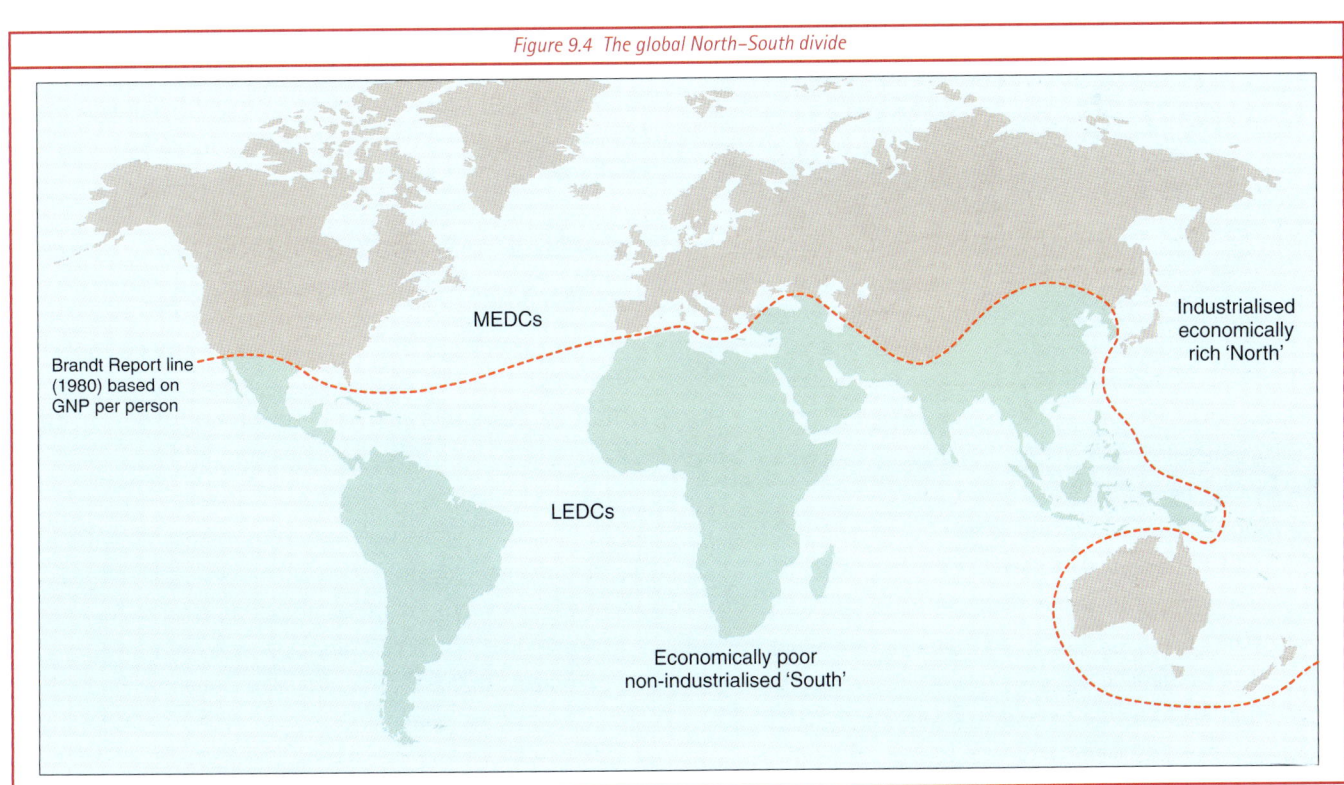

Figure 9.4 The global North–South divide

Section 3: Managing Economic Development

- east Asia, e.g. China, Philippines
- Middle East and north Africa, e.g. Egypt, Algeria
- Latin America and the Caribbean, e.g. Mexico, Jamaica

Around a quarter of the world's population (1.5 billion people) live in these five areas in conditions of absolute poverty. Their numbers are growing and the contrast between their conditions of life and those in the UK can be drawn from the map below showing average food intake in Africa, Europe and the Middle East.

The mushroom-shaped diagram below shows how unequal the distribution of economic development is. The richest 20% of the world's people benefit from around 85% of its economic activity; the poorest 20% being responsible for only 1% of world economic activity. This massive gap between the world's richest and poorest has more than doubled over the past thirty years.

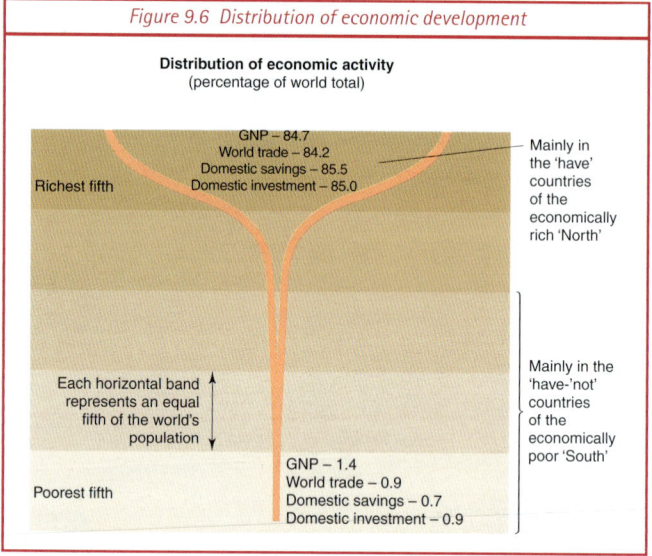

Figure 9.6 Distribution of economic development

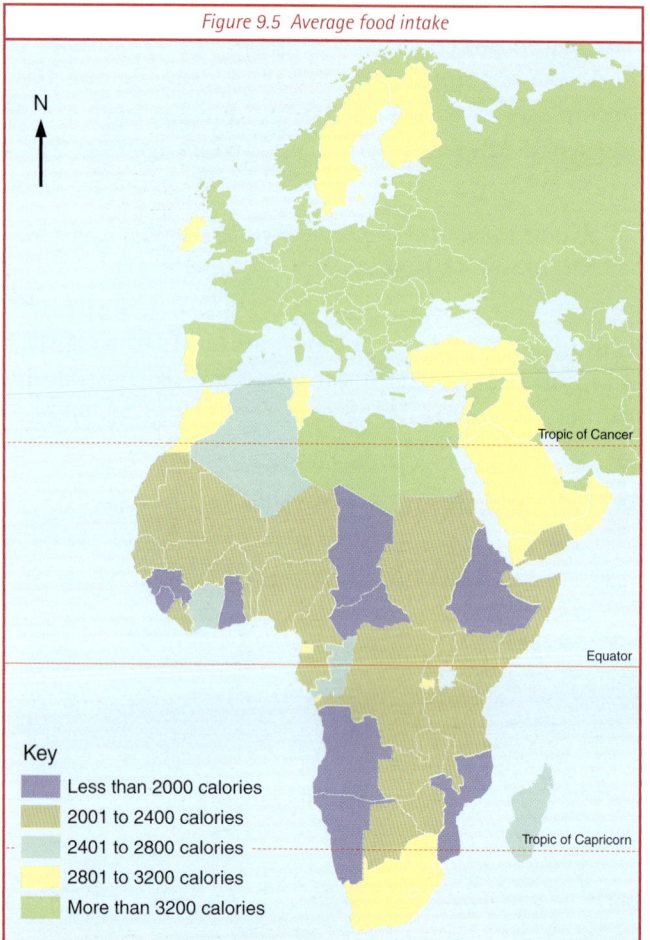

Figure 9.5 Average food intake

Some countries have in recent times genuinely joined or are on track to join the group of MEDCs. The Newly Industrialising Countries (NICs) in South-East Asia have successfully been through an economic miracle since the 1960s. The map of Asia and data file on Taiwan opposite describe how multinational manufacturing companies and cheap exports have played a big part in the rapidly rising GDPs per person in this area.

Levels of development can vary within a country

A development gap has become pronounced in many countries. It can be almost as pronounced as that which we have seen exists between countries. In the

ACTIVITIES

1. Suggest why Taiwan and some of the other countries in east Asia have been so successful in raising the level of their GDP per person.
2. Environmental pollution has been one consequence of this economic success. Explain why and suggest what might be done about it. (PS1.1, PS1.2, PS2.1)
3. The term 'global shift' has been used to describe the rapid industrialisation in east Asian countries. Discuss why it might be a suitable term to describe this process, and how as a result of the process the distribution pattern shown by Figure 9.4 might change.

Chapter 9: Contrasting levels of development

Figure 9.7 NICs of S.E. Asia and Taiwan datafile

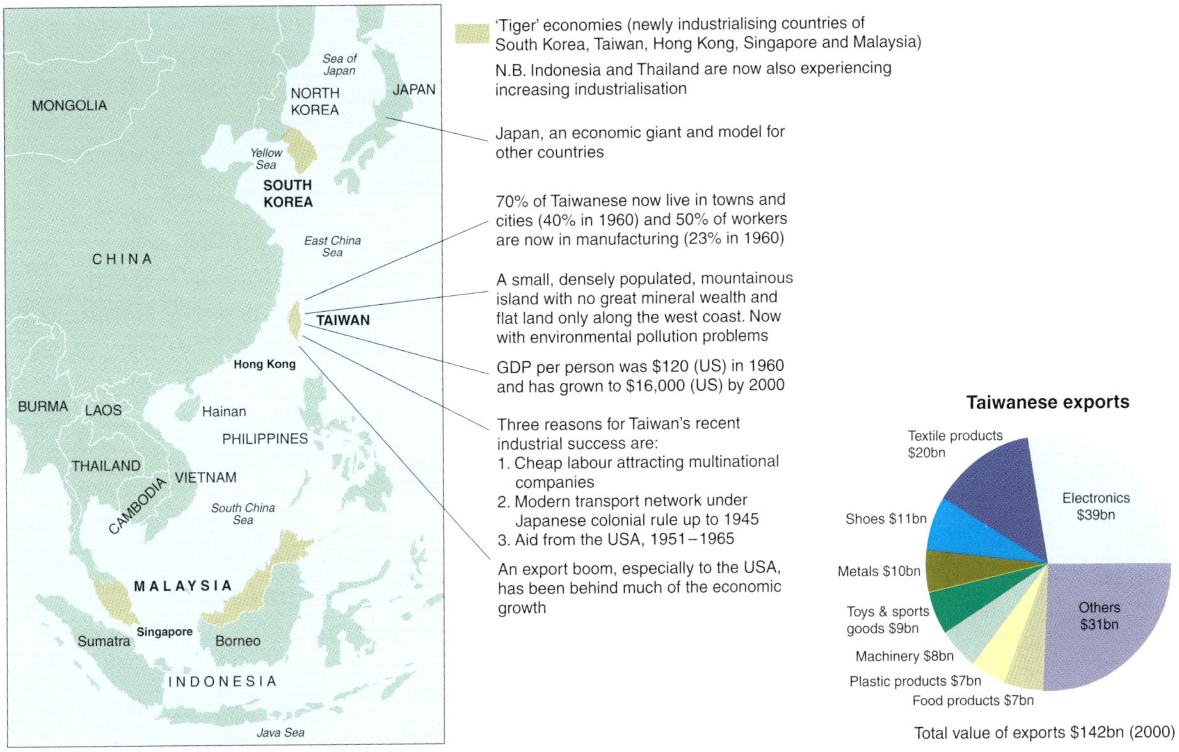

'Tiger' economies (newly industrialising countries of South Korea, Taiwan, Hong Kong, Singapore and Malaysia)
N.B. Indonesia and Thailand are now also experiencing increasing industrialisation

Japan, an economic giant and model for other countries

70% of Taiwanese now live in towns and cities (40% in 1960) and 50% of workers are now in manufacturing (23% in 1960)

A small, densely populated, mountainous island with no great mineral wealth and flat land only along the west coast. Now with environmental pollution problems

GDP per person was $120 (US) in 1960 and has grown to $16,000 (US) by 2000

Three reasons for Taiwan's recent industrial success are:
1. Cheap labour attracting multinational companies
2. Modern transport network under Japanese colonial rule up to 1945
3. Aid from the USA, 1951–1965

An export boom, especially to the USA, has been behind much of the economic growth

Taiwanese exports
- Textile products $20bn
- Electronics $39bn
- Shoes $11bn
- Metals $10bn
- Others $31bn
- Toys & sports goods $9bn
- Machinery $8bn
- Plastic products $7bn
- Food products $7bn

Total value of exports $142bn (2000)

UK the poorest fifth of the population have incomes which are less than a quarter of the national average: about £2 700 in 1999 instead of £12 480. The traditional picture is that this inequality is distributed so as to give on a map of the UK a North–South divide like the cartoon (Figure 9.8) shows.

Figure 9.8

"TRAGICALLY DIVIDED FOR YEARS...THE SOUTH BOOMING, THE NORTH POOR."

It is true that at a regional scale, development indicators such as GDP per person can be used to show this pattern. The GDP per person in 1999 for each of the eleven standard regions of the UK was:

Figure 9.9 GDP per person, 1999

	£1000s
South East (including Greater London)	16.82
Scotland (including North Sea)	14.19
East Anglia	12.52
South West	11.72
West Midlands	11.26
North West	10.87
East Midlands	10.78
Yorkshire & Humberside	10.73
Wales	9.90
North East	9.83
Northern Ireland	9.39

With the exception of Scotland the pattern at this scale is clearly one of higher incomes in southern England and lower incomes in northern England, Wales and Northern Ireland. Other indicators such as the unemployment rate often show a similar pattern of regional inequality.

Some geographers prefer to look at the variations in development within the UK at a smaller scale than the regional. A more accurate picture may be gained if the scale of study is more local and the size of a town or an area of a city or county. At this scale development in the UK looks on a map like a patchwork of booming, prosperous areas, and areas

111

Section 3: Managing Economic Development

ACTIVITIES

1. **a** On an outline map of the UK mark the eleven standard regions, their GDP per person data given in Figure 9.9 and their unemployment rates (which you can obtain from your teacher or the Internet).
    b Explain why this map can give a misleading impression of differences in living standards and quality of life in the UK.
2. Suggest why infant mortality rates and life expectancy vary so much in a UK city like Sheffield.

of poverty and high unemployment (or bust areas). There are boom and bust areas in both:

- The North, e.g. Macclesfield and Warrington are booming while Mansfield and St Helens are poorer areas.
- The South, e.g. Newbury and St Albans are booming while Gosport and Hastings are poorer areas.

Prosperous towns with low unemployment and rising incomes can be very close. Booming Warrington is only 5 miles from struggling St Helens. This pattern of rich and poor alongside each other has long been a feature of our large cities. The map of Sheffield (Figure 9.10) shows how infant mortality rates vary between the poor inner city and eastern wards, and those more prosperous ones towards the outer western edge of the city. There is a difference of eight years between the life expectancy of men living in the most prosperous wards and those in the most deprived inner city wards.

Development must be sustainable

It is important that any present development does not ruin prospects for future generations. In the past much of the development in the UK and other countries has taken a short-term view of life; changes have been for the 'here and now' and resources have been allowed to run out and environments damaged. Sustainable development takes a long-term view and looks for developments which can be kept up and improve life for everyone now and in generations to come.

Protecting the environment and using natural resources carefully at the same time as improving living standards is a key feature of sustainable development. Development which leads to the pollution of air, sea, rivers and groundwater, to deforestation, to reduction in the variety of plants and animals (biodiversity), and to global warming, acid rain and ozone layer depletion is not sustainable development. Fairness or equity is another important aspect of sustainable development. Sustainability requires international cooperation, with everyone pulling their weight. Fish stocks cannot be preserved for future generations if one or two countries continue to overfish. NICs cannot be expected to

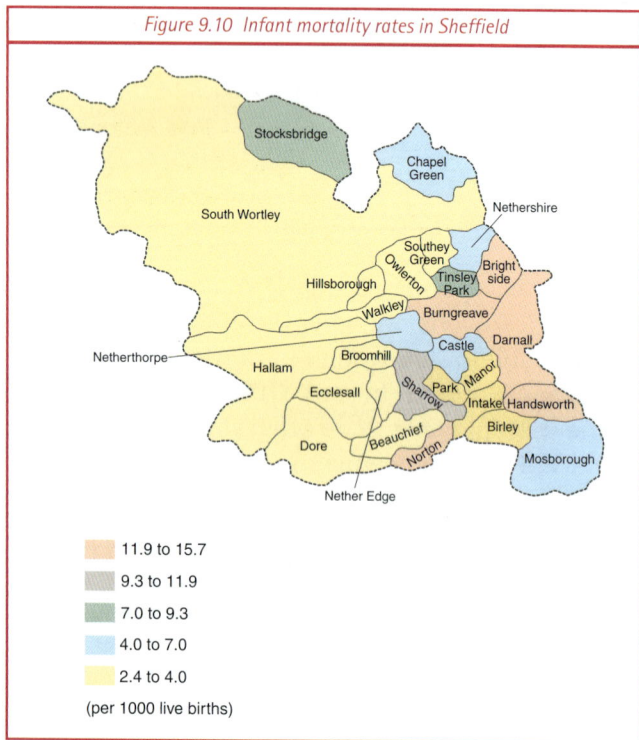

Figure 9.10 Infant mortality rates in Sheffield

11.9 to 15.7
9.3 to 11.9
7.0 to 9.3
4.0 to 7.0
2.4 to 4.0
(per 1000 live births)

ACTIVITIES

1. Imagine that you investigated production and economic activity in five countries, A to E, and found the following were dominant:
   - country A – production of capital goods, e.g. machinery and machine tools in polluting factories
   - country B – import of consumption goods, e.g. cars, luxury foods
   - country C – provision of education and health care, e.g. schools, clinics
   - country D – production of military equipment, e.g. tanks, fighter jets
   - country E – development of infrastructure, e.g. roads, electricity grid with little countryside damage

   How sustainable is development in each country? Give your reasons. (PS1.1, WO1.3)

clean up air pollution while car use continues to grow in MEDCs. Everyone needs to benefit from development, including the poor and LEDCs, if development is to be sustainable. Some of the broader social welfare ways of measuring development such as the Human Development Index and the Skylark Index give a better indication of sustainability than GDP/GNP per person. Fairness and the environment can be found in these indexes.

Measuring development is not easy

We have so far raised a number of difficulties in defining and measuring development. Summarising these:

- GDP/GNP per person gives only a partial view of life, only of things that can be measured in terms of money. It is also only an average and tells you nothing about how GDP/GNP is shared out among the population.
- Single indicators alone are of very limited use in measuring development. Indexes combining a number of indicators give a fuller picture but a problem is, which indicators to include?
- The idea of quality of life is both difficult to define precisely and to measure. Development that does not incorporate quality of life and human welfare is incomplete.
- Meeting the needs of ordinary people and assessing whether it is they who have gained from economic development creates measuring difficulties.
- We can only be clear about full sustainability of present development in the future. Future generations will know whether present development has harmed their prospects and the environment.

Reasons for development differing from place to place

The causes of areas and countries developing or not developing economically are various. The best way to think of these various causes is in terms of resources which are factors of production. There are four basic factors of production:

- natural resources, including land, climate, minerals
- labour, including enterprise
- capital and technology
- knowledge and education

Greater production and economic development comes about when one or more of these factors becomes more productive. The curve showing possible production in the country (Figure 9.12) moves out to the right. There are more goods and services for the country's people to consume. Their living standards could rise.

It is the supply of these resources that enables areas and countries to break out of a cycle of poverty into a cycle of growth and development (Figure 9.14). Some economically successful countries like Japan only have a few natural resources but they may have developed large supplies of **entrepreneurship** (enterprise and high business skills) and capital which has been invested in better education, technology, health care and nutrition. A fitter, well-educated and enterprising labour force using **hi-tech** equipment can now be found in Japan. Other countries might be lacking in natural resources because of their location in a hostile environment where the climate is harsh and natural hazards occur. Hazards like earthquakes, volcanoes, drought, flooding and soil erosion mean that nature can become more of a threat and constraint on people than a resource to be used by them and an opportunity for development. The development

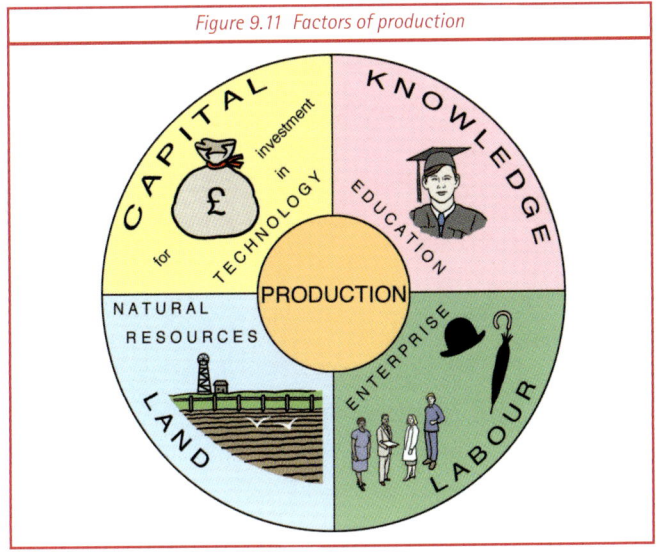

Figure 9.11 Factors of production

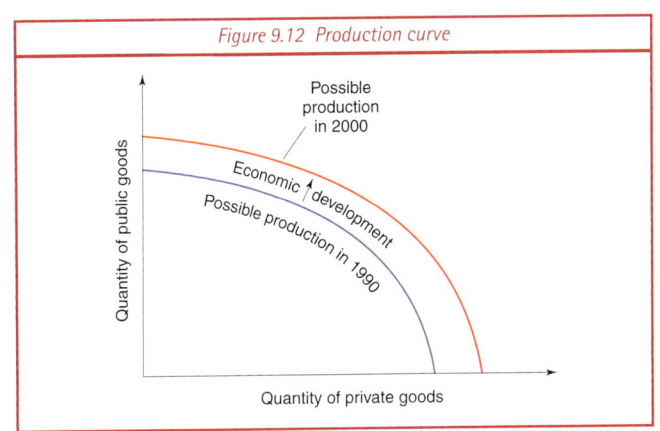

Figure 9.12 Production curve

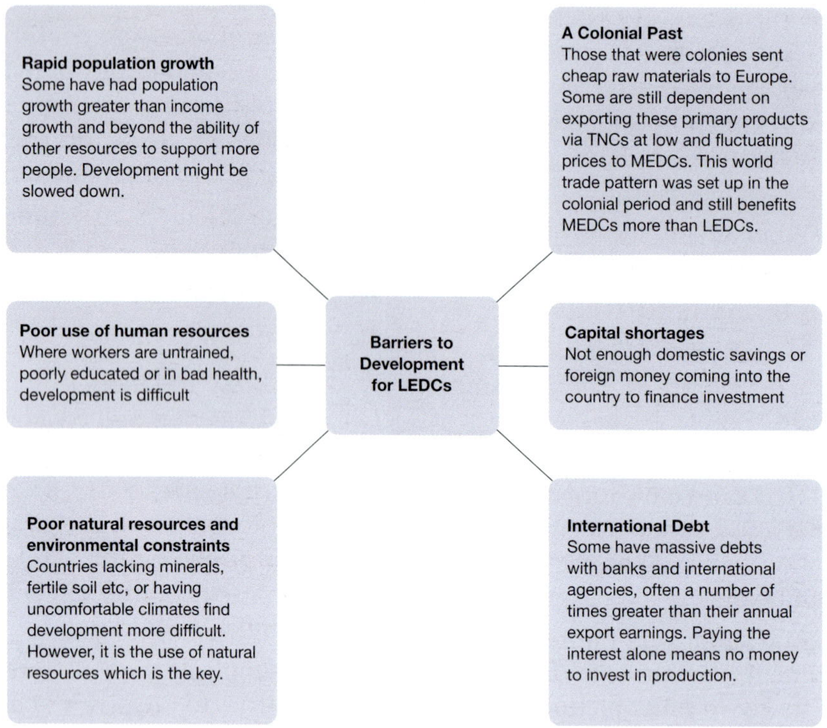

Section 3: Managing Economic Development

> **Figure 9.13 Barriers to development for LEDCs**
>
> **Rapid population growth**
> Some have had population growth greater than income growth and beyond the ability of other resources to support more people. Development might be slowed down.
>
> **A Colonial Past**
> Those that were colonies sent cheap raw materials to Europe. Some are still dependent on exporting these primary products via TNCs at low and fluctuating prices to MEDCs. This world trade pattern was set up in the colonial period and still benefits MEDCs more than LEDCs.
>
> **Poor use of human resources**
> Where workers are untrained, poorly educated or in bad health, development is difficult
>
> **Capital shortages**
> Not enough domestic savings or foreign money coming into the country to finance investment
>
> **Poor natural resources and environmental constraints**
> Countries lacking minerals, fertile soil etc, or having uncomfortable climates find development more difficult. However, it is the use of natural resources which is the key.
>
> **International Debt**
> Some have massive debts with banks and international agencies, often a number of times greater than their annual export earnings. Paying the interest alone means no money to invest in production.

process can be slowed down in these environments. Some of the other resource problems which can act as barriers slowing down development in LEDCs are shown on Figure 9.13.

Improvements in agriculture, advances in technology, both agricultural technology and technology generally, and the discovery and use of mineral resources can lead to a country breaking into the cycle of growth and development shown on Figure 9.14b. The introduction of hi-tech methods in industry has been one of the causes of a cycle of growth setting up in Newly Industrialising Countries (NICs). New, more efficient machinery speeds up production and leads to greater output. This releases workers for retraining in new lines of production and new jobs. The area or country generally develops as it goes round and round the successful spiral shown on Figure 9.14b. Through history new technology, whether the latest hi-tech or only **intermediate technology** which is appropriate to where it is being introduced, has generally been a cause of general economic development. Technology drives

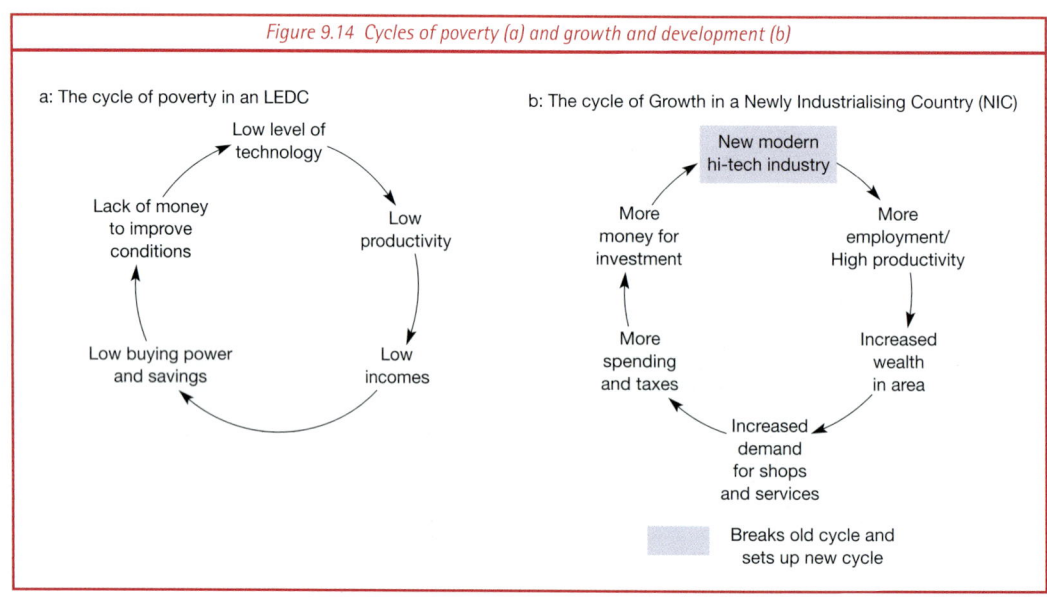

Figure 9.14 Cycles of poverty (a) and growth and development (b)

a: The cycle of poverty in an LEDC

Low level of technology → Low productivity → Low incomes → Low buying power and savings → Lack of money to improve conditions → Low level of technology

b: The cycle of Growth in a Newly Industrialising Country (NIC)

New modern hi-tech industry → More employment/High productivity → Increased wealth in area → Increased demand for shops and services → More spending and taxes → More money for investment → New modern hi-tech industry

Breaks old cycle and sets up new cycle

development. Look back at the Taiwan case study page 111: and see how the **industrialisation** and development of this country and other NICs in Pacific Asia over the past 30–40 years has partly followed the example of Japan. There, industry has become very reliant on new technology with very **capital-intensive** and **knowledge-based** methods of manufacturing becoming common.

Agricultural change, like technological advances, can also lead to general economic development. In fact, technological advances can be in agriculture as well as manufacturing. The mechanisation of agriculture, the use of fertilisers and pesticides, the introduction of higher-yielding varieties of crops and animals and generally more scientific farming has the following four beneficial effects as far as general development is concerned:

- It boosts industry by giving demand to farm machinery makers and agricultural chemical companies. New jobs are created.
- More and better food improves general health, making people able to work harder. General production should rise.
- Labour can transfer from farm work to more productive employment in factories and offices as machinery replaces manual labour in agriculture. Again, general production should rise.
- Rising agricultural yields may give a food surplus so food exports become possible. Income from abroad may flow into the country.

One of the causes of some countries being MEDCs is that they have a generally high level of technology, including in agriculture where output is high but with only few inputs of labour. The UK could be broadly **self-sufficient** in food with only 2–3% of the labour force working in agriculture. Mineral wealth or lack of it can also be an explanation of different levels of development. In 1973 11 countries led by Saudi Arabia formed the OPEC (Organization of Petroleum Exporting Countries) cartel. Enormous oil price rises between 1973 and 1980 enabled Saudi Arabia to finance considerable economic development. During this period its GDP roughly doubled and in 2000 was approximately 150% higher than in 1973. Mineral wealth in the form of oil has transformed the country. Oil

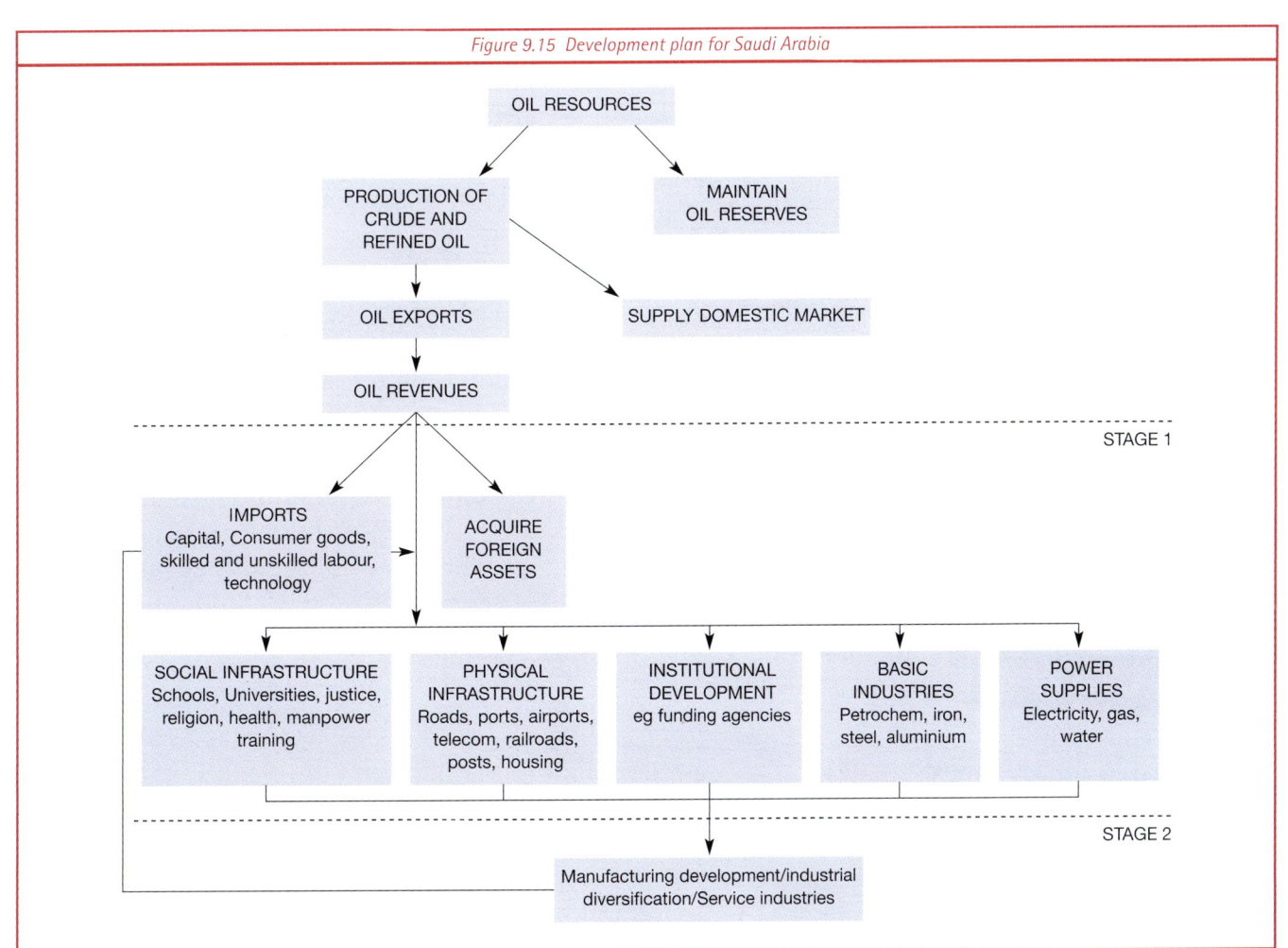

Figure 9.15 Development plan for Saudi Arabia

Section 3: Managing Economic Development

Figure 9.16 Modern developments in Saudi Arabia

Effects and consequences of the level of development on a country's people and economy

We might think in terms of positive effects and negative effects. Economic development is an opportunity but it can also bring threats. One of the great threats facing NICs today is that of environmental damage brought about by industrial pollution. Air, water and soil pollution in Taiwan have reached critical levels. The need for **government regulation** to clean up polluting industries is great. However, the opportunity that economic development provides for a country is of enormous benefit. It provides the opportunity for:

1. improvements in quality of life and human welfare via:
   - more consumption of food
   - better medical and educational services
   - improved housing

2. greater and more varied employment which:
   - gives people more job choices
   - raises job satisfaction
   - gives people chances to be more **mobile geographically** and between occupations

3. greater personal freedoms and opportunities in the form of:
   - democracy and social justice – **freedom of speech**, of voting and from discrimination are necessary features of a well developed country and associated with MEDCs but not all LEDCs.
   - individuals fulfilling their potential – Equal Opportunities policies common in many MEDCs aim to encourage people to feel a sense of their own identity and raise their self-esteem.

revenues mainly from Europe, North America and Japan have given it the ability to quickly create a physical and social infrastructure of roads, airports, hospitals, schools, other modern city buildings, machinery etc. Figure 9.16 shows some of these modern developments. The country has been careful to plan its future development as Figure 9.15 – Saudi Arabia's Development Process – shows. Some of the key features of this development plan are:

- **sustainability** – they recognise that they must try not to deplete their oil resources too quickly
- **industrialisation** – during the 1980s and 1990s they have been trying to reduce their dependence on oil by investing oil money in setting up productive manufacturing industries
- **diversification** – they realise that development must be balanced and spread across all aspects of life in Saudi Arabia.

ACTIVITIES

1. Suggest in what ways:
   i. simple water pump in a Kenyan village
   ii. a high-speed train linking all the islands of Japan might lead to economic development in that country.
2. Do a web or CD-Rom (e.g. Encarta) search on OPEC. List the OPEC member countries. What is a cartel? How has the cartel been able to raise oil prices since 1973? (IT1.1 – IT2.3)
3. Saudi Arabia has a fairly high level of GNP per person but only a medium level of quality of life/human welfare. Can you suggest why this might be so?

Clearly, quality of life, employment and opportunity are linked. Improved employment opportunities leads to greater personal freedom and improved quality of life. Economic development tends to bring all three in time. All three are not fully present in all NICs at present; there is scope for development in the areas of civil rights and quality of life. Development is about meeting human needs. As economies change and develop, which they are presently doing rapidly in the NICs of Pacific Asia, the range of human needs that can be met increases.

ACTIVITIES

1. Do a brainstorm. Think of how employment, opportunity and quality of life are linked for people living in a shanty town (squatter settlement) on the edge of an LEDC city. How does the low level of development affect these three aspects of life for residents of the settlement? (WO1.1, WO1.2, WO2.1, WO2.2)

2. Do a web or CD-Rom search on political freedom and civil rights in one of the NICs of Pacific Asia (e.g. Singapore, Hong Kong, Taiwan). How might this be a source of human suffering in the country? How does this compare with the opportunities available to people in the UK? Political changes often follow economic development; why and what might the political future look like in this NIC? (IT1.1 – IT2.3)

3. After working through this chapter so far, try to summarise how you would attempt to measure development in a country. Do not forget to include the so-called 'greening' of development! (WO1.1 – WO2.3. LP2.3)

Strategies used by LEDCs to try to raise their level of development

The effects of low levels of development, as we have just seen, are so depressing for people affected by them, and the gap between their lives and those of most people in MEDCs is so enormous, that the need for LEDCs to develop economically is one of the most urgent that the world faces. This gap continues to grow despite efforts over many years to reduce it. The efforts to narrow the gap by boosting development in the economically poorer countries involve people and policies in:

- LEDCs
- MEDCs
- international agencies such as the United Nations and World Bank.

A system of **foreign aid** has grown up since the Second World War in which governments and people in the more economically developed world give resources to the LEDCs either:

- directly from government to government – this is known as bilateral aid – or through private charities (e.g. Oxfam) where the resources go directly to people in need
- through the international agencies set up for this purpose. This is known as **multilateral aid**.

This aid can be in the form of:

- money
- technology
- goods (e.g. food) and services (e.g. know-how)

and can be given for:

- Emergencies – this **short-term aid** is likely to be in the form of food, shelter and medical supplies after a disaster. Figure 9.17, a cartoon drawn after the Honduras floods in 1998, shows short-term aid being given in response to the disaster.
- Real development – this is referred to as **long-term aid** and is likely to involve money and technology (capital) which can be invested in long-term development.

Governments in MEDCs recognise that giving aid helps to protect their interests abroad, often by boosting industries in their own country. It can

Figure 9.17

create markets to export to now or in the future. This is why some bilateral aid is tied, which means that the aid-giving country puts a condition on the aid. It has to be used to buy goods and services from industries in the aid-giving country. Foreign aid as a way of encouraging development has both advantages and disadvantages. Here are two advantages and two disadvantages.

Advantages of Foreign Aid:
- It can lead to improvements in technology which raise the output of agriculture and industry.
- It can provide better health know-how which can improve and save lives.

Disadvantages of Foreign Aid:
- It can develop a dependency culture in LEDCs with them becoming too dependent on aid from abroad.
- It can be mis-spent, perhaps on prestige projects in the capital city.

ACTIVITIES

1. Try to list some other advantages and disadvantages for LEDCs of them receiving aid from MEDCs
2. Suggest why Figure 9.17 shows Honduras in need of emergency aid as well as in need of help with debt (World Bank and IMF).

Aid can play a part in development, especially where it is properly directed but LEDCs themselves often prefer other strategies or any aid to be used to set up these other strategies. Strategies which involve small-scale projects are often preferred. These can lead to local development, often in rural areas and are more likely to be sustainable. Introducing water pumps to many village wells can be more appropriate than building one major hydro-electric power scheme. Small, self-help and sustainable – the three Ss – are thought to be the best way to a fourth S – success !

Appropriate (or intermediate) technology is often lacking from LEDCs. There may be no technology or low-tech available as part of the low level of development of the country. Hi-tech exists in MEDCs and some can be transferred to LEDCs but this may be quite inappropriate for countries with shortages of skilled labour, spare parts and capital for investment. Basic pieces of equipment which are straightforward to use and cheap to mend, e.g. wind-powered water pumps, fuel-efficient wood-burning ovens, are far more suitable and exactly what these countries need to improve their technology.

The Development Puzzle

Just which is the best way forward for LEDCs? Opinions vary from country to country and between different people in every LEDC. Some will argue, for example, that resources need putting into the country's infrastructure of roads, railways, water and electricity supplies.

Figure 9.18 Putting resources into a LEDC's infrastructure

Others will favour a green revolution in agriculture to increase the supply of food or putting more resources into education and training.

The general approach is often to encourage globalisation within the LEDC through:
- fair trade agreements and more exports
- more tourism into the country

Transnational companies can play a big part in developing both of these.

For a long time now the key question in the development puzzle has been 'aid or trade?' Many people have believed that trade is much more valuable than aid to an LEDC trying to develop. The problem has been that LEDCs have not had the opportunity to increase their export trade by opening up markets in MEDCs. Figure 9.19 shows some of the reasons why this has happened.

Transnational companies (TNCs) have, as Figure 9.19 shows, played a part in this problem.

The largest TNCs are found in the motor vehicle and oil industries, and are based in the USA, European Union or Japan. General Motors, Ford, Mitsubishi, Toyota, Volkswagen, Shell, Exxon,

Chapter 9: Contrasting levels of development

Figure 9.19 Fair trade for LEDCs?

Globalisation brings more exports. Exports earn income for a country abroad. This is known as **export-led growth**. We saw earlier (page 111) how export-led growth has been crucial to the economic success of NICs such as Taiwan over the past thirty or so years.

ACTIVITIES

1. One positive effect of TNCs operating in LEDCs is that they create employment. Try to think of more positive effects of TNCs. Figure 9.19 may help.
2. One negative effect of TNCs operating in LEDCs is environmental damage because health and safety regulations are limited in the country or relaxed for the TNC. Try to think of more negative effects of TNCs. Figure 9.19 may help.
3. Referring to Figure 9.19, suggest what you think LEDCs would regard to be fair trade.

BP and Mobil operate in many countries around the world. Some LEDCs, e.g. Brazil, have been keen to attract them to their country as a way of encouraging economic development. Inward investment by Japanese and Korean TNCs have also been encouraged by the UK government. Despite bringing some benefits to the host country it is often the case that the home base country and the TNC itself benefit most from these international operations and locations.

Globalisation is about the economic, social and cultural integration of countries. The economies and cultures of some countries are becoming more integrated into a wider **global economy** and society. Some people argue that globalisation is about sameness across national boundaries: MacDonalds and Coca-Cola everywhere! TNCs are heavily involved in the process. They link countries in the developing global economy. Those countries not linked into this connected, interdependent and 'smaller' world will find development more difficult to achieve. **Globalisation** provides opportunities for development for included LEDCs. India through its adoption of new information technology and telecommunications is one of the LEDCs to be included in and benefit from globalisation.

LEDCs can also take advantage of globalisation by encouraging the development of a tourist industry. Aeroplanes have shrunk the world in terms of travel times. Leisure and service sector industries are of major importance in MEDCs. The economic importance of earnings from tourism is very significant in some LEDCs. Since the development of a noticeable tourist industry in The Gambia, West Africa, over the past 20–25 years, the country has added some $30 million to its GNP per year. All other sources of income and production in the country have added around half of this amount during this period. Tourism now accounts for 17% of the GNP of The Gambia. A successful tourist industry requires improvements in a country's infrastructure. Airports, roads, electricity and drainage systems have to be put in place. This is to the general benefit of the country.

A recent newspaper headline read: 'THAILAND PINS HOPE ON TOURISM FOR ECONOMIC REVIVAL.' For ten years from 1985 to 1996, Thailand was the world's fastest growing economy. It was one of the 'tiger' economies or NICs (Newly Industrialising Countries) of south-east Asia. The whole shape of its economy changed from one based on agriculture to one where 80% of exports were manufactured goods. After 1996 the economy went into recession and exports declined. This is where the headline above fits in. A sharp rise in tourist arrivals is playing an important role in Thailand's economic recovery. It is earning the country more than any other industry and is growing. The Tourism Authority of Thailand says that tourism is as important to Thailand as the City of London is to the UK. Tourists are attracted to Thailand, especially from the EU by:

- the opportunities for jungle treks in northern Thailand and beach holidays in southern Thailand
- the fact that they do not have cold winters but warmth all year round
- the regular direct flights to Bangkok
- the low value of the Thai currency which means that prices for tourists seem cheap
- the friendly, smiling people

The urban population of Bangkok doubled after 1985. Some people believe that such urbanisation is not sustainable in the long run. Big cities, especially in LEDCs, seem to be growing uncontrollably, unable to support their populations, and to be causing damage to the environment. Whether this is true or not is a matter for discussion. Cities can accommodate large numbers of people on quite small areas of land, and urban poverty may be less than rural poverty. Urbanisation may be more sustainable than some people believe, and provide solutions to environmental problems.

Generally, some progress is being made towards making development sustainable. It is increasingly realised that there are no quick fixes to development. Developers must take a global view and consider the long-term sustainability of all development. Development that leaves urban poverty or leads to environmental destruction is not global sustainability.

Exam-style questions with answer comments for Chapter 9

Question 1 – F Tier

Study Figure 1 which shows types of aid that are sent from More Economically Developed Countries (MEDCs) to Less Economically Developed Countries (LEDCs).

Figure 1	
SHORT TERM AID	LONG TERM AID
Food	Education and Training
Medicines	Transport improvement schemes

(a) (i) Which type of aid – short-term or long-term – provides for real development? **(1)**

Long-term aid is the answer.

(ii) Give a reason for your answer to (i) **(1)**

The idea is that education, training and transport enable an economy to develop further.

(iii) Which type of aid – short-term or long-term – is a response to an emergency? **(1)**

Short-term aid is the answer.

(iv) Give another example of emergency aid. **(1)**

There are many examples, e.g. tents.

(b) (i) What term is used to describe aid that has conditions linked to it? **(1)**

The term is tied aid.

(ii) Explain why this type of aid may be more of an advantage to the MEDC donor than the LEDC recipient. **(3)**

Tied aid has conditions linked to it, often that it is spend on goods/services from the country providing the aid.

(c) Suggest why LEDCs now often prefer aid from MEDCs to be put into low-cost, self-help rural development programmes. **(4)**

These are more sustainable, hit the grassroots and slow migration out of the countryside.

(d) In what other ways can the governments of MEDCs assist the development of LEDCs? **(3)**

The best answer will focus on trade. LEDCs need access for their exported goods into MEDC markets.

(15 marks)

Question 1 – H Tier

Study Figure 2 which lists several economic development indicators.

Figure 2		
Indicator	More Economically Developed Country	Less Economically Developed Country
Birthrate	Low	High
Life expectancy	High	Low
Literacy rate	High	Low
Energy consumption	High	Low
Primary sector employment	Low	High
GDP per person	High	Low
Car ownership rates	High	Low

(a) Give the meaning of each of the following terms:
1. Life expectancy
2. Literacy rate
3. Primary sector employment **(6)**

For a full definition worth two marks in each case, check the terms out in the glossary. Answers with only some truth in them would receive one mark.

(b) Why are birth rates in LEDCs high? **(3)**

There are various reasons. You can list three (e.g. large families are an asset in LEDCs for older parents) or develop two for full marks.

(c) What are the positive effects on a country's economy of having a high literacy rate? **(3)**

Democracy, hi-tec advances, etc. require people to read and write well.

(d) Why do geographers often regard a country's GDP per person as its single most important economic indicator? **(3)**

The basic idea is that GDP measures a country's wealth and its level of wealth determines how many resources can be put into all the other indicators we use to raise their level.

(15 marks)

Chapter 9: Exam-style questions

Question 3 – F Tier

Study Figure 3 which gives information about four countries at different levels of economic development.

(a) (i) Which one of the four countries shown is a:

1. Least Less Economically Developed Country (LLEDC)

   Tanzania is the best choice because of its very low GNP per capita. Peru is really LEDC not LLEDC.

2. More Economically Developed Country (MEDC)

   Netherlands is the only answer. It has a high GNP per capita and its largest industrial sector is tertiary.

3. Newly Industrialising Country (NIC). **(2)**

   South Korea with its 7% rate of economic growth is the answer.

(ii) Give two reasons for each of your choices in (a)(i). **(6)**

In each case there is a mark for naming the right country and a mark for giving a good reason from the data in Figure 3.

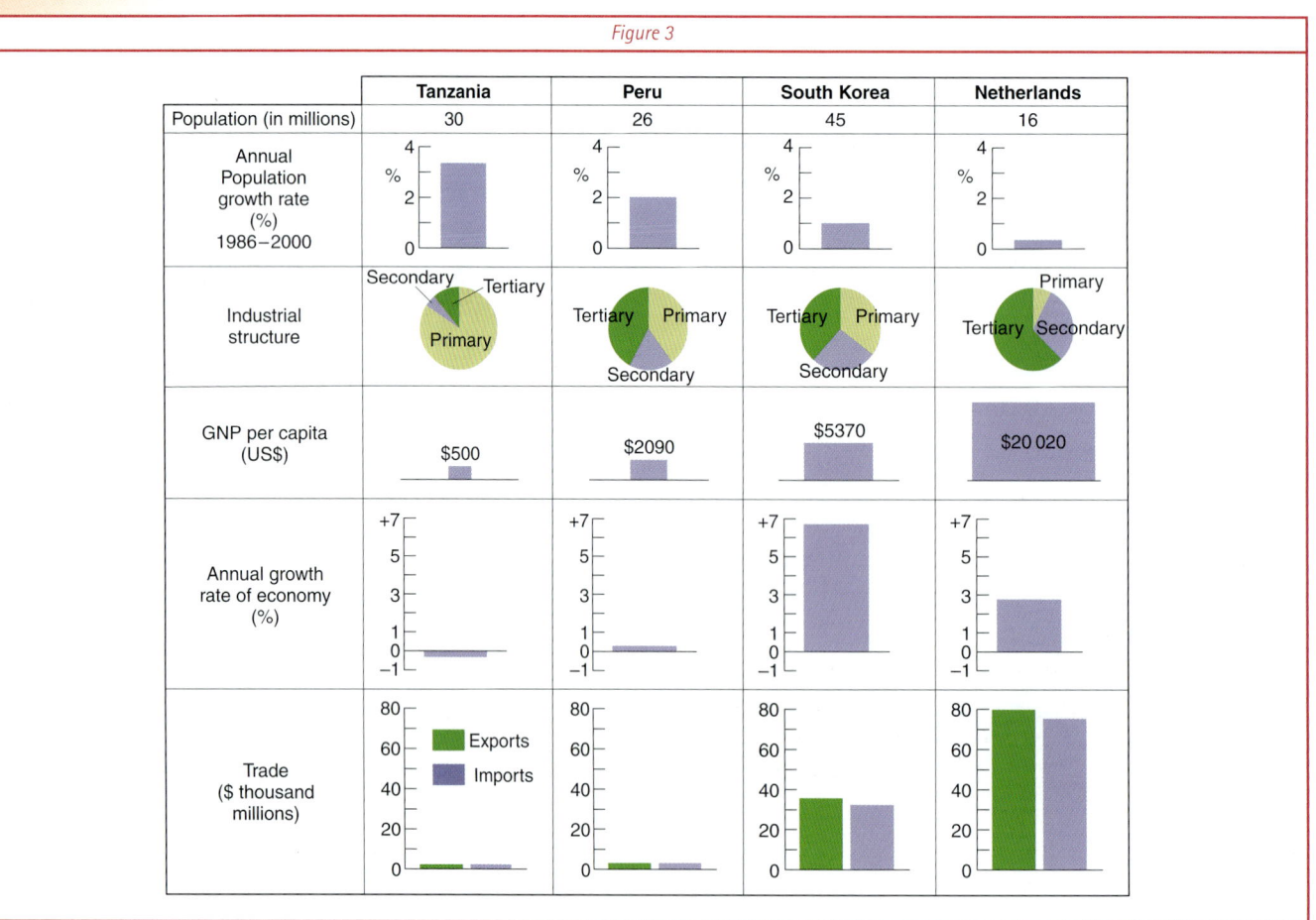

Figure 3

122

Chapter 9: Exam style questions

(b) Study Figure 4 which shows how development in countries with lower levels of economic development can create problems for those who want to protect and sustain world environments.

  (i) Describe the sources of danger to the environment shown. **(4)**

   There are various dangers shown. Identifying two (e.g. deforestation, air pollution) and describing them will get high marks.

  (ii) Comment on how suitable the title is for the cartoon **(3)**

   You will need to explain what sustainable means, and then point out that the title is suitable because deforestation or air pollution ar not sustainable developments.

  (15 marks)

Figure 4 Developing an unsustainable future

Question 4 – H Tier

Study Figure 5.

(a) (i) State how the five basic indicators on the map show that Africa is very badly off compared with other parts of the World.

  1. Wealth
     Africa has the lowest GNP per capita shown.
  2. Birth rate
     Africa has the highest birth rate shown.
  3. Life expectancy
     Africa has the lowest life expectancy shown.
  4. Infant mortality
     Africa has the highest infant mortality rate shown.
  5. Literacy levels **(5)**
     Africa has the lowest literacy rate shown. Better answers might give numbers and/or say what the indicators mean.

 (ii) Suggest **one** further indicator and show how you could use it to measure living standards. **(2)**

   There are various e.g. calorie intake per person per day, numbers of doctors per 1000 people. Remember to show how, for instance, your daily diet helps to tell us something about your living standards.

(b) Study Figure 6 which shows population and food output changes, 1980–2000.

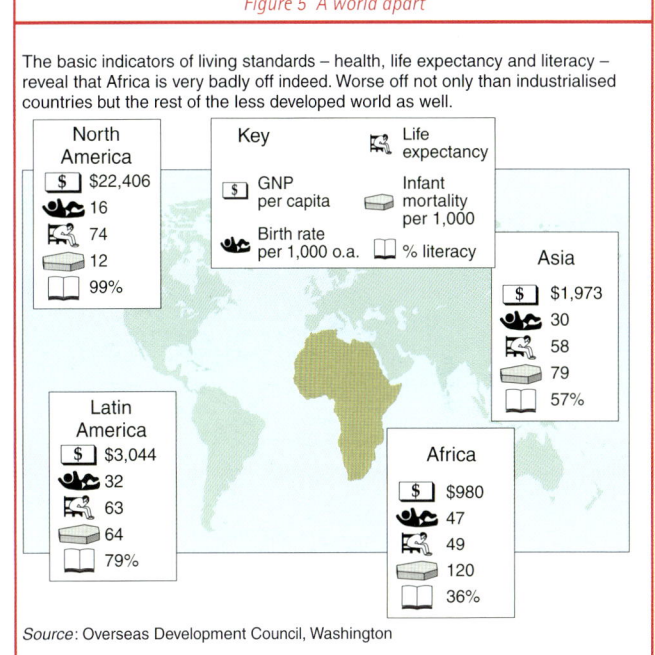

Figure 5 A world apart

Figure 6 Population and food output changes, 1980–2000			
	% population change	% increase in food output	% change in food output per person
Average for MEDCs	+10%	+21%	+11%
Average for all LEDCs	+27%	+40%	+10%
Average for Africa	+37%	+23%	−10%

123

Chapter 9: Exam-style questions

(i) What does the table show about population change in Africa compared to the rest of the world? **(1)**

It is increasing far faster than anywhere else.

(ii) Suggest reasons for the difference. **(2)**

The gap between birth and death rates is positive and wider than in the other two regions. High birth rates in Africa result from high death rates, poor family planning, no state provision for the elderly, etc.

(iii) What does the table show about food output in Africa compared to the rest of the world? **(1)**

Increasing but less slowly than in other LEDCs. The change in MEDCs is not as important.

(iv) Explain why Africa's food output per person has declined when in the rest of the World it has increased. **(2)**

Population growth in Africa has outstripped the growth of food output in that continent. Elsewhere it is the other way around.

(v) What has been a result of this decline in food output per person? **(1)**

Malnutrition, starvation and famine are a result.

(c) Study Figure 7 which shows examples of appropriate technologies which are needed so that LEDCs can develop their rural areas.

Describe the main features of appropriate technology and show how they suit rural development. **(6)**

Appropriate technology is equipment, methods of production, etc., that is suitable for local conditions. To suit rural areas in LEDCs it needs to be simple, inexpensive, easy to use and repair, time-saving, etc. This way it is likely to sustain development.

(20 marks)

Figure 7 What is needed?

WHAT IS NEEDED?

AGRICULTURAL TECHNOLOGIES

COMMUNICATIONS TECHNOLOGIES

HEALTH AND SANITATION TECHNOLOGIES

ENERGY TECHNOLOGIES (MAKING AND SAVING)

INCOME-MAKING TECHNOLOGIES

Resource depletion — Chapter 10

Natural and non-renewable resources

Features of the **natural environment** such as water, climate, soils and raw materials which can be used to benefit people are described as natural resources. Figure 10.1 shows a classification of resources. A basic distinction is usually made between **natural resources** and **human resources** such as machinery and knowledge. Resources are of use to people in meeting their needs of life. Raw materials such as minerals (e.g. iron ore) and fuels (e.g. crude oil) are important to industry but are found only in some places. Other natural resources such as water can be found almost everywhere. As Figure 10.1 shows some natural resources are renewable, others are non-renewable. **Renewable resources** can be used over and over again. Non-renewable resources are fixed or finite in supply. They are exhaustible, will eventually run out and cannot be replaced. **Fossil fuels** are a good example of a natural and non-renewable resource.

Coal, oil and natural gas were formed in past geological periods, and are burned as a source of fuel and power. The bulk of the world's energy is produced from fossil fuels. Figure 10.2 shows how most of India's commercial energy and about half of its total energy now comes from fossil fuels. Generally, in MEDCs most energy consumption is commercial and mostly still from fossil fuels.

Why natural, non-renewable resources are being depleted

Continued reliance on fossil fuels will lead to an **'energy crisis'** in the future. There is estimated to be about 400 times as much oil and coal in the Earth as the world now consumes in a year. Coal is more plentiful than oil. The trouble with non-renewable resources is that use reduces the supply. The supply of fossil fuels can only reduce if we continue to extract and burn them as a source of electricity. The discovered and commercially available supplies of a non-renewable resource are known as the reserves. There may be as yet undiscovered supplies of coal and/or oil which, once discovered and made accessible to extracting companies, will add to the world reserves of fossil fuels. This would extend the 'shelf life' of coal and oil as energy sources. Concerns that any global warming due to atmospheric pollution from coal and oil burning should be slowed could have the same effect. Fewer fossil fuels would be consumed.

Mineral mining will also eventually lead to the exhaustion of mineral reserves. Figure 10.3 shows how the reserves of copper, lead and tin are limited, and how continued consumption of them will lead to their exhaustion. Discoveries of new supplies of these minerals in the Earth's crust will put back the date when the world runs out of copper, lead and tin.

The world's copper reserves in 2000 will be depleted at present rates of consumption by about 2017. The 2000 lead reserves will have run out at present consumption rates a year or so later. Without new discoveries of tin deposits tin reserves will be 100% depleted in about five year's time.

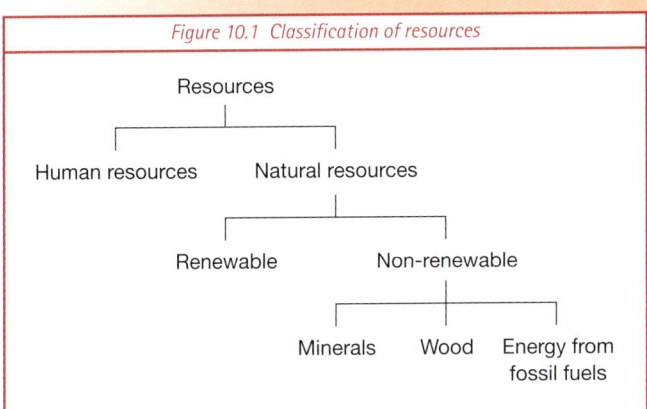

Figure 10.1 Classification of resources

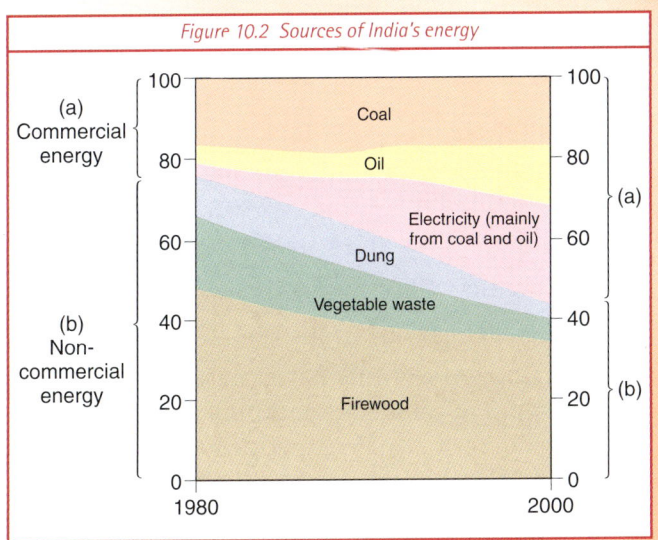

Figure 10.2 Sources of India's energy

Figure 10.3 Table of mineral reserves

	Proven reserves (m. tonnes)		Annual consumption (m. tonnes)	
	1975	2000	1975	2000
Copper	409.5	165.5	7.3	9.5
Lead	150.1	69.8	3.6	3.8
Tin	3.7	0.9	0.2	0.2

Section 3: Managing Economic Development

The causes of mineral and timber resources depletion

The demand for natural, non-renewable resources has continued to grow. The two basic reasons for this are:

- economic development – more industrialisation and greater wealth around the world increases the demand for these raw materials
- population growth – the more people there are in the world, the greater the demand

The combined effect of economic development and population growth is to increase the demand for both minerals and timber significantly.

The increased demand for hardwoods in MEDCs and for local firewood is a key reason for the large-scale deforestation of tropical rainforests. More than half of the original rainforest has disappeared, most of it in the last 50 years. Much of this has been felled for timber. **Hardwood** such as teak, mahogany and rosewood has for a long time been in big demand for making furniture. Countries such as Malaysia export hardwood logs to MEDCs such as the UK and Japan. Tropical rainforests are often found in LEDCs whose need to trade with MEDCs is immense. **Commercial logging** is one of three broad reasons for tropical deforestation. The other two are:

- firewood – wood supplies a large proportion of energy for many of the world's poorest people. Figure 10.2 shows that it still supplies 36% of India's energy. Trees are felled to burn.
- land clearance – trees are felled so that the land can be used in various ways: farming, cattle ranching, mining, oil exploration, **hydro-electric** projects and road building.

The importance of these three broad reasons does vary from place to place. Much forest clearance in Sarawak state, northern Borneo, Malaysia (Figure 10.4) has been for logging and timber exports. The timber industry is the state's biggest employer. Large-scale logging projects can be found throughout the state. Malaysia as a whole has 47% of the world's tropical hardwood exports (see Figure 10.5). Japan is the main importing country for Sarawak timber exports with Japanese transnational companies heavily involved in the trade. The rate of deforestation is so fast that at present rates of logging, Sarawak's rainforests will have completely disappeared in about 20 years' time. Today there are relatively few patches of untouched rainforest left. This 'chainsaw massacre' of the forest ecosystem resource for its hardwood logs may lead to the cleared land being diverted to other uses, especially farmland.

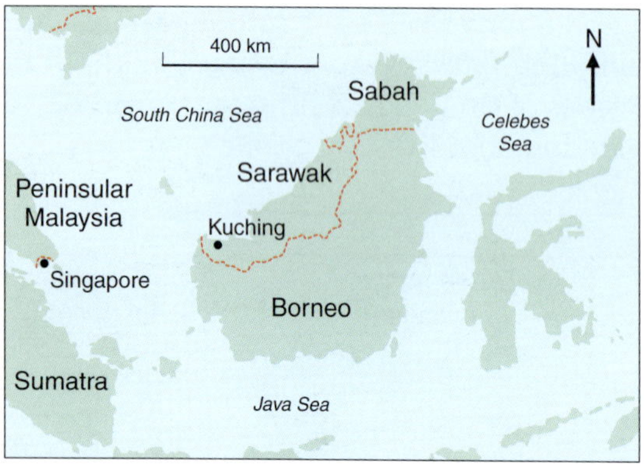

Figure 10.5 Hardwood logs for export from Sarawak

Some minerals are another depleting resource at present. Large-scale mined areas can be found in Australia. There is a great variety of rich mineral reserves scattered throughout the old rocks of that continent with their high mineral content. It is one of the world's largest mineral producing areas, and each year more deposits are discovered. The island state of Tasmania is a major mining region. As Figure 10.6 shows a range of minerals are mined in western Tasmania. The settlement pattern of north-west Tasmania largely results from the possibilities for mining that this area offers. The area is mountainous and isolated. Mining settlements and transport links to serve them were built. As Figure 10.7 suggests, providing transport in isolated areas becomes worthwhile if deposits are:

- large, high-grade and have a long life
- not too difficult to mine
- in large demand by industry

Most production is in the hands of large transnational mining corporations such as BHP Billiton, Rio Tinto, Anglo-American and Alcoa. Some are foreign-based but Australian control of mining interests has been increasing. Mining is a major

Figure 10.4 Location of States of Sarawak and Sabah, Malaysia

Chapter 10: Resource depletion

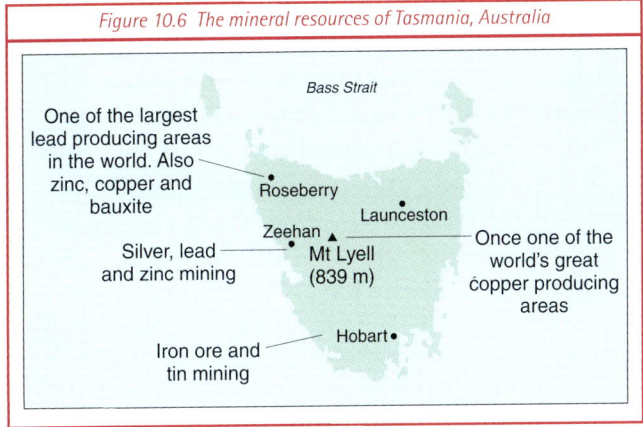

Figure 10.6 The mineral resources of Tasmania, Australia

Figure 10.7 Australian iron ore mining

source of wealth for Tasmania. Minerals are mainly exported for processing elsewhere.

The Mount Lyell copper mine closed December 15, 1994. The 101-year-old mine had come to the end of its life, during which time it had produced 1.1 million tonnes of pure copper. This involved the mining of around a 1 000 million tonnes of copper ore by open-cast quarrying. The heavy surface mining of the area for a century has left environmental damage. Underground mining of minerals occurs elsewhere in western Tasmania.

ACTIVITIES

1. Copper is a good conductor of electricity. Hardwood is tough and durable.
   List as many uses as you can for:
   a. Copper – think of the electricity, plumbing and brewing/catering industries.
   b. Hardwood – think of the building, furniture and transport industries. (WO1.1 – WO2.2)
2. Suggest why the demand for minerals and hardwoods remain high.
   Think in terms of wealth and industrialisation.

The effects of the continued depletion of these resources

People, economies and the environment are affected by the continued extraction of non-renewable resources. The eventual running out of their supply will have serious effects on people, economies and the environment. Careless use or over-use of these resources, as is the case with deforestation, threatens their existence. Supply crises become real possibilities in the future. As supplies run down resource prices are likely to rise. Shortages of these resources will then put up costs of production and so the prices of many goods. Hardships may occur as people cut back on using these resources.

The government of **Saudi Arabia** is very aware that it should not deplete its oil reserves too rapidly. It prefers to try to stop world oil demand from rising too much. With large reserves as Figure 10.12 shows, it does not want to be pressed into increasing supply too much to meet a rapidly growing world demand for oil. Efficiency in the use of oil is encouraged. The Saudi Arabian economy is very dependent on oil exports. 40% of the country's GDP comes from these exports. Future shortages of oil and a cut-back in its consumption may lower living standards around the world. It would also spell disaster for the economy and people of Saudi Arabia. Its policy of conserving reserves is a sustainable strategy.

Figure 10.8 shows some effects of deforestation in tropical rainforests. The environmental impacts associated with the removal of these timber resources include:

- exhausted, infertile soils or areas of bare soil in which few plants will grow
- rivers which flood more because, as Figure 10.9 shows, more rain hits the soil and more soil ends up in the river

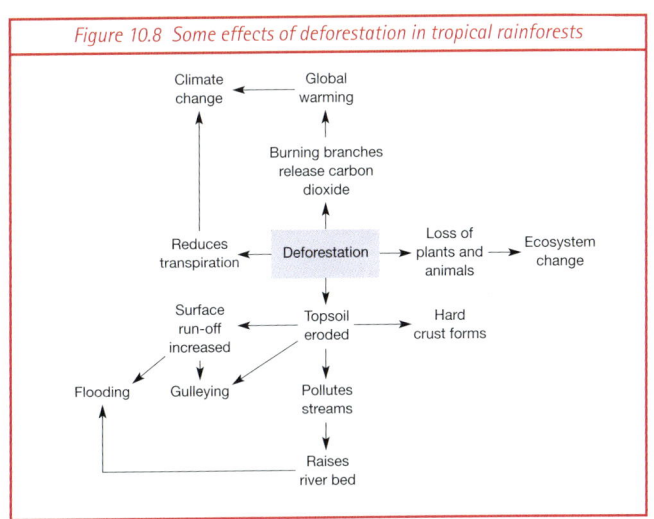

Figure 10.8 Some effects of deforestation in tropical rainforests

127

Section 3: Managing Economic Development

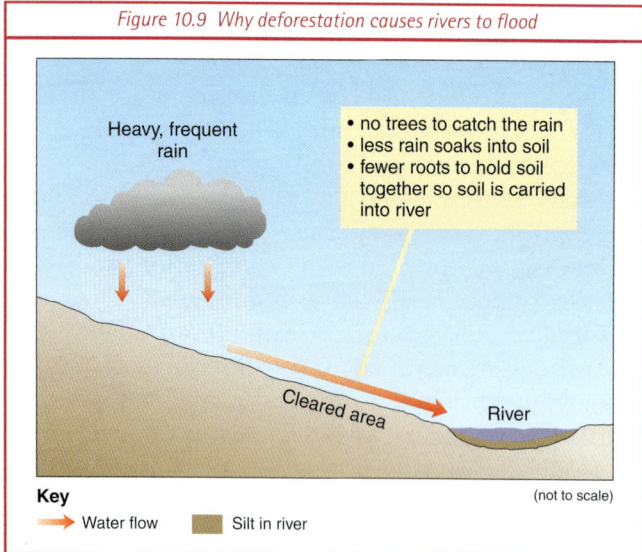

Figure 10.9 Why deforestation causes rivers to flood

- the loss of plants and animals, some of which may provide medical cures in the future and/or improved crop farming
- adding to the Greenhouse Effect which may increase global temperatures and change the world's climate.

The whole rainforest **ecosystem** is a resource which is in a very delicate ecological balance. Figure 10.10 shows what might happen to this ecosystem when it is disturbed by large-scale deforestation. Higher temperatures and lower rainfall could see rainforest turning into desert.

Mining and the closure of a mine when the minerals run out can have serious economic effects. Some of these effects as well as some of the causes of a copper mine closing in Zambia are shown by Figure 10.11.

Loss of exports and unemployment are a real threat to the Zambian economy. The national economy might collapse without copper mining. It provides 98% of the country's export earnings and 15% of its jobs. Mining 'though' does destroy natural environments.

One major cause of environmental damage has come from the transport of oil. There have been many examples of oil spills into the sea after tanker accidents. In 1989 the Exxon Valdez tanker ran aground in Alaska. The escape of more than 11 million gallons of crude oil into the Pacific Ocean cost £1.25 billion to clean up. Early in 2001 the Jessica tanker also ran aground but in the Galapagos islands off Ecuador. The islands are a National Park and World Heritage Site, have a unique ecosystem, and are where Charles Darwin developed his theory of evolution. 5 000 species of animals and birds live on the islands; 40% of them cannot be found anywhere else. Figure 10.13 shows some of the rare wildlife threatened by poisoning from the spillage of diesel from the Jessica. Some of the impact will be long-term as the effects are passed down the **food chain**.

Extracting resources from the Earth provides opportunities but also brings threats. It can meet the needs of both local people (e.g. jobs in the local mine) and of the economy (e.g. export earnings from a valuable mineral), but it can fail to meet the need to conserve landscapes and natural environments. Damaged environments can be the price, future generations especially, pay for the economic opportunities brought by, say, mining. The effects of these activities can conflict.

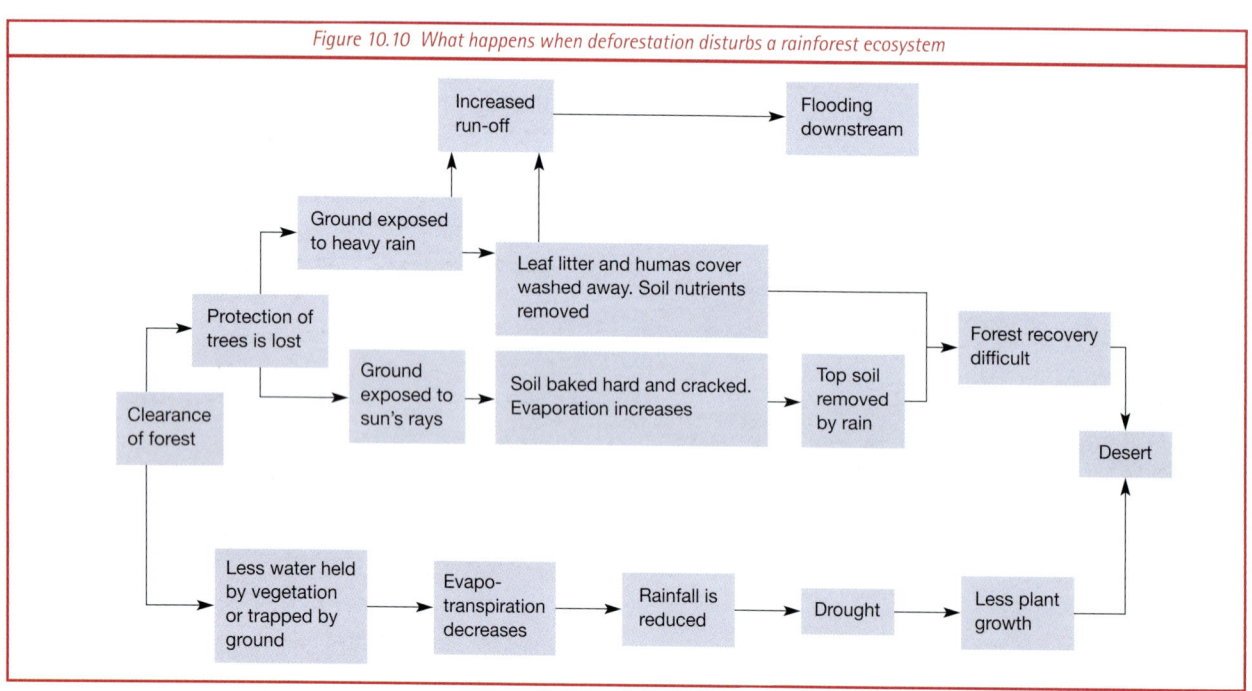

Figure 10.10 What happens when deforestation disturbs a rainforest ecosystem

Chapter 10: Resource depletion

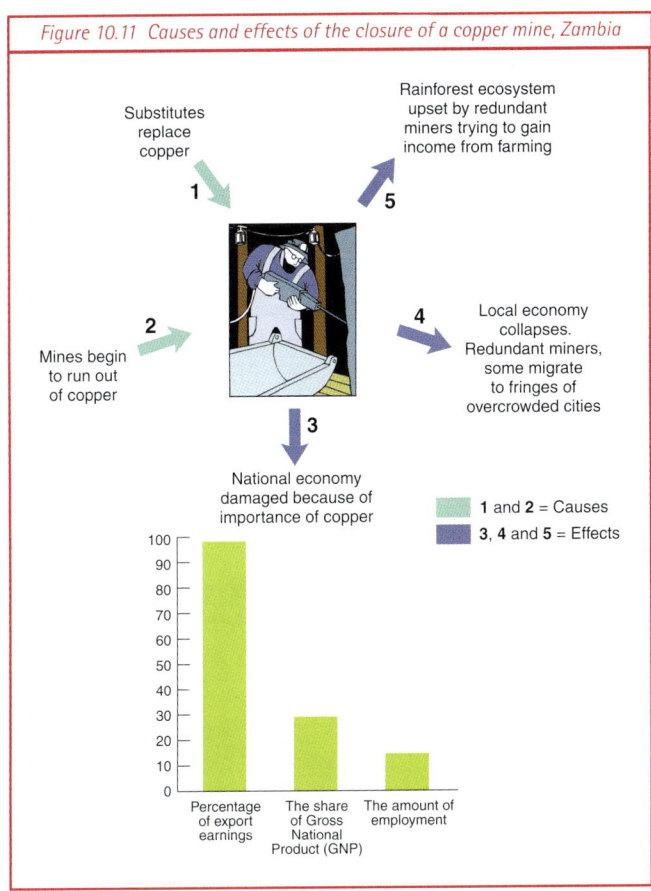

Figure 10.11 Causes and effects of the closure of a copper mine, Zambia

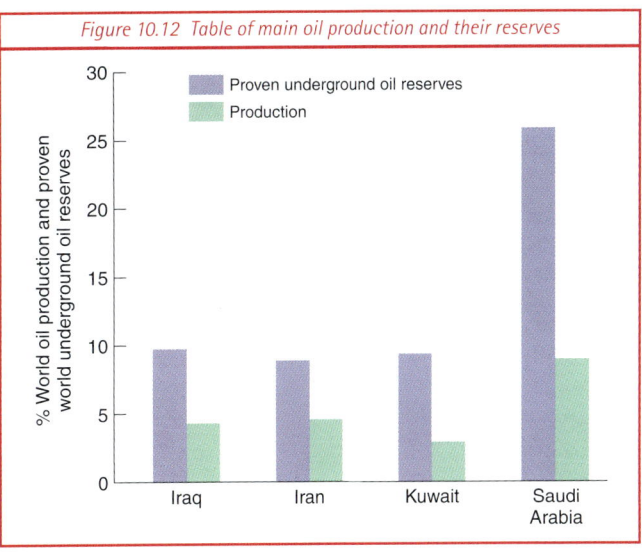

Figure 10.12 Table of main oil production and their reserves

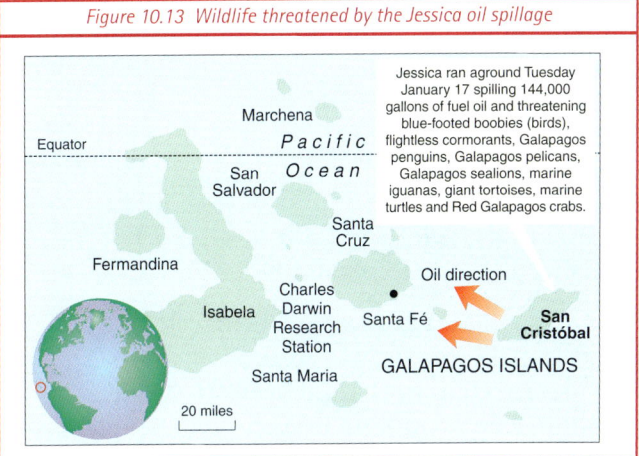

Figure 10.13 Wildlife threatened by the Jessica oil spillage

ACTIVITIES

1. List three ways in which people will be affected by the continued depletion of natural resources.
2. Do an Internet search on mining in National Parks (e.g. limestone quarrying at Castleton in the Peak District National Park – www.peakdistrict.org.uk). What are the conflicts of interest in this issue? (IT1.1 – IT2.3)
3. Draw a food chain for a coastal ecosystem (e.g. seagulls eat sand eels, etc). Show how long-term poisoning from an oil spillage can be passed on to a number of creatures.
4. Study Figure 10.14 which contains information about the importance of oil extraction to Saudi Arabia.
   a. How has oil extraction since the 1970s benefited Saudi Arabia and other OPEC countries?
   b. When oil prices rise significantly, who and what are the gainers and losers?
   c. Suggest why oil prices in the future might follow the price trend of the 1970s. (C1.2, C2.2, N1.1, N1.3, N2.1, N2.3)

Managing the extraction of resources

It should now be clear that there is a growing need to manage resources, especially non-renewable ones. Unless we do so properly a future generation may reach the limits of economic growth on the Earth. Their living standards and quality of life would then fall. With present rates of growth in world population, industrialisation, pollution, food production and resource depletion some geographers believe that this point could be reached about the year 2100. Future generations need access to natural, non-renewable resources. Resource management so that this is achieved might be through a range of approaches:

- Substitute renewable for non-renewable resources where possible.
- Develop appropriate technology which alters methods of production and living in a way which saves on resources. Greater efficiency in the use of existing resources is often possible.
- Recycling of materials using natural resources.

Section 3: Managing Economic Development

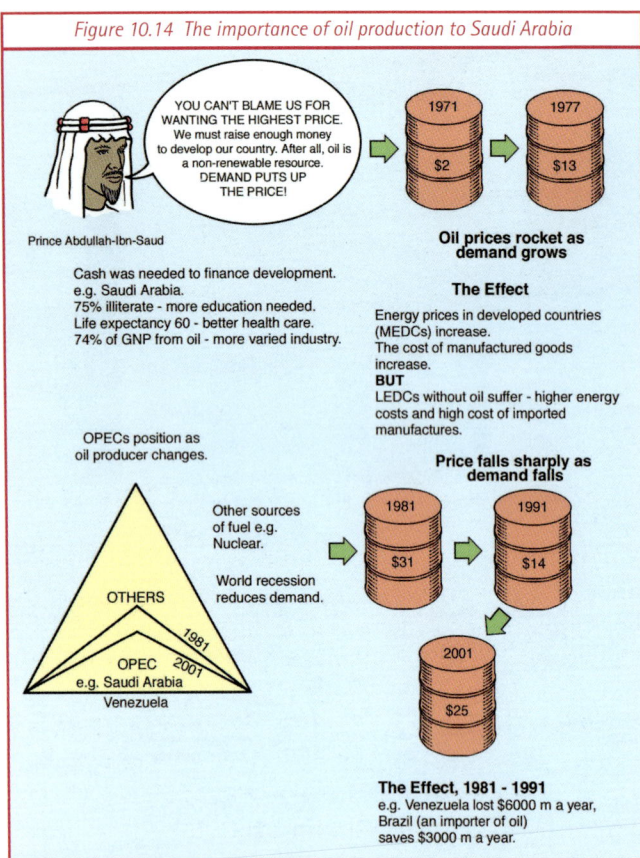

Figure 10.14 The importance of oil production to Saudi Arabia

These approaches are sustainable developments. Other approaches include creating alternative resources such as **artificial substitutes** (e.g. plastics), using natural substances that people have not yet begun to develop and see as resources, conserving existing natural resources, exploring for new supplies of natural resources and extracting previously uneconomic reserves of these resources.

This range of management strategies available to us now and in the future may mean that resource depletion will be less of a problem than many people think. Local governments, national governments, the **European Union** (EU), non-government organisations (NGOs) such as the United Nations and **Greenpeace** and many transnational companies (TNCs), such as Rio Tinto, have environmental policies. The EU takes environment or 'green' issues seriously, encourages sustainable development and favours tight controls on pollution. Rio Tinto's environmental policy includes the rehabilitation of old mining sites. Greenpeace reports highlight the threats if present trends in resource use continue unchanged. Their campaigners sometimes take direct action to conserve natural resources and promote alternatives to their use. There are other NGOs (e.g. WWF – World Wildlife Fund) involved in peaceful, creative environmental protest. The UN and its various agencies encourage sustainable approaches, especially in LEDCs where the UN often funds development projects. Hydro-electric power schemes involving damming rivers are generally preferred on environmental grounds to schemes based on either nuclear power or fossil fuels (coal and oil).

Developing renewable energy sources is a key part of the UK government's energy policy; its present target is to make sure that at least 20% of the country's electricity comes from non-fossil fuels. Leaving aside the controversial supply of electricity from nuclear power stations, this means developing:

- wind power
- tidal and wave power
- hydro-electric power

Solar power and geothermal power are less suitable sources in the UK. Tropical locations and plate margin locations are really needed if these two sources are to be of commercial use. Figure 10.15 shows the two sources of renewable energy suited to the UK, they cannot be exhausted and are environmentally clean. They do not contribute to atmospheric pollution though they may have a small environmental impact locally. Wind turbines can be noisy and an 'eyesore' – noise and visual pollution! Dams and **tidal barrages** can disturb local ecosystems. A barrage across the Mersey is likely to change the nature of the saltmarsh ecosystem on the south bank of the estuary. One further disadvantage of renewable energy sources is that they are limited in their ability to generate electricity. At present they do not have the capacity to replace non-renewable sources but are a useful way of supplementing them to slow down the eventual exhaustion of coal and oil

Figure 10.15 Sources of renewable energy

supplies. The development of fast-growing 'energy crops' that can be processed after harvesting into a substitute fuel like **bio-diesel** will also ease the pressure on non-renewable energy supplies. 5% of Brazil's energy needs are supplied by biomass. 75% of its cars run on ethanol produced from sugar cane, cassava and maize.

Appropriate technology can be designed to minimise the use of natural resources. The EU encourages the making of homes, cars and factories which are energy-efficient. Less coal and oil will be needed if homes are well insulated, use low-energy light bulbs and have solar panels fitted for summer use. The government and car industry can also help to slow down the depletion of non-renewable oil supplies by:

- developing more fuel-efficient vehicle engines
- trying to cut car journeys
- discouraging the driving of bigger cars

The plan to charge motorists to drive into London may be one of achieving the second of the three points above.

The **appropriate technology** idea also applies to factories. Appropriate technology makes the most of natural resources and reduces the use of scarce **factors of production** (land, labour, capital and knowledge). In LEDCs simple, **low-tech** and cheap machines to transport water, peel crops or bake bread save working time, limit environmental damage, use local materials and skills, are affordable and generally appropriate to the community. Where they use much less firewood there is a clear advantage. Traditional methods of manufacturing in MEDCs use lots of energy, generally non-renewable. Machinery which uses less energy and renewable energy may better meet the needs of today and the future. For LEDCs to transfer the traditional energy-inefficient technology of MEDCs would perhaps be an environmental mistake. The United Nations Development Programme in LEDCs generally supports the idea of these countries introducing industrial technology which is efficient and inexpensive in its use of energy resources. The NGO Intermediate Technology Development Group set up in the UK promotes small-scale appropriate technology in India, often low-cost machinery in rural areas.

The **recycling** of materials is encouraged throughout the EU. Recycling is the re-use of resources. It involves the processing for re-use of chiefly waste glass, paper and some scrap metals from factories and homes. It is estimated that only 8% of all household waste in the UK is recycled.

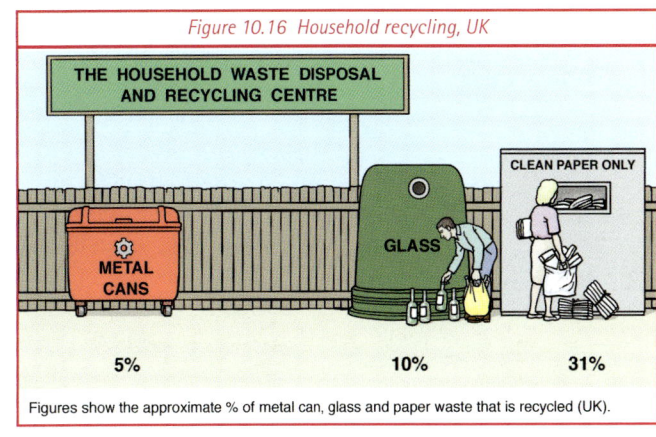

Figure 10.16 Household recycling, UK

Figures show the approximate % of metal can, glass and paper waste that is recycled (UK).

Other EU countries recycle more. The present target set by the government is that we recycle, compost and incinerate (burn with the energy or heat used) 45% of waste by 2010. Recycling in the UK needs encouraging if this target is to be met. Figure 10.16 shows the percentage of waste paper, waste glass and used metal cans from households recycled in the UK. There are natural resources in all three of these items so recycling should slow down the rate at which the resources (raw materials) run out. Recycling is one method of helping the environment. It helps to save scarce resources, especially non-renewable ones, and does reduce environmental pollution. It should cut down on the amount of waste disposed of in landfill sites though this does require there to be a market for recycled materials and goods. The lack of firms willing to re-process waste has dogged the success of many of the voluntary recycling schemes run by **local authorities** in the past. Around a half of the commercial and industrial waste produced in the UK is recycled. The amount of recycling in the UK is set to rise towards European levels. Roughly 20% of a new BMW car is made of recycled resources.

ACTIVITIES

1. List what the UK government might do to increase the amount of waste that is recycled. (WO1.1, WO1.2, WO2.1, WO2.2)
2. Study Figure 10.15.
   a. Name two of the sources of energy shown.
   b. For each source list as many advantages or benefits, and as many disadvantages or costs as you can think of. You should refer to the idea that these are renewable energy sources which are alternative to non-renewable sources. (C1.3, C2.3)

Section 3: Managing Economic Development

Figure 10.17 Model for a sustainable scheme of rainforest-based industries

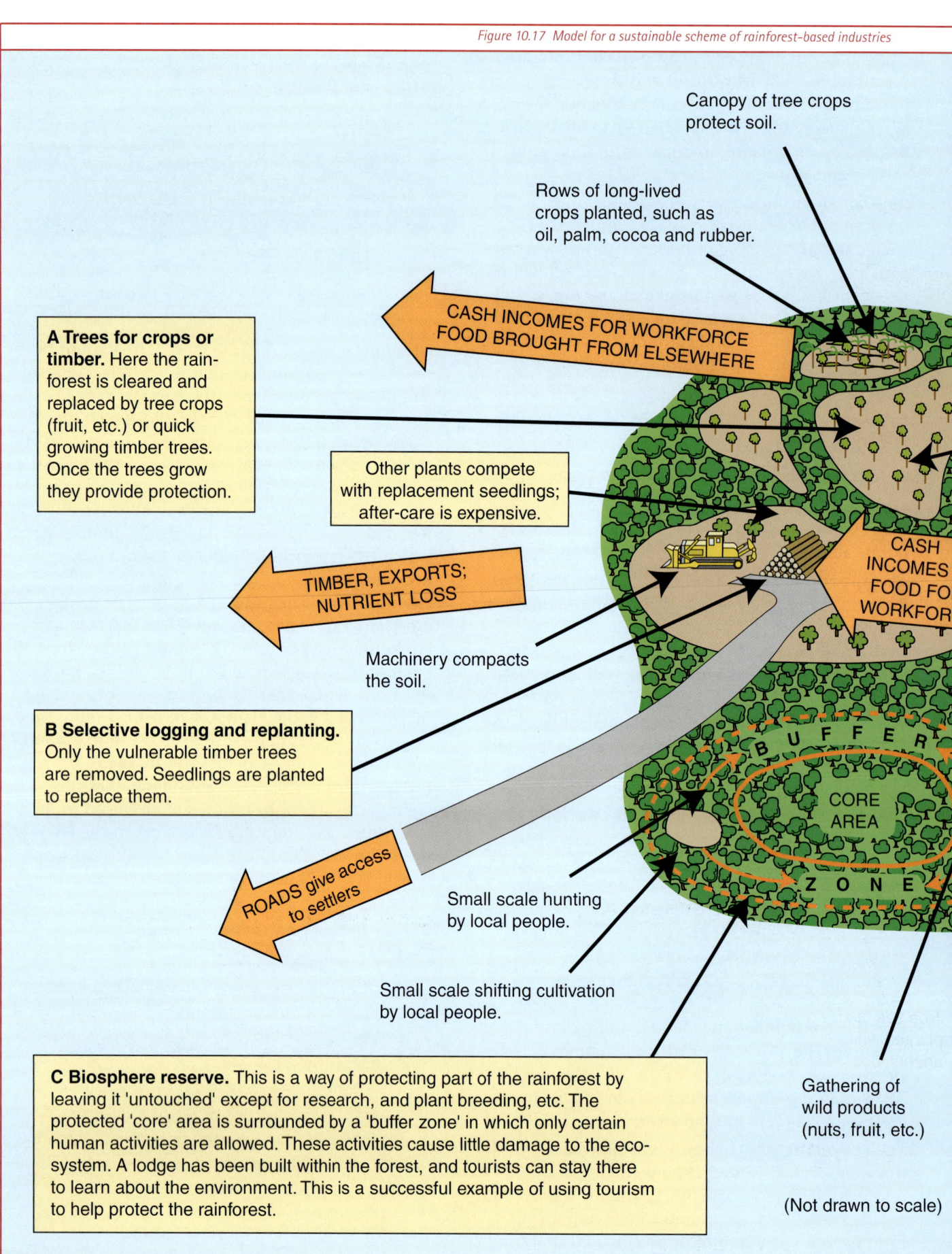

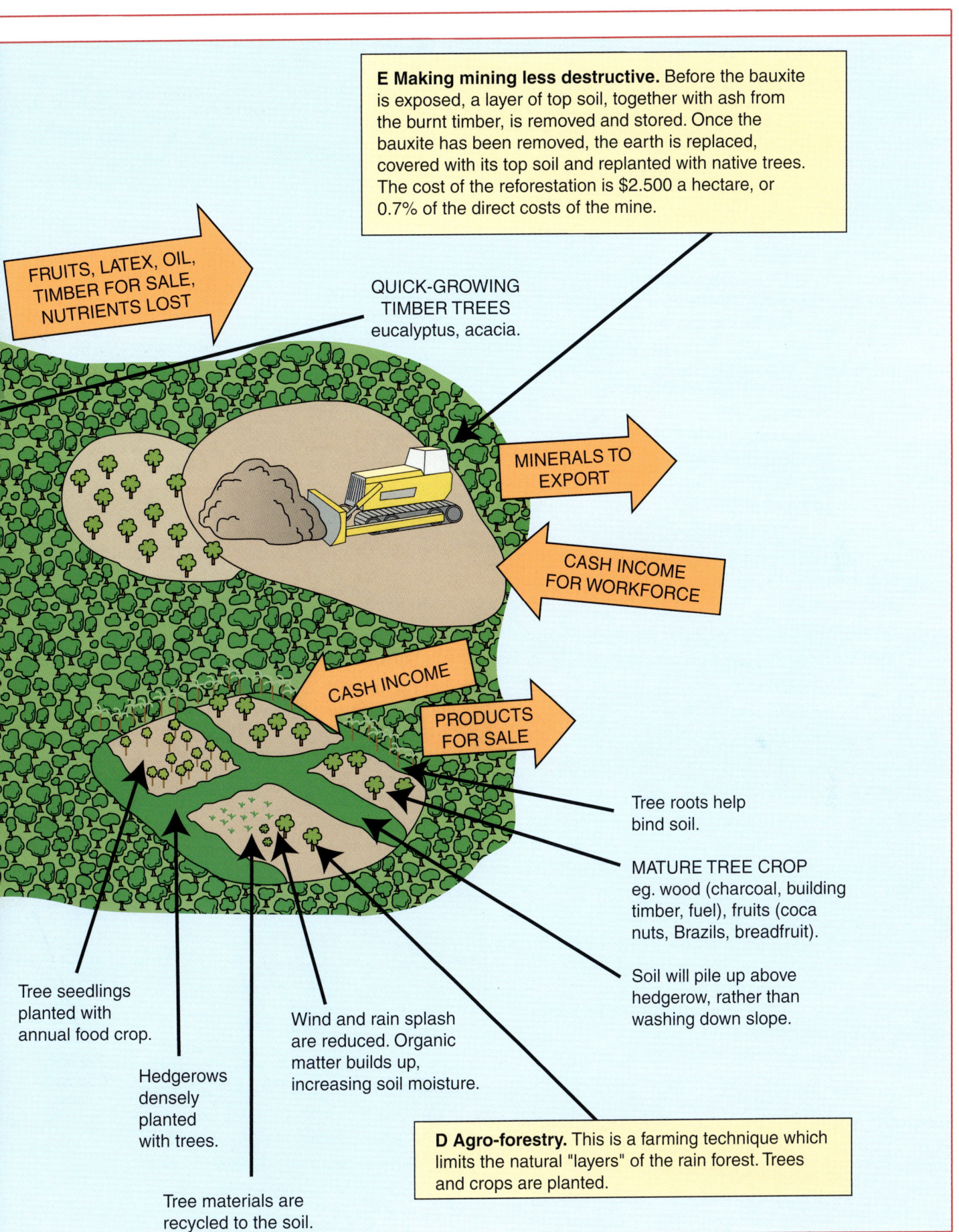

Section 3: Managing Economic Development

Sustainable forestry: the way forward for all depleting resources

Deforestation causes serious environmental problems and in the future will bring economic difficulties, but the timber is a valuable resource which LEDCs feel the need to exploit at present for economic development. Sustainable forestry may enable both **conservation** of the forest and development to take place. Only some trees are removed. It involves managing rather than destroying the forest. It depends on a partnership between governments, developers and local people. Trees can be removed and timber cut provided **over-felling** does not occur, and new young trees, often fast-growing species, are planted in their place. In Sabah state, Malaysia, which now practises sustainable forestry, only 7 to 12 trees per hectare are felled. Government rules require at least 30 substantial trees to be left in each hectare. The Malaysian government plays a large role in trying to sustain the country's rainforest:

- It issues **licences** and **quotas** (Malaysian Timber Council) to harvest logs to logging companies.
- It sets regulations and standards for these companies and tries to police them.
- It co-operates with western NGOs (Non-Government Organisations) like the WWF and other conservation groups.
- It supports forestry research such as the Reduced Impact Logging (RIL) scheme.
- It negotiates with licensed logging companies in the belief that to work, sustainability requires their commitment. They favour dealing with large companies committed to long-term profit from large forestry projects over a long period.

The Sabah forests are logged but the forest is still there and is expected to be into the future. Government rules ensure that tree cover and **biodiversity** are maintained, and that the impact of forestry on the ecosystem and local people is kept to a minimum. Trees are cut so that they fall where they do least damage to other trees. Roads are designed so that they do not become gullied rivers of mud and the rivers clogged with soil eroded from these roads. Logging companies and conservationists now co-exist side by side in these rainforests.

NGOs also play a big part in the sustainable development of these rainforests. The United Nations Commission for Sustainable Development encourages countries like Malaysia with threatened rainforests to work towards a Forest Convention designed to safeguard the future of this type of environment and its timber. The Forest Stewardship Council (FSC) is a voluntary organisation involving NGOs (e.g. WWF, Friends of the Earth) and companies involved in the timber market (e.g. B&Q) committed to sustainable forestry. The idea is to award levels of certificate to forests according to how well managed they are. The hope is that the most sustainable forests will find it easier to sell their timber to increasingly environmentally-conscious customers, especially in the European Union.

Sustainable schemes such as that shown by Figure 10.17, help local people by offering a chance of long-term income without clearing large areas of their trees. A range of sustainable rainforest-based industries are possible, e.g.:

- carving of craft items out of bark from trees selected for sustainable logging
- nut oil production
- farming of valuable butterflies in their natural habitats
- ecotourism – jobs as guides, rangers, caterers etc.
- **agro-forestry** – where trees for timber are combined with food crops and/or animals

Natural resources should be able to be exported for generations to come. Conserving forests also preserves animal and wildlife habitats, the natural beauty of this type of environment, and may limit global climate change. Forests act as so-called 'carbon sinks'. They remove as part of the carbon cycle carbon dioxide emitted into the atmosphere by human activities such as power stations. More forests may mean a weaker Greenhouse Effect and less global warming.

ACTIVITIES

1. Study Figure 10.17. Write down all the ways in which this patch of forest is being managed for future sustainability. Make sure that the sustainability of each way you identify is explained. (WO1.1, WO1.2, WO2.1, WO2.2)
2. What are the lessons that managers of minerals and fossil fuels can learn from the way in which some forests are being managed? Suggest sustainable strategies for mineral and fossil fuel management. (PS1.1, PS1.2, PS2.1)

Chapter 10: Exam-style questions

Exam-style questions with answer comments for Chapter 10

Question 1: F Tier

Study Figure 1 which shows eight ways of providing energy: coal, hydroelectric power, nuclear, oil, solar, wave, wind and wood, Each is labelled with a letter from A to H.

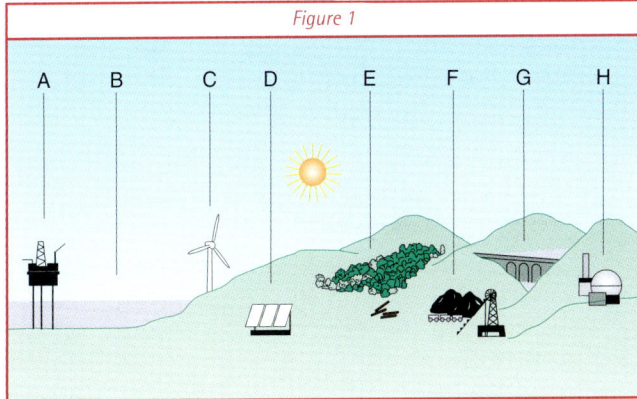

Figure 1

(a) Name any two ways by choosing from the list of forms of energy printed above Figure 1. **(2)**

A = oil; B = wave; C = wind; D = solar; E = wood; F = coal; G = hydro-electric; H = nuclear.

(b) Choose one of the two ways named in (a). Give two advantages and two disadvantages of using this way as a source of energy. **(4)**

Advantages might concentrate on 'renewability' or 'non-polluting' or 'efficient at producing energy'. Exhaustion, pollution or danger are disadvantages.

(c) Some sources of energy are renewable, others non-renewable.
  (i) Complete the table below, by giving the meaning of these two terms, and by ticking the appropriate box to indicate whether each of the four sources of energy is renewable or non-renewable. **(6)**

(ii) What is the evidence from your own knowledge and the table below that the world may be facing an 'energy crisis'? **(3)**

The 'energy crisis' is about a future world which is too dependent on non-renewable fossil fuels whose supply is running out and too low. The table shows two such sources of energy that may become a future problem.

(15 marks)

Question 2: H Tier

Study Figure 2, which shows the changing production of a non-renewable resource.

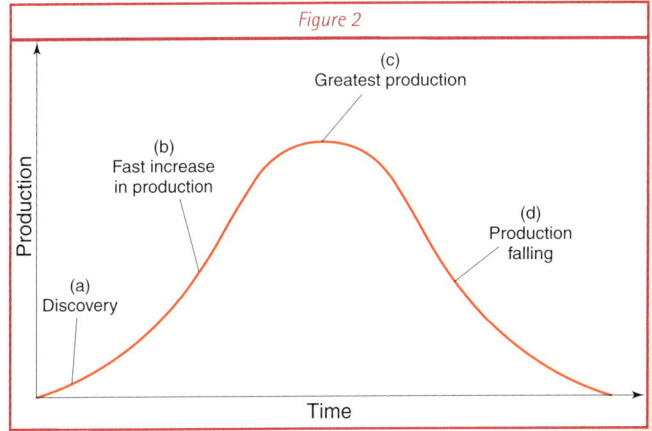

Figure 2

(a) (i) At which point on Figure 1 is the price of the resource most likely to be rising, B or D? **(1)**
  B
  (ii) Give two reasons why production of many of the world's natural resources is likely to fall during this century. **(2)**
  Dwindling supplies and falling demand because artificial substitutes are now available would be a good answer.

Term	Meaning	Source of Energy			
		Petroleum	Natural Gas	Hydro-electricity	Nuclear electricity
Renewable	Supply will not run out or become exhausted			✓	✓
Non-renewable	Resources which once used cannot be replaced	✓	✓		✓

135

Chapter 10: Exam-style questions

(b) List three ways in which countries will be affected by the continued depletion of these resources. **(3)**

Loss of export earnings and unemployment in producing countries, more research, development and discovery in using countries.

(c) (i) State briefly the meaning of the term 'recycling'? **(1)**

The re-use of resources would be fine.

(ii) State two materials very suitable for recycling **(2)**

Glass, paper or scrap metal.

(iii) Describe two other ways in which people can conserve limited resources. **(4)**

Two ways might be to switch to renewable resources or to put legal bans on using limited resources.

(d) Explain how prolonging the life of a non-renewable resource helps to protect and conserve the environment. **(2)**

One way to answer this might be to pick a non-renewable resource like oil. Say what non-renewable means and explain how, by making the oilfield last longer, the environment is less damaged.

(15 marks)

Question 3: F Tier

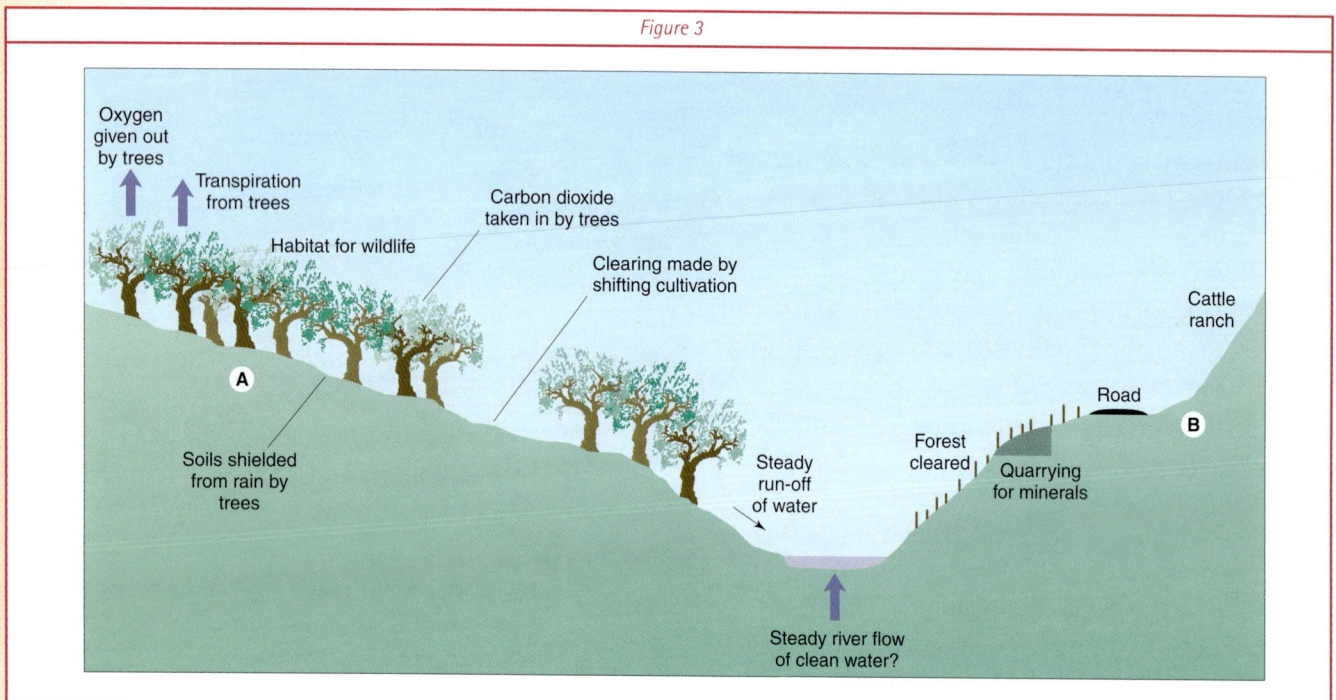

Figure 3

Study the field sketch (Figure 3) which shows a river valley in the Amazon rainforest of Brazil. There is traditional rainforest life on one side of the valley. Large-scale deforestation (clearance of all the trees) has occurred on the other side.

(a) Which side of the valley – A or B – has been deforested? **(1)**

(b) (i) Describe the natural vegetation in a tropical rainforest under the following three headings:

1 layers of vegetation
   Rainforests can be split into three or four distinct vertical layers of vegetation.

2 vegetation on the forest floor
   There is generally little but it can be patchy.

3 the quantity of vegetation **(3)**
   Rainforests are the richest ecosystem on the planet with lots of vegetation and lots of variety.

(ii) Choose **one** of the headings in (b) (i) above, and explain how its features are adapted to the climate **(2)**

Obviously, this depends on which one you choose but, for example, the great quantity of vegetation is due to the year-round hot, wet climate in which plants grow quickly all year.

Chapter 10: Exam-style questions

(c) Explain how shifting cultivators (farmers) live in harmony with and cause no permanent damage to the rainforest. **(2)**

You would need to explain that because shifting cultivators only slash-and-burn small patches of forest at a time before moving on, the forest can quickly recover in the hot, wet conditions that encourage rapid growth.

(d) Draw a bar graph to represent this information about deforestation in the Amazon rainforest. **(3)**

One mark for a decent horizontal axis showing the years involved; one mark for a vertical axis showing area in thousand hectares drawn to scale; one mark for accurate plotting of the six bars.

Year	Area of forest clearance (in thousand hectares)
1970	5
1975	10
1980	15
1985	145
1990	240
1995	327

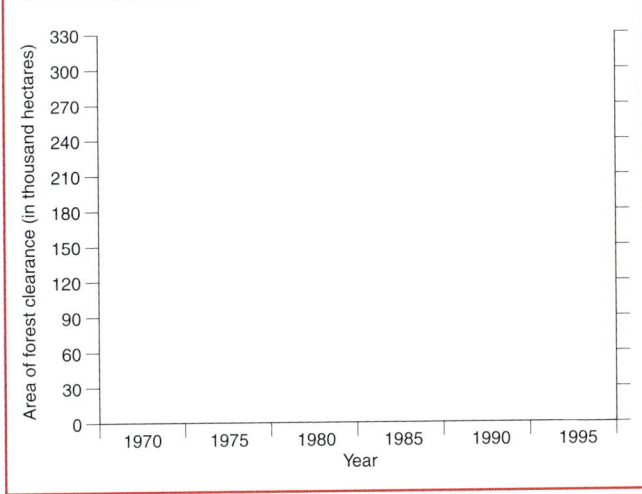

(e) Why are some people concerned about the deforestation of this and other tropical rainforests? **(4)**

Make sure that you make clear that deforestation means cutting down forests and that this leads to more carbon dioxide being put into the atmosphere, fewer 'carbon sinks' to absorb the extra carbon dioxide being put in by industries and vehicles around the world, and special ecosystems with rare plants and habitats being gradually lost.

(15 marks)

Question 4: H Tier

Study Figure 4 which shows world energy production and use.

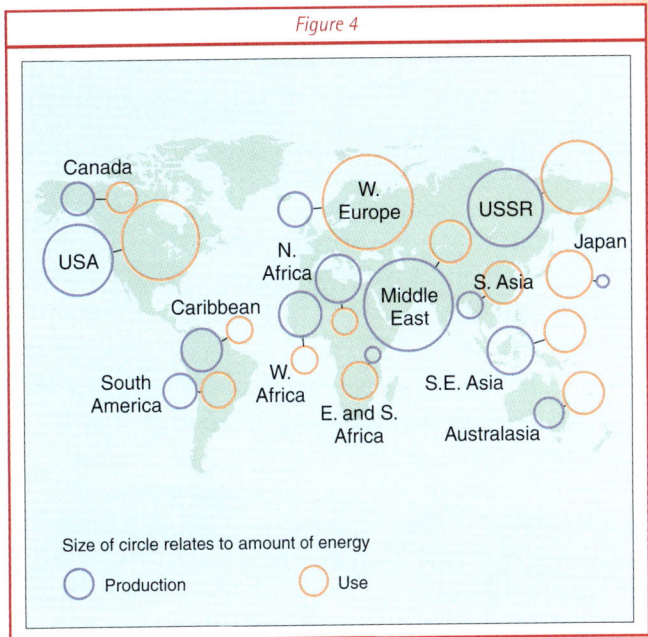

Figure 4

(a) (i) Name the area in which the amount produced and the amount used are almost the same. **(1)**

The USA or South America or Russia and former Soviet Republics.

(ii) Give two reasons why some areas are large producers of energy. **(2)**

You need to give natural supply reasons like large oilfields, large coalfields, etc.

(iii) Explain two reasons which might explain why Western Europe and North America are large users of energy. **(4)**

When it asks you to explain a reason there will be one mark for giving it and a second mark for developing it, e.g. lots of industry, high living standards which mean … .

(b) (i) Some areas use a lot more energy than they produce. Name the area shown on Figure 3 which has the largest energy gap. **(1)**

Western Europe.

(ii) What action can areas such as these adopt to overcome their energy gap? **(3)**

There are various possibilities from importing oil; searching for gasfields of your own; developing alternative sources of energy, etc.

137

Chapter 10: Exam-style questions

Study Figure 5, a newspaper extract about the UK's North Sea oil fields.

(c) What does Figure 5 suggest about North Sea oil as a Future source of energy for the UK? **(3)**

The answer should be about its future uncertainty. There are few new fields and existing fields are producing less. The UK is no longer self-sufficient in oil.

(d) Uncertainty over oil supplies from the Middle East is used to justify drilling for oil in the Alaskan wilderness with its special environment. Explain why some people object to extracting resources in special environments. **(6)**

This is a question about environmental damage caused by drilling, quarrying, mining, etc. you need to write about wildlife being disturbed, habitats being destroyed, 'eyesores' spoiling attractive environments, etc., with an example. Do an Internet search for material on how oil extraction in Alaska has changed parts of that state's environment. Add it to your answer.

(20 marks)

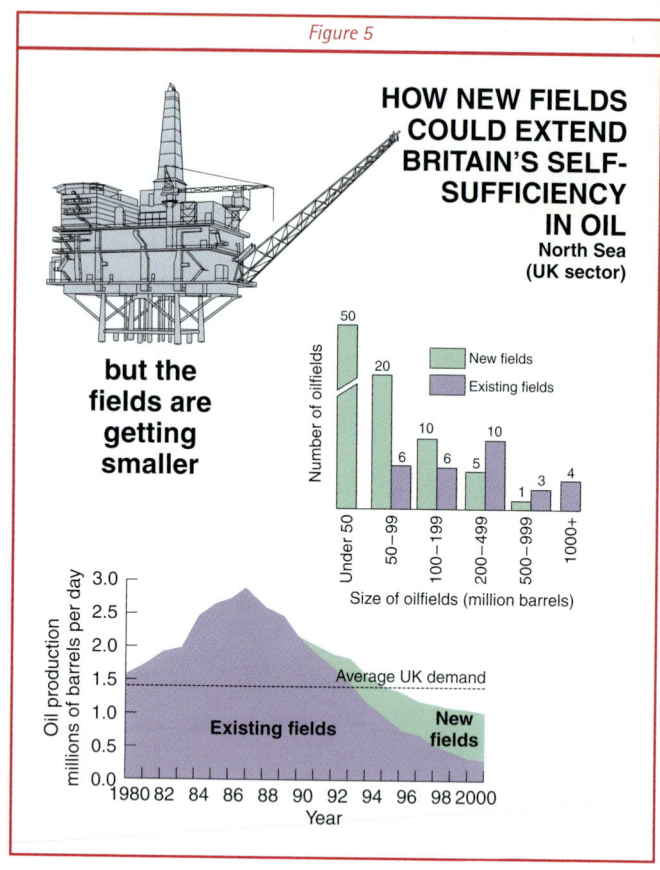

Figure 5

HOW NEW FIELDS COULD EXTEND BRITAIN'S SELF-SUFFICIENCY IN OIL
North Sea (UK sector)

but the fields are getting smaller

Economic development and the global environment

Chapter 11

Threats to the global environment from economic development

The cartoon (Figure 11.1) shows a knight fighting a many-headed dragon in order to protect a princess. Each of the many heads represents an environmental threat. The heads or threats have grown from the dragon's body which represents economic development. Our environment is represented by the princess. Her advice to the knight, who represents those who try to care for our environment, is that the dragon will go on sprouting more and more heads so that it is its body that he must go for. Economic development can create more and more environmental threats. According to the cartoon, the problem is economic development.

Figure 11.1 Cartoon from the Ecologist

Three of the heads or threats shown on Figure 11.1 – acid rain, ozone layer damage and CO_2 (global warming) – are brought by a polluted atmosphere. Economic development can bring pollution. This chapter looks at these three atmospheric threats.

Concern about the state of the global environment grows. As the recent newspaper headline (Figure 11.2) shows the United Nations sees signs of doomsday from these environmental threats. They warn of economic, social and political disaster around the world during the next century if economic development continues as it has done.

The link between these three threats to the global environment

They are all the result of massive pollution of the atmosphere from industry, transport, farming, burning forests and generally modern life, especially in the more economically developed areas of the world. The effects of polluted air are global. The **Industrial Revolution**, starting in 18th-century Britain, was the beginning of the large-scale pollution of the atmosphere. Over the last two centuries air polluting gases have been emitted into the atmosphere at an increasing rate, especially since 1950. The main air-polluting gases are:

- CO^2 (carbon dioxide) emitted from burning fossil fuels, e.g. **car exhausts**, power station chimneys
- carbon monoxide emitted from similar sources to CO^2
- **CFCs (chlorofluorocarbons)** emitted from cleaning solvents, **aerosol** sprays, fast-food packaging, manufacture and the disposal of old refrigerators and air conditioning systems
- methane emitted from biological decay (e.g. animal waste) and fossil fuel burning
- nitrous oxide emitted from agricultural fertilizers and fossil fuel burning
- sulphur dioxide emitted from fossil fuel burning and various industrial processes
- surface ozone emitted as a result of a reaction between some of the above pollutants (e.g. carbon monoxide and methane) and sunshine.

The burning of **fossil fuels** such as coal, oil and wood is clearly the major cause of polluted air. The petrol burned in an average car (without a **catalytic converter**) each year emits:

- four times the car's weight of CO^2
- the car's weight in carbon monoxide
- 20% of the car's weight in nitrous oxide
- 1% of the car's weight in sulphur dioxide

Figure 11.2

139

Section 3: Managing Economic Development

CO_2 is generally regarded as the main polluting gas in global warming; CFCs in ozone layer damage; and sulphur dioxide in acid rain. In each case, other gases are also to blame. For example, CFCs destroy high-level ozone as well as act as a greenhouse gas which, molecule for molecule, is a far stronger warmer of the atmosphere than CO_2. Ozone is also a greenhouse gas. **Greenhouse gases** strengthen the natural **Greenhouse Effect** in the atmosphere and may be causing any global warming. But any global warming may be being delayed by other types of air pollution. The thinning of the high-level Ozone Layer might be causing cooling at the same time as CO_2 and CFCs are causing warming. Sulphur dioxide, a major gas contributing to acid rain, may also be slowing down any global warming. Some scientists believe that the gas is forming a thin haze around the Earth which, like the clouds and dust (including volcanic dust), reflects some sunlight back into space. People are altering the weather and climate of the globe in these three ways: strengthening the Greenhouse Effect, thinning the Ozone Layer and making rainfall more acid. They are linked by a common cause – pollution.

ACTIVITIES

Complete the three following passages below. Find suitable words to fill in the blank spaces by studying the diagrams, Figures 11.3 and 11.4.

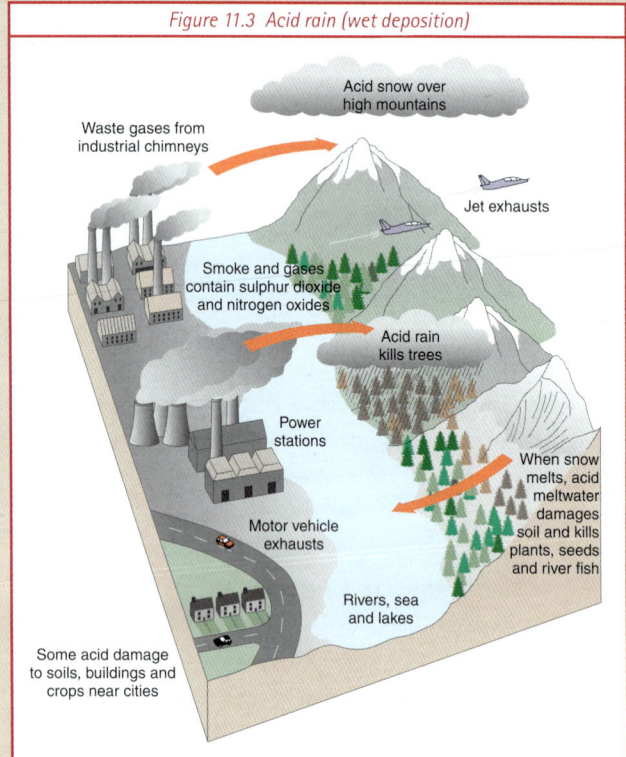

Figure 11.3 Acid rain (wet deposition)

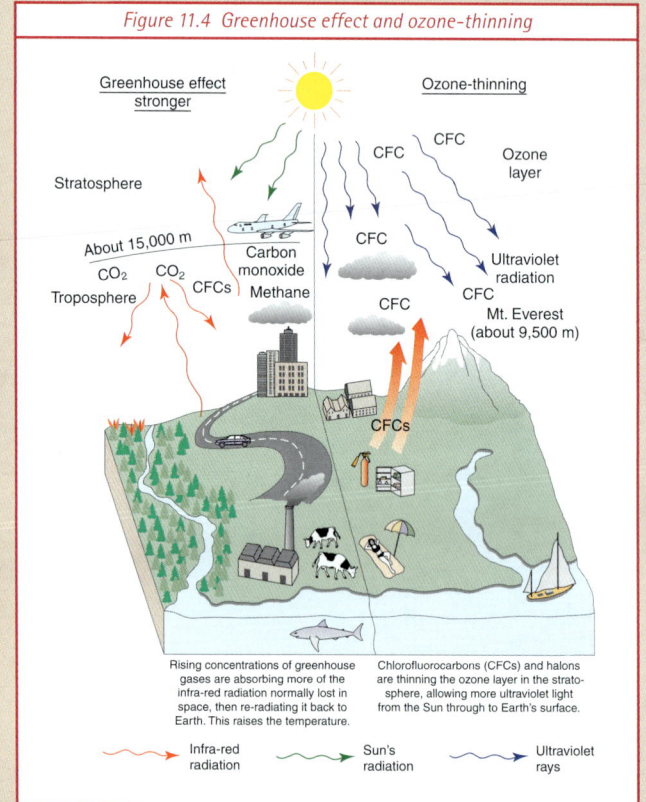

Figure 11.4 Greenhouse effect and ozone-thinning

1 **Acid rain**
   Coal, oil and natural gas are called _____ fuels. They are burnt in _____, _____ and _____. When they burn they give off _____ and _____ such as sulphur dioxide. These emissions mix with moisture in the _____ and _____ to make _____. These acids fall back to earth when it rains and snows. Acid rain and acid snow are known as _____ deposition. Dry deposition on the Earth of these pollutants in the air also occurs. Deposition of acids pollutes _____, _____ and _____, and can kill _____, _____, _____ and other plants. The acids also damage _____ and _____.

2 **Ozone Layer thinning**
   Ozone forms a layer of _____ in the upper atmosphere. It is a kind of oxygen that absorbs ultra-violet radiation from the _____ . It is these ultra-violet rays that _____ our skin. In recent years _____ have been found in the Ozone Layer over the Arctic and Antarctica.

140

This high-level ozone is being destroyed mainly because we use chemicals called _____. These are found in _____, refrigerators and beefburger containers.

3 CO_2 and a stronger Greenhouse Effect
The gases that exist naturally in the Earth's atmosphere let some of the sun's rays through to warm us. They also trap some of this heat in the lower atmosphere rather like the glass in a _____ does. If they did not, the Earth would be a _____ planet. However, the amount of some of these gases in the atmosphere, such as _____ has been rising. Forest _____, vehicle _____ and coal/oil/gas-fired _____ produce this extra gas. This creates a kind of blanket in the atmosphere which _____ heat from escaping back into space. World temperatures have begun to generally _____. Each of these three climate changes seems to be the result of people _____ the atmosphere

Acid rain

Rainwater can be acidic. Increased acidity is shown by a drop in **pH value**. Since the Industrial Revolution, especially in 19th-century-Britain, the pH value of our rainfall has fallen. Our rainfall was becoming more acidic. Acid rain is a weak sulphuric and nitric acid, resulting from the mixing of rainwater with sulphur dioxide and nitrogen oxide. The cleanest rain in Britain with pH values of around 5.6 tends to be brought by Atlantic westerly winds. In eastern England pH values of 4.6 and below can be found in the rainfall. This is ten times dirtier than a pH of 5.6. Acid rain and acid snow in which acid substances are mixed with water is often known as **wet deposition**. **Dry deposition** describes when these acid substances fall to the ground dry. Figure 11.5 shows these two types of acid deposition.

Causes of acid rain

Acid rain is caused by the burning of fossil fuels – coal, oil and natural gas – in industry, power stations and by road traffic. Sulphur dioxide, nitrogen oxides and hydrocarbons are emitted into the atmosphere. The link between sulphur dioxide and acid rain is well recognised. This is the main polluting gas in the case of acid rain. Over half of sulphur dioxide pollution in Britain comes from power stations. As Figure 11.6 shows, it is easy to see why British power stations are widely blamed for causing acid rain in Denmark and Scandinavia. Countries downwind of Britain will receive British air pollution.

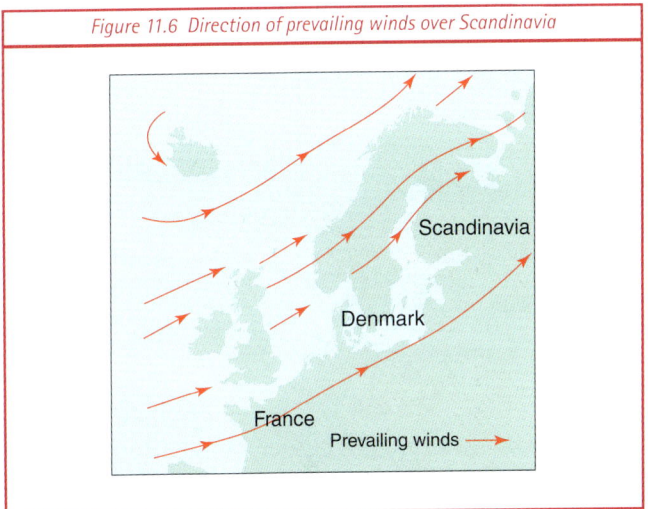

Figure 11.6 Direction of prevailing winds over Scandinavia

ACTIVITIES

1 Study Figure 11.7. Explain how people introduce acid substances into the atmosphere. Refer to the photographs, and suggest how these emissions are linked to a country's level of economic development.

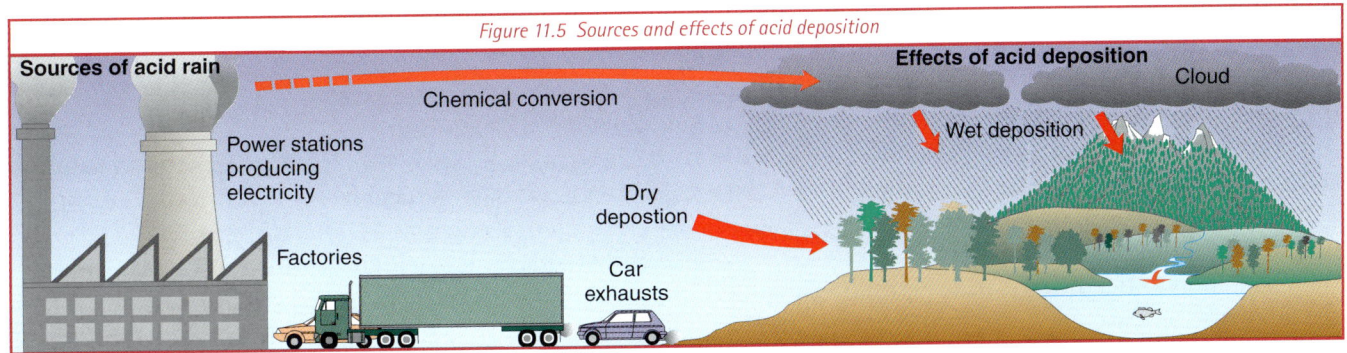

Figure 11.5 Sources and effects of acid deposition

Section 3: Managing Economic Development

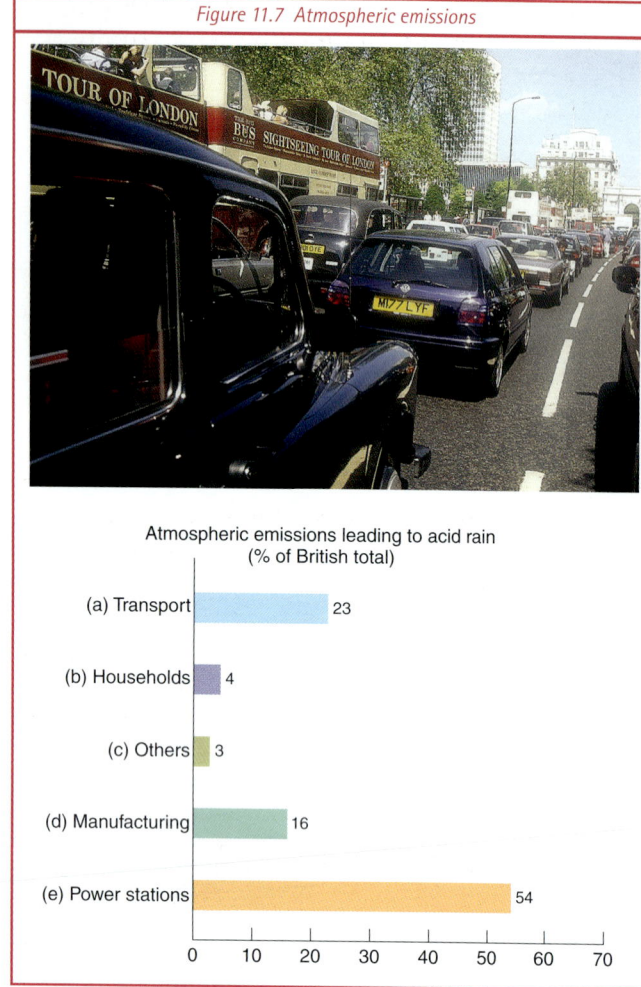

Figure 11.7 Atmospheric emissions

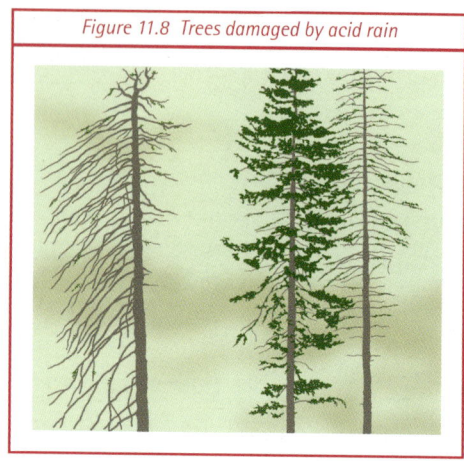

Figure 11.8 Trees damaged by acid rain

Breathing acidified air and drinking water from acidified lakes and rivers can bring health problems for some people. Cases of bronchitis are greater but the effects of acid rain on human health may be indirect. Poisonous acids can contaminate water supply and cause disease which eventually might lead to death.

Acid rain is slowly dissolving the external stonework of many historic buildings and monuments. This problem seems to be worse in more polluted urban areas. It has been estimated that the most affected buildings have lost up to 4% of their weight in a year. The Acropolis in Athens has crumbled more in the last 30 years than in the previous 2000 years.

Figure 11.9 shows some effects of acid rain on trees, lakes, people and buildings.

Effects of acid rain

Scientists, including geographers, have become concerned about the effects that present levels of acidity have on:

- natural environments
- human health
- buildings

It has been estimated that the annual cost of acid deposition in the European Union amounts to around £3 billion. Soils, rivers and lakes can become acidified and poisoned. Toad and frog populations drop. Some fish such as tench, bream and carp have reproduction difficulties. Trees can die or be weakened. Leaves can drop off deciduous trees. Fir tree needles can yellow. Damage to trees, including fruit trees, is a major effect of acid rain. Crown thinning has been getting worse in many areas as Figure 11.8 suggests. Scandinavian forests have been damaged by acid rain, usually attributed to air polluted by British **power stations**. Some plants such as lichen and sundew can also disappear from an area experiencing very acid rain.

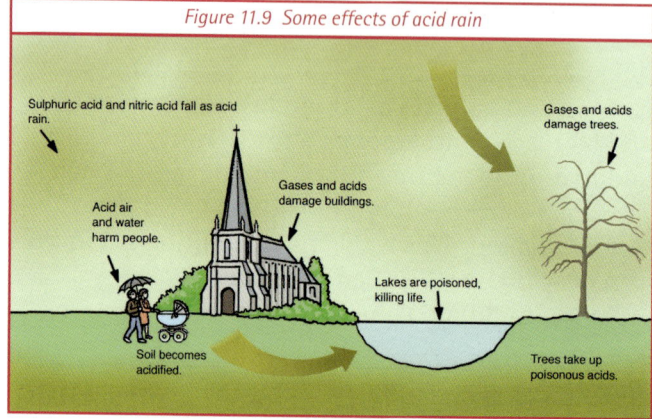

Figure 11.9 Some effects of acid rain

Action against acid rain

Since the late 1980s various measures to cut sulphur dioxide emissions have been introduced. Cleaning equipment known as **FGD (flue-gas desulphurisation)** has been fitted to coal-fired power stations in Britain. FGD requires large amounts of limestone to trap the sulphur. Sulphur dioxide emissions have been falling but the FGD process is expensive and the process reduces the electricity output of power stations. Modern power stations, whether gas-fired or nuclear, and alternative, renewable sources of energy help to cut sulphur dioxide concentrations. Cars with catalytic converters fitted make less contribution to air pollution. Some countries, e.g. Japan, have strict emission controls. Acidified soils need alkaline fertilizers like lime adding to them. Stone buildings can be sprayed with a protective coating. These are examples of action to treat directly the effects of acid rain. International and bilateral agreements between countries to curb sulphur dioxide pollution do exist. There are a range of United Nations regional air pollution treaties.

The Ozone Layer

Ozone is a type of oxygen. Most of the ozone in the atmosphere is found in a layer between about 15 and 40 kilometres above the Earth's surface. This layer is within the **stratosphere**, a zone of the upper atmosphere. Here, as Figure 11.10 shows, **ultra-violet (UV)** sunlight reacts with oxygen to form ozone. This ozone is not very stable and naturally breaks down as well as it naturally forms. UV radiation from the sun gets absorbed by the upper atmosphere in this way. Ozone is continuously created and destroyed by the UV radiation from which it protects life on Earth.

Ozone is also formed close to the ground. Again, it is formed by a chemical reaction between sunlight, oxygen and the sort of polluting gases that cause acid rain, which are released from car exhausts and factories. Ground-level ozone levels are highest in summer and sunny areas, and in areas busy with traffic, e.g. **tourist honeypot areas** on sunny summer days in rural Britain, and the **petro-chemical smogs** of sunny, urban Los Angeles.

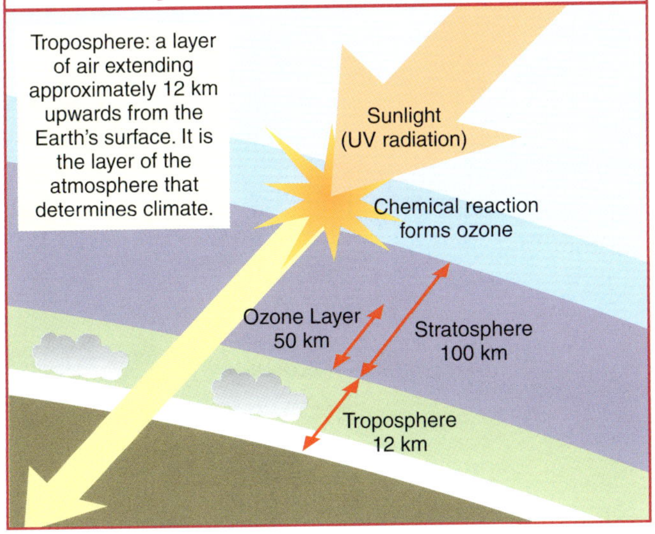

Reasons for the disappearing Ozone Layer

A hole in the Ozone Layer over Antarctica was first discovered in 1979, (see Figure 11.11). A similar hole was identified over Arctic areas in the early 1990s. Since then it has been feared that these holes would continue to get bigger and that the Ozone Layer would thin over the whole world. The holes have grown. By 2001 the Antarctic hole covered 15 million square kilometres and reached the southern tip of South America. General thinning has also been occurring. The concentration of ozone in the stratosphere over Britain is thought to have diluted by over 10% over the past century. However, recently some scientists have thought that the rate of destruction of the Ozone Layer has slowed down, and that some of the holes have even begun to fill in.

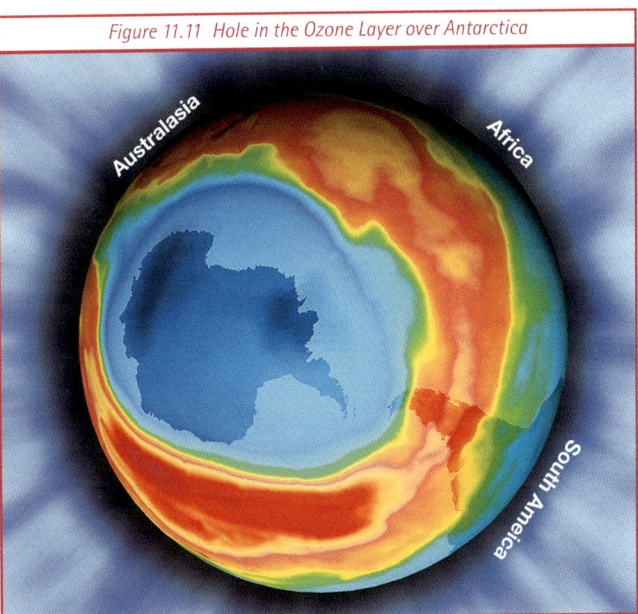

Figure 11.11 Hole in the Ozone Layer over Antarctica

The main cause of ozone destruction in the stratosphere is pollution of the atmosphere below. The main culprit is the family of cheap, cooling chemicals known as CFCs (chlorofluorocarbons) which have been increasingly produced during the past 50 years. As Figure 11.12 shows, their use ranges from refrigeration and air conditioning systems and industrial solvents, to the gases that provide pressure in aerosols, and are used in some foam furniture, packaging and plastics such as burger cartons. The CFCs rise into the stratosphere

Figure 11.10 Where and how ozone is formed

Troposphere: a layer of air extending approximately 12 km upwards from the Earth's surface. It is the layer of the atmosphere that determines climate.

Sunlight (UV radiation)
Chemical reaction forms ozone
Ozone Layer 50 km
Stratosphere 100 km
Troposphere 12 km

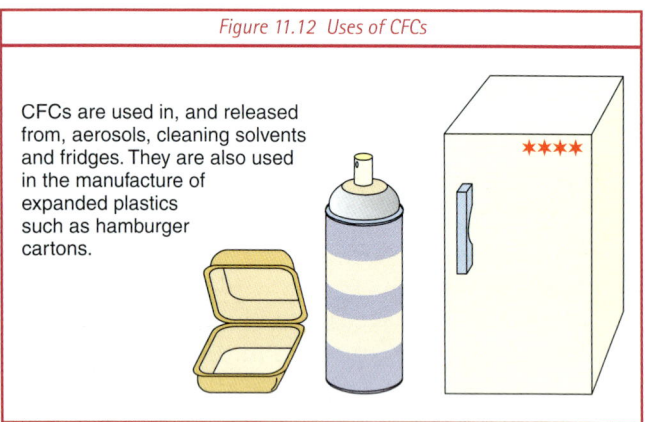

Figure 11.12 Uses of CFCs

CFCs are used in, and released from, aerosols, cleaning solvents and fridges. They are also used in the manufacture of expanded plastics such as hamburger cartons.

where they break down under UV radiation to give off chlorine monoxide, especially in the very low temperatures above the Poles in winter. In spring when longer periods of sunshine return, the chlorine attacks ozone and the Ozone Layer is eroded. Very high levels of chlorine monoxide can now be recorded in the **stratosphere**.

Effects of Ozone Layer depletion

As Figure 11.13 shows the Ozone Layer acts as a shield against the sun's lethal UV radiation. It filters out much of this dangerous radiation and has allowed life as we know it to develop on the Earth. More UV rays would harm people, plants and animals. Loss of the ozone shield is likely to cause an increase in sunburn, skin cancers and eye problems such as cataracts. White-skinned people are thought to be more vulnerable, and some doctors talk of 'sunbathing deaths' that is, death from skin cancer, especially the often fatal melanoma type caused by exposure to UV radiation. People's immune systems may be depressed by the rise in UV radiation with the large-scale health problems which that would bring as infectious diseases spread more easily. The threat to human health may be vast.

Food supplies may also be put at risk. Many crops such as maize, wheat and rice give lower and poorer quality yields if too much UV radiation reaches them. Malnutrition may rise. Forests and tree crops, e.g. fruits, may also be harmed. Equally, marine life may be damaged. Plankton, vital to food chains in the sea, may grow less well.

Some scientists wonder whether life on Earth would be possible without an Ozone Layer.

Efforts to repair the Ozone Layer

There is some evidence that the Ozone Layer issue will turn out to be an environmental success story. Some scientists report that the rate of ozone creation in the stratosphere is no longer slower than its rate of destruction by chlorine monoxide from pollution. They estimate that the polar holes will last for most of this century, but that the Ozone Layer over Antarctica will recover to 1979 levels by 2060.

Ozone Layer damage is a global problem and international cooperation was needed in order for the downward trend in CFC emissions, shown in Figure 11.15, to be achieved. Under the **Montreal Protocol** on Substances that Deplete the Ozone Layer, 1987, most countries agreed to cut emissions of CFCs by 50% by 1991. This protocol was part of the United Nations Environment Programme. The 1991 target was met by the USA, Western Europe and Japan who had produced 38%, 36% and 12% of world emission respectively. The decision by the US Government to ban CFC use in aerosols in 1988 did much to reduce their 38% of world production. Public pressure to ban CFCs has also been strong in Western Europe as the advert in Figure 11.14 suggests. CFC production largely stopped in these three areas by 1995. A total phasing out of these gases had been planned at the Montreal conference

Figure 11.13 Ozone depletion

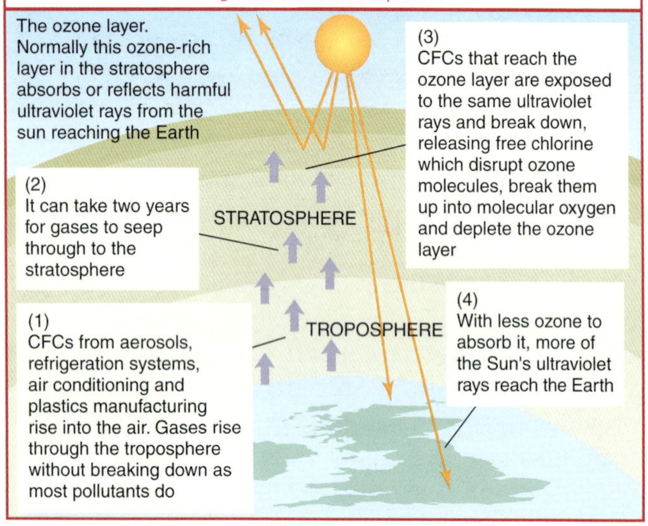

The ozone layer. Normally this ozone-rich layer in the stratosphere absorbs or reflects harmful ultraviolet rays from the sun reaching the Earth

(2) It can take two years for gases to seep through to the stratosphere

(1) CFCs from aerosols, refrigeration systems, air conditioning and plastics manufacturing rise into the air. Gases rise through the troposphere without breaking down as most pollutants do

(3) CFCs that reach the ozone layer are exposed to the same ultraviolet rays and break down, releasing free chlorine which disrupt ozone molecules, break them up into molecular oxygen and deplete the ozone layer

(4) With less ozone to absorb it, more of the Sun's ultraviolet rays reach the Earth

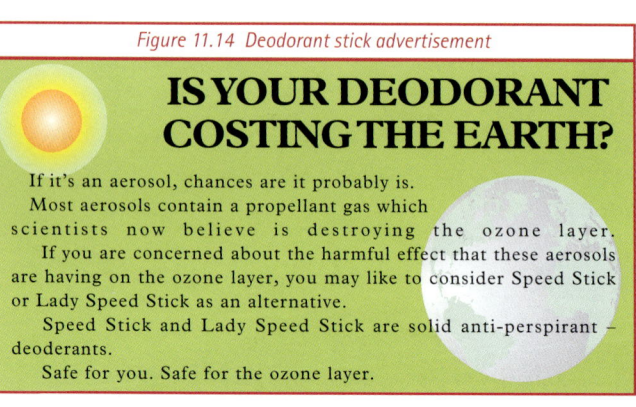

Figure 11.14 Deodorant stick advertisement

IS YOUR DEODORANT COSTING THE EARTH?

If it's an aerosol, chances are it probably is. Most aerosols contain a propellant gas which scientists now believe is destroying the ozone layer.

If you are concerned about the harmful effect that these aerosols are having on the ozone layer, you may like to consider Speed Stick or Lady Speed Stick as an alternative.

Speed Stick and Lady Speed Stick are solid anti-perspirant – deodorants.

Safe for you. Safe for the ozone layer.

Chapter 11: Economic development and the global environment

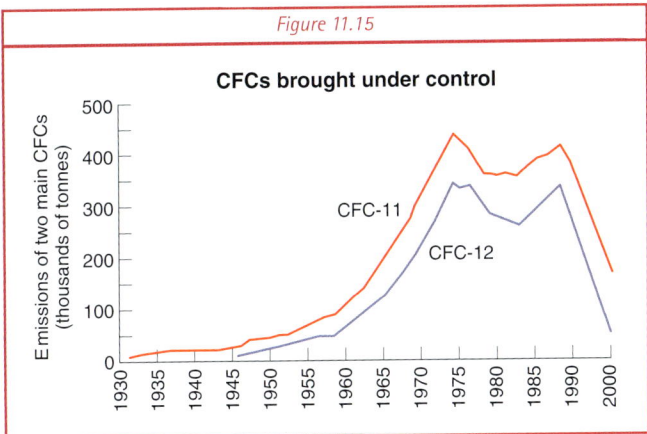

Figure 11.15

and this was almost achieved by MEDCs by 2000. Manufacturing industry has developed less harmful but more expensive alternatives. Reducing CFC pollution in the rest of the world is proving more difficult. The plan is that LEDCs phase out their use by 2010. The Montreal Protocol includes an aid scheme under which MEDCs who have produced and benefitted from CFCs pay LEDCs not to use them. India and China have joined the protocol and now benefit from the aid scheme.

There are a number of natural causes of ozone destruction, including volcanic eruptions, but steps to reduce CFC production do seem to be having a positive effect on the Ozone Layer.

ACTIVITIES

1. Explain how CFCs are depleting the Ozone Layer.
2. Why is it that in spring ozone levels drop over the Antarctic and Arctic regions?
3. Study Figure 11.14, the deodorant stick advertisement. How effective do you think it was in persuading people? Why do you think that?
4. Why was it important that India and China joined the Montreal Protocol? (C1.3, C2.3)

Global Warming

The average temperature across the world as a whole seems to have risen by about 1°C over the past 50 years. The worry is that this slight warming is the beginning of a future trend. The term 'global warming' is meant to describe three other features as well as rising temperatures:

- a world average
- a human activities cause
- a change in climate generally

A warmer atmosphere with human activities to blame is likely to cause:

- aspects of climate other than temperature to change, e.g. rainfall
- some places to warm more than others, and some perhaps to cool

One prediction is for the world as a whole to have warmed by 1.5°C. by 2050 with the deserts of North Africa 2.5°C warmer but Britain beginning to cool. A warmer atmosphere could mean that the **North Atlantic Drift (Gulf Stream)** decreases in size, bringing much colder winters to north western Europe.

Climate change is a fact of life on Earth and nothing new. We know that climate does and has changed from time to time for natural reasons. Until about 20 000 years ago an Ice Age with average temperatures 5–6°C cooler than today affected Britain. Short-term climate change, say over a few hundred years, has also happened naturally. It is now generally agreed that temperature change is happening and that it is happening at a faster rate than we have ever known before. The question is whether it is global warming, that is people-made, or natural variability as in the past. It cannot yet be fully proven but the evidence for global warming is strong and gets stronger year by year.

Evidence for global warming

The evidence so far is of a slight upward trend during the 20th century as Figure 11.16 shows. The run of warmer years in the 1980s and 1990s can be seen. The temperature curve for the world rises sharply during the 1990s. Temperatures in the UK have also risen and are now about 1°C above the country's long-term average. It is believed that the 1990s were the warmest decade in the UK on record over the past thousand years. The rise in temperatures has not been uniform across the world. Siberian temperatures are now 3°C higher than their long-term average. Antarctica has warmed

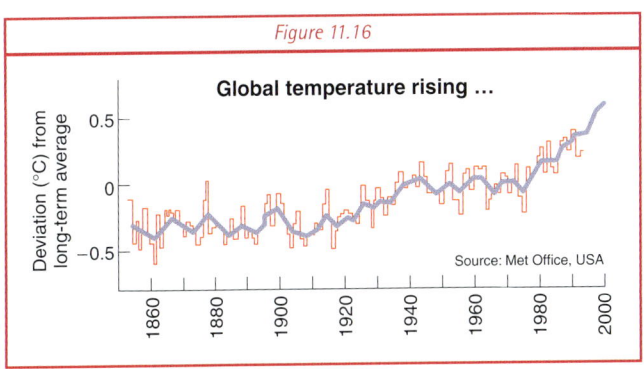

Figure 11.16

145

more slowly over the past 50 years and there is at present even some suggestion that parts of it may cool as other continents warm. The plant life and Siberian ecosystem is showing clear signs of this warming just as icebergs continue to break off parts of the Antarctic ice cap.

The widening of differences in temperature between regions has helped to generate more and stronger storms. There has been more extreme weather during this warmer period. 2001 was the wettest year in the UK since records began. 1995–97 had been a period of drought. Floods and droughts have become more frequent around the world. **Hurricanes** have also been more common and more severe. Higher sea temperatures are thought to be the explanation.

This is exactly how scientists predict that the world's climate system might behave in a warmer world. There is no firm proof yet that pollution from human activities is warming up the world for a long time to come, but it is strongly suspected that this is happening. The United Nation's IPCC (International Panel on Climatic Change) fear that global warming is occurring and that a 2–3°C hotter world by 2100 is likely. They fear the rise could be as much as 5°C. This is unless urgent action during this century manages to slow down the rise. Their view is that unlike climatic changes in history, this change is not due to the natural variability of temperature over time.

The UK's climate in the future

The UK seems to be facing more unstable weather patterns. The seasons have been changing. Winters have become much milder and summers warmer and wetter. The usual prediction for the UK is that the climate will continue to get warmer, and that southern England will become more like the Mediterranean with hot, dry summers and wet winters. Some predict that Bournemouth will have similar weather to the Loire Valley in France by 2040; others say it will be more like the French Riviera. Whatever the prediction for the speed and size of the warming, many scientists expect a lot of warmer summers and a generally balmy and frost-free UK. The predictions for rainfall do vary from a drier England but wetter Scotland to wetter everywhere.

If global warming is a reality, its impact will be uneven. Another possibility for the UK is that the heat it gets from the north Atlantic Ocean decreases in a warmer world and the UK might actually get colder. The fear is that the melting of Arctic ice could cause the Gulf Stream that warms the UK, especially in winter, to fail. As more freshwater flows into the north Atlantic between Greenland and Norway the cold, salty water found there now will weaken. It is the sinking of this heavy water in the north Atlantic which helps to drive the whole ocean current circulation and pumps the Gulf Stream northwards from the Caribbean Sea. Should the Gulf Stream stop flowing or alter its course average winter temperatures in the UK could fall to the level now found in the Czech Republic and Canada. In this picture of our future, the UK could face a 'big freeze' as the rest of the world warms up because the cooling from the weakening of the Gulf Stream might be stronger than the temperature rises from global warming.

Figure 11.17 shows these two futures for the UK. What seems certain is climatic uncertainty. The worry is that we might not be absolutely certain which future we could be facing until it is too late to do anything about it.

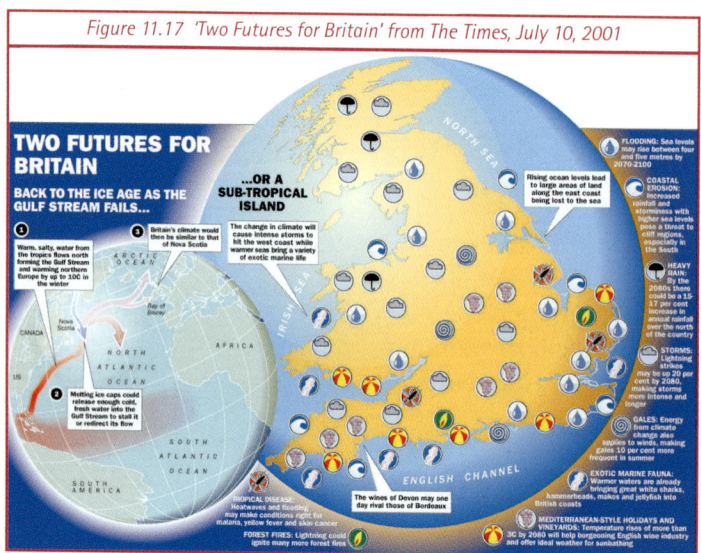

Figure 11.17 'Two Futures for Britain' from The Times, July 10, 2001

Chapter 11: Economic development and the global environment

The Greenhouse Effect and how it is being strengthened

Global warming would be responsible for either of the two futures just described. Global warming is the result of an enhanced or stronger Greenhouse Effect. It is well known that the Greenhouse Effect traps heat in the lower atmosphere, and that one of the main gases responsible for the Effect is carbon dioxide (CO_2). Figure 11.18 shows how there has long been a natural Greenhouse Effect which has kept the Earth warmer than it would otherwise have been. The Earth's atmosphere naturally acts like the glass of a greenhouse, letting shorter wave radiation from the sun through to the surface but preventing longer wave (**infra-red**) from radiating back into space. Several gases, known as greenhouse gases, contribute to the Greenhouse Effect. Scientists have measured increases in the amounts of these gases in the atmosphere, notably the **Keeling Curve** (see Figure 11.19) plotted to show the changes in the amount of carbon dioxide in the atmosphere over Hawaii since 1958. Carbon dioxide levels in the Northern Hemisphere are now thought to be about 25% higher than they were 200 years ago. Other greenhouse gases include methane, nitrous oxide, ozone and CFCs (chlorofluorocarbons). The scientific effects of these increased gas concentrations are believed to be a strengthening of the natural Greenhouse Effect and global warming. Carbon

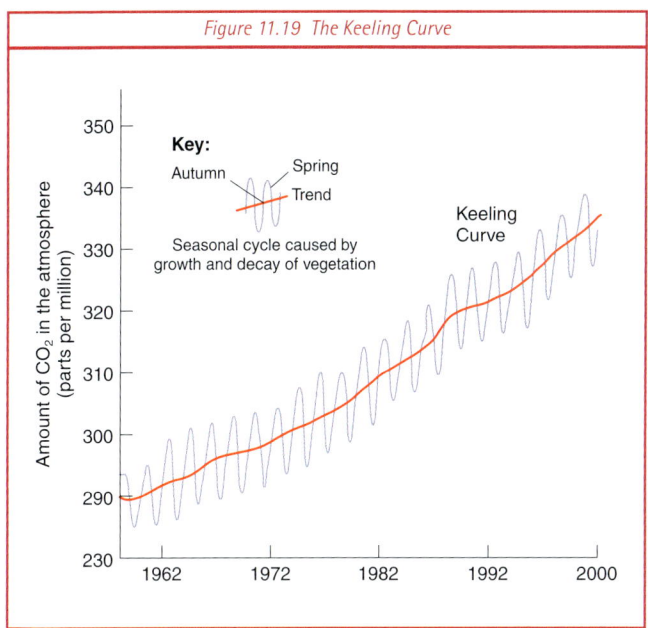

Figure 11.19 The Keeling Curve

dioxide makes the biggest contribution to any global warming because of its far higher concentrations in the atmosphere. As Figure 11.20 shows, other greenhouse gases, especially methane and CFCs make a smaller contribution because there is less of them but they are more effective as trappers of heat (infra-red radiation) in the lower atmosphere.

These increased concentrations of greenhouse gases are the result of pollution. People and human activities have polluted the atmosphere as country after country has industrialised and urbanised over the past 200 years. Modern society's burning of fossil fuels – coal, natural gas, oil and timber – seems to be changing the global climate by enhancing the Greenhouse Effect. In the UK we emit about 2.5 tonnes of atmospheric carbon for each person per year. About half a tonne comes from road transport and another half a tonne from domestic energy use. Most of the rest comes from businesses and factories.

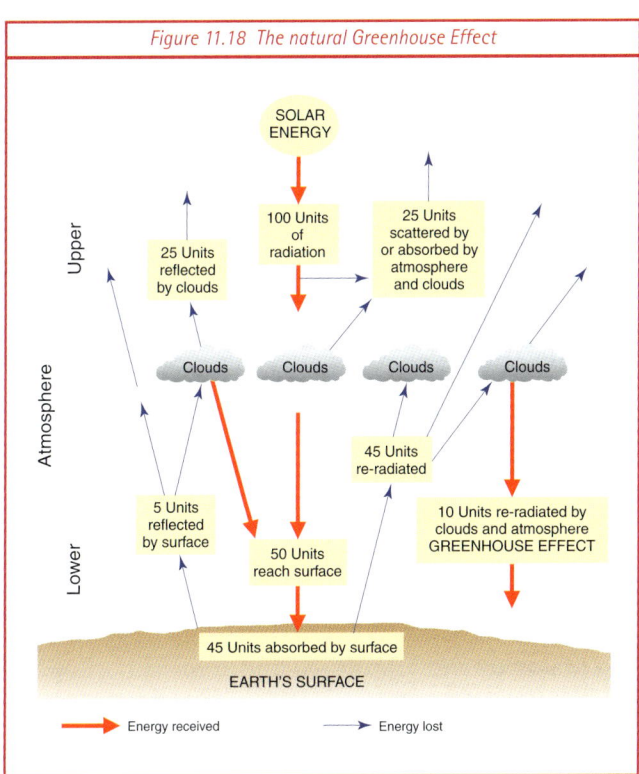

Figure 11.18 The natural Greenhouse Effect

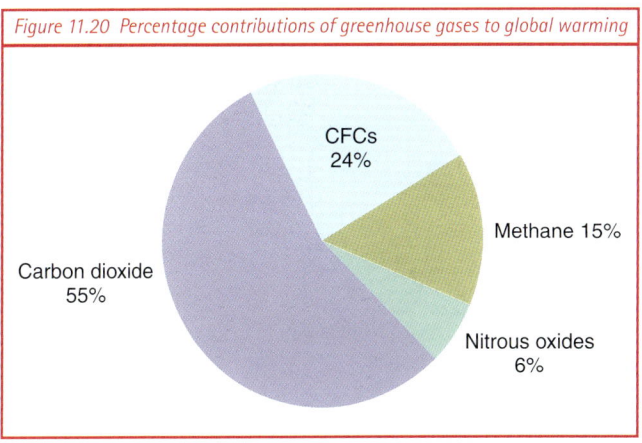

Figure 11.20 Percentage contributions of greenhouse gases to global warming

Section 3: Managing Economic Development

Rising temperatures and global warming

Global warming refers to rising temperatures caused by people and their activities. Temperatures are warming and climate is changing; the question is, whether it is people-made or not? There are various natural causes of climate change. These include:

- a stronger or weaker sun
- slight wobbles in the Earth's orbit
- dust from volcanic eruptions

The present climatic changes are faster than any known of in the past when only natural causes must have been at work.

ACTIVITIES

1. Find out by how much average winter and summer temperatures in London must change for them to be like: **a** Nice, southern France **b** Prague, Czech Republic ? (N1.2, N2.2)
2. Study Figure 11.17. Which climatic future for the UK would you prefer? Give reasons for your choice.
3. Write about six specific ways in which modern society burns fossil fuels. (C1.1, C2.1a, C2.1b)
4. Study Figure 11.21. Describe how some areas of the world emitted more carbon dioxide into the atmosphere than others in 1985. Suggest reasons why this was so. How and why are these contributions expected to change by 2025? (C1.3, C2.3, N1.3, N2.3)

Some likely Impacts from global warming

Continued changing climate from global warming during this century will alter the face of the world. People and aspects of their natural environment have adjusted to the climates in which they live. A new climate will mean adjustments in people's way of life and in ecosystems. The economy (e.g. food supply) and the environment (e.g. plant and animal habitats) will be severely affected. The nature of the impacts will depend on how a warmer globe actually affects the climate of each part of the world. We have already seen that some areas will be heated more than others, and that others may actually cool. As Figure 11.17 has shown, there is uncertainty about the UK's climatic future. One country may become wetter, another region drier. It is unknown whether the UK will be wetter or drier.

The picture will be complicated with each part of the world experiencing costs and benefits, and having winners and losers. Some people, some industries, some plants etc. will face costly threats; other people, other industries, other plants, etc. will benefit. In the warmer UK future shown by Figure 11.17, hoteliers in Cornwall will gain from the warmer South Coast weather while those on the low-lying East Anglian coast may find their hotels flooded by the rising sea. Figure 11.22 suggests some of the impacts in the UK should the climate of the country warm during the present century.

A warmer world is likely to be a wetter and a greener world. Higher temperatures and rainfall encourage lusher growth of vegetation, though some crops such as potatoes may struggle to grow. Warmer temperatures, especially milder winters, lead to fewer deaths and to energy costs falling. Weather changes affect the hydrological (water) cycle. More frequent and more intense floods and droughts are likely. Dwindling water supplies could lead to war in some parts of the world, especially the Middle East.

Rising sea levels are often associated with global warming. A combination of the thermal expansion of warmer seawater and some melting of Antarctic and high mountain ice will raise sea levels. Some islands in the West Indies and Pacific Ocean (e.g. the Maldives and Mauritius) and some other low-lying areas might disappear. Figure 11.22 shows vulnerable areas around the UK coast. There would

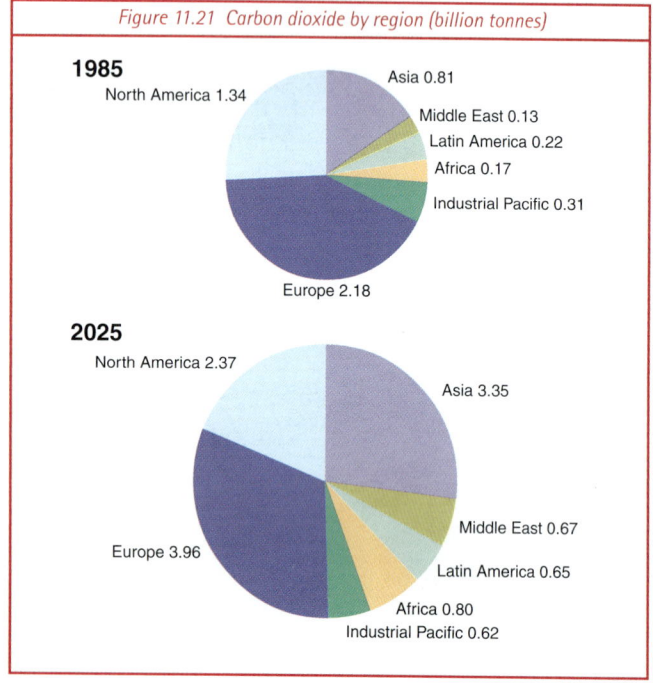

Figure 11.21 Carbon dioxide by region (billion tonnes)

1985
- North America 1.34
- Asia 0.81
- Middle East 0.13
- Latin America 0.22
- Africa 0.17
- Industrial Pacific 0.31
- Europe 2.18

2025
- North America 2.37
- Asia 3.35
- Middle East 0.67
- Latin America 0.65
- Africa 0.80
- Industrial Pacific 0.62
- Europe 3.96

Chapter 11: Economic development and the global environment

Figure 11.22 How warmer weather could affect the UK

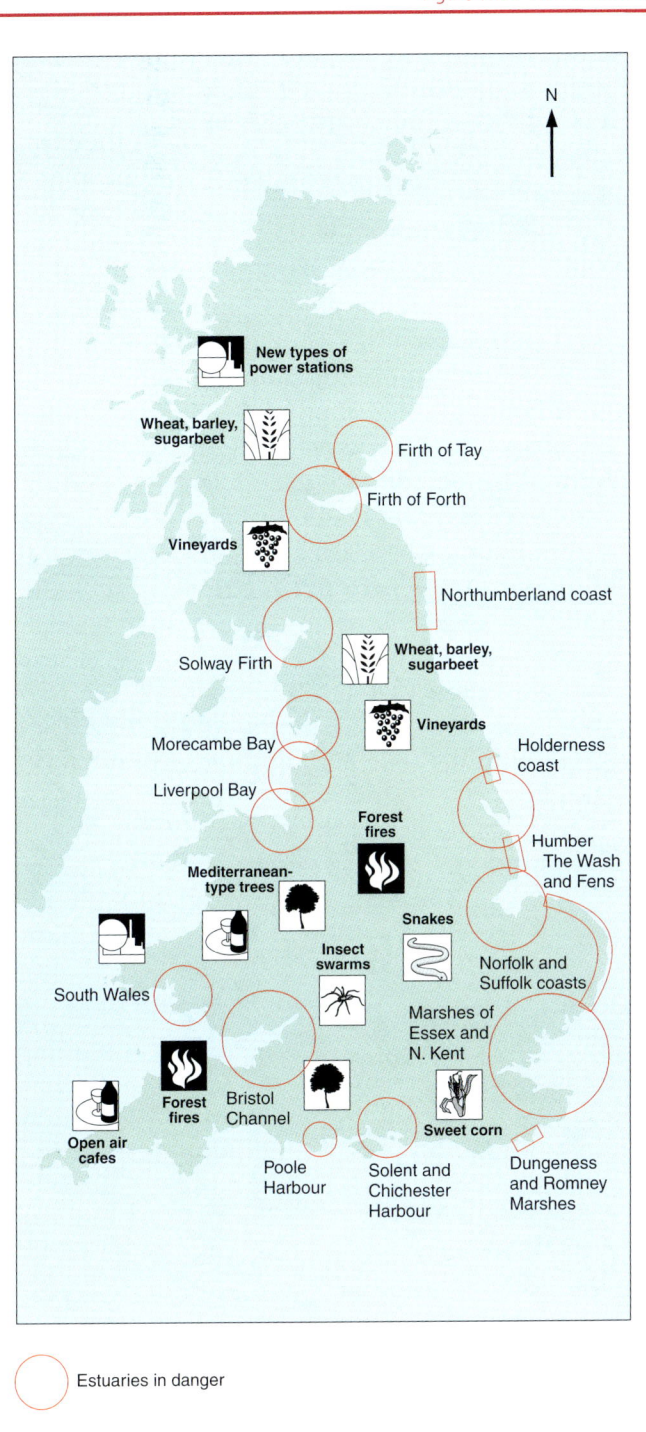
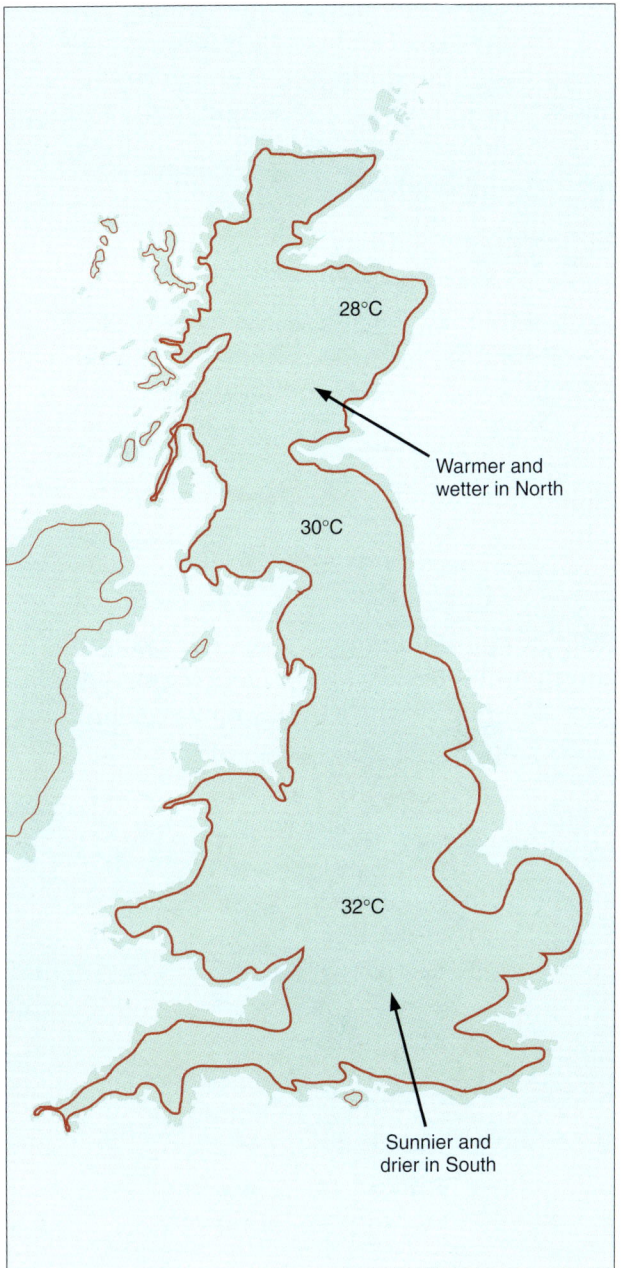

- ◯ Estuaries in danger
- ▭ Coast under threat

— Possible coastline in 2080

30°C Possible average summer temperature in 2080

Some views on how warmer weather could affect the UK

- Lots of lovely sunshine?
- New crops for farmers?
- Problems for our forests?
- Changes in our power supplies?
- Swarms of insects?
- Possibility of widespread coastal flooding?

149

be major problems in many coastal regions around the world. Millions of people would find themselves displaced. An enormous refugee crisis would result. Figure 11.23 shows this impact in various areas of the world, and the high costs of trying to protect land against a 1-metre-rise in sea level that could result from a 1.5°C rise in world temperature by 2050.

ACTIVITIES

1. From Figure 11.22 for the UK and Figure 11.23 for the world, list some regions, industries and groups of people who are likely to: **a** gain and **b** lose from the climate changes shown by the information. (C1.2, C2.2)

People's response to the challenge of global warming

On balance global warming is seen as a threat. The negative impacts on some people and regions and the costs of adjusting everywhere may outweigh any benefits. A significant cut in greenhouse gas emissions is thought to be needed. A drop in pollution will lessen the climate changes but the world's population needs to start cutting gas emissions now before it is too late to stop the really major impacts. Some climate change may now be irreversible because of gases already in the atmosphere. The effect of these will take years to work out.

The response to global warming has to be a global and international agreement. Since 1992 there have been efforts, under the United Nations Climate Change Convention, to stabilise and then cut greenhouse gas emissions. That year many of the countries of the world met at the so-called **Earth Summit** in Rio de Janeiro and agreed to curb emissions. Five years later, in 1997, 160 countries met in Kyoto, Japan, as part of this UN Convention and signed the **Kyoto Protocol** which took the policy of cutting emissions further. 38 industrialised countries agreed to cut their emissions by at least 5% below their 1990 levels by 2008–12. The UK government agreed to a 12.5% cut by 2010. These targets were set only for countries, mainly MEDCs,

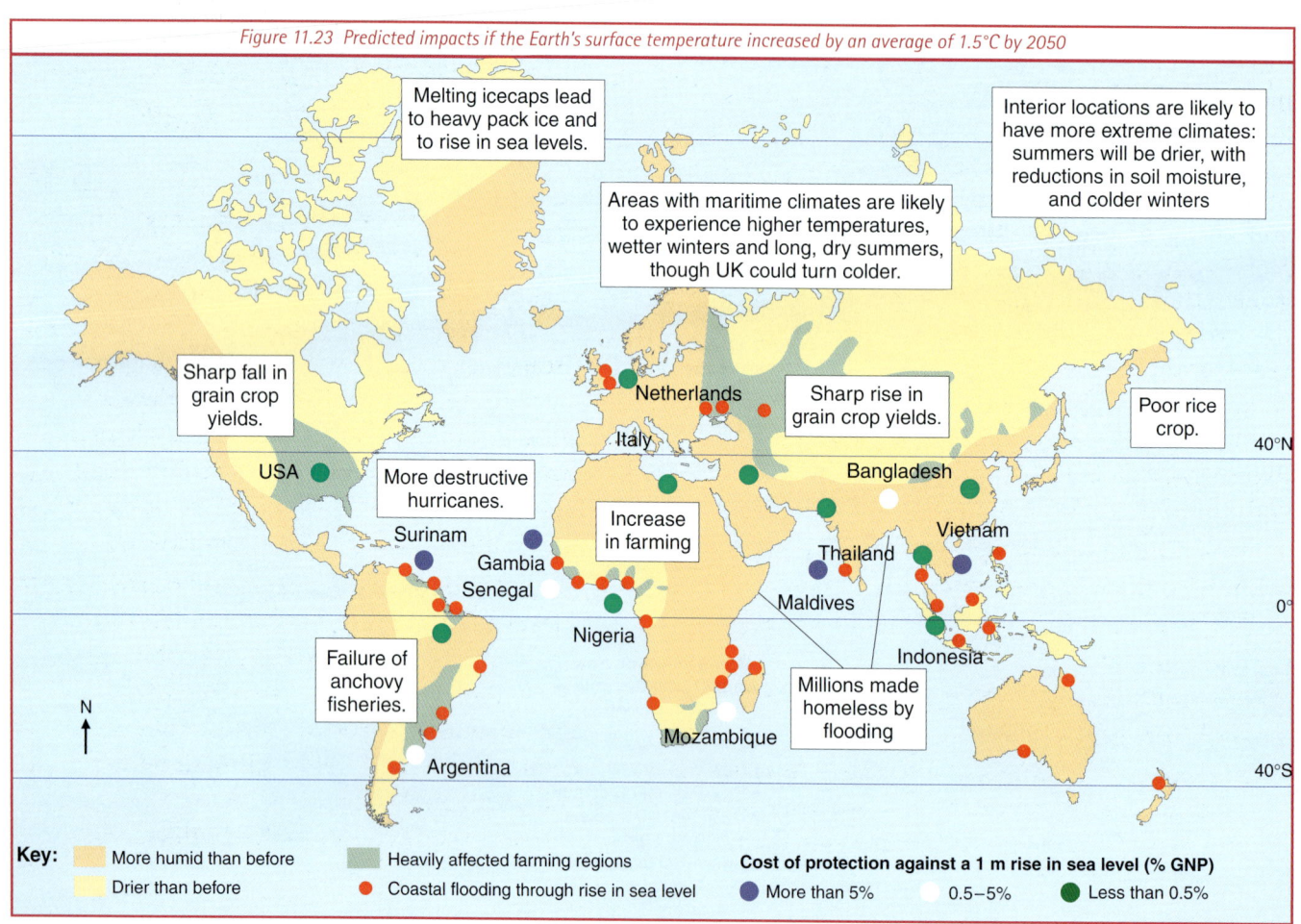

Figure 11.23 Predicted impacts if the Earth's surface temperature increased by an average of 1.5°C by 2050

who believed themselves to have the greatest responsibility for the stronger Greenhouse Effect. They agreed that they should be first to clean up their air. MEDCs emit more polluting gases per person than LEDCs. It was also thought to be unfair to try to prevent LEDCs industrialising. Had India, for instance, agreed to cut back on these emissions, its economic development may have virtually stopped. LEDCs generally accepted the principle of cutting emissions but were not set targets. Various factors were used when each country's target was set. These included its degree of development, wealth per person, level of energy efficiency and rate of population growth.

Figure 11.24

% contribution to world CO_2 emissions.	
US	22%
Russia*	18.4%
China	5.0%
Japan	4.4%
Germany	3.2%
UK	2.8%

*including former USSR states.

Some countries (e.g. New Zealand and Germany) have now reduced their emissions below previous levels. Other countries such as the UK and Japan have managed to stabilise at earlier levels. There is little or no control of emissions still in the USA and Russia. Controlling these emissions does mean sacrificing some short-term economic growth. It has been estimated that implementing the Kyoto Protocol will cost over $1 000 billion a year worldwide.

So far people have responded to the threat of global warming by adopting preventive measures like the Kyoto Protocol. But how will we actually cut emissions ? There are a number of possible ways:

- We need to use less energy and use what we need to use more responsibly. For example, drive less so using less petrol, switch off some street lights at night.
- We need to use more energy from renewable sources. Wind turbines and solar power panels do not require the burning of fossil fuels. As Figure 11.25 suggests, the greater use of **nuclear power** also causes greenhouse gas emissions to drop.

Figure 11.25

An advertisement placed in newspapers by Nuclear Electric.

- We need to improve energy efficiency so that we get the same amount of heat or light from less fuel. Homes, cars and factories can be made more energy-efficient, e.g. homes from which less heat escapes need to use less gas or electricity to heat them. Gas-fired power stations emit less air pollution than coal-fired stations. **Unleaded and lead-free petrol**, non-petrol cars and petrol or engines which offer more miles per litre are helpful in trying to limit global warming.
- We need to develop the so-called '**carbon sink**' idea. Forests, farmland and oceans absorb greenhouse gases as Figure 11.26 shows. Slowing down the rate of tropical **deforestation** and encouraging **afforestation** (e.g. the planting of trees in the English Midlands to develop the National Forest) might be effective in lowering carbon dioxide concentrations sufficiently.

Section 3: Managing Economic Development

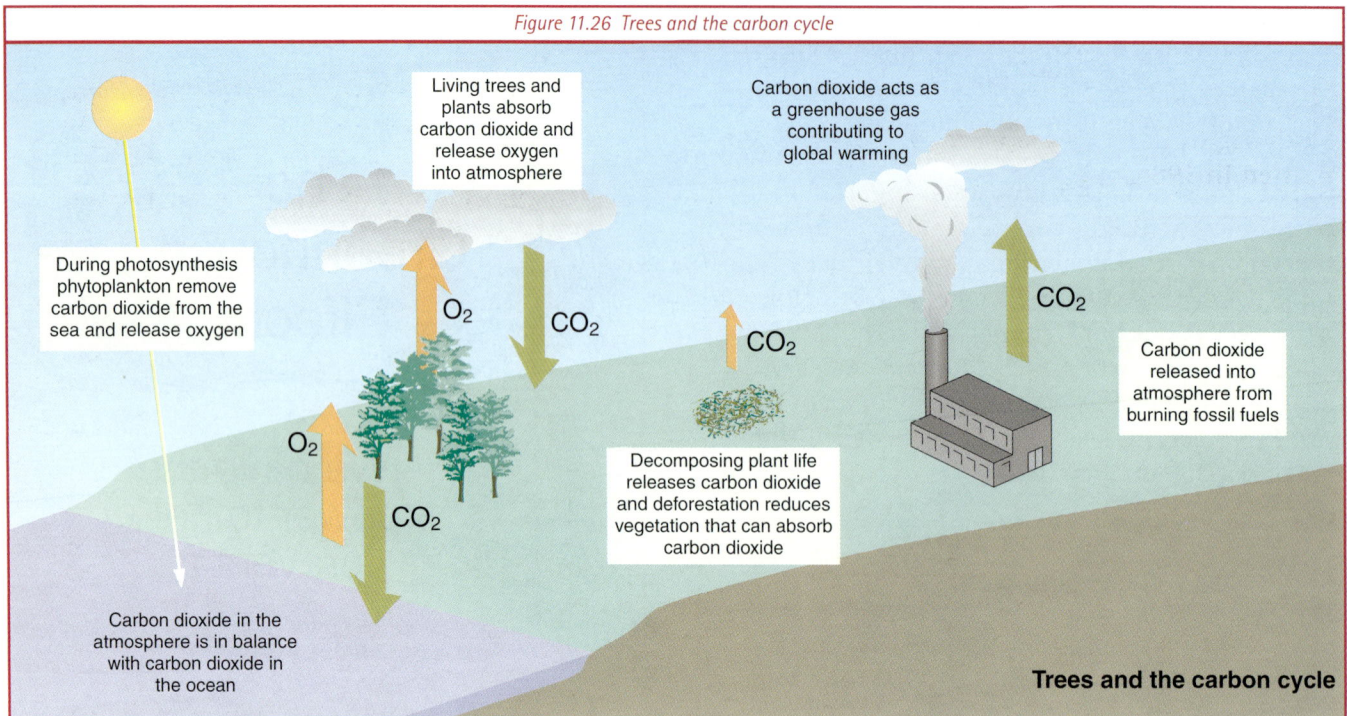

Figure 11.26 Trees and the carbon cycle

The alternatives to this preventive approach are:

1. Do nothing and learn to adapt to a different climate and environment when it comes.
2. Invest now in protective measures, e.g. start building coastal **dykes** to try to hold back rising sea levels that are predicted to come in the future.

ACTIVITIES

1. The US Government claims that the 7% cut in greenhouse gas emissions it agreed to at Kyoto is now not achievable and will damage the US economy. What harm would it do? Use Figure 11.24 to argue why some countries are disappointed by the US claim.
2. Study Figure 11.27 which shows how action to try to curb global warming is also being taken at a small, local scale. Devise a local action plan that might help with the issue in your area. (PS1.1, Ps1.2, PS2.1, PS2.2)
3. Action to deal with the threat of global warming is part of sustainable development. Explain how.

Figure 11.27 Examples of local action

COUNCILS WARM TO CONTROLLING CLIMATE CHANGE

Chesterfield Borough Council has joined with four other East Midlands' local authorities to work in a unique partnership and help pilot a world-wide initiative that aims to tackle climate change and air pollution.

With increasing concerns about the environmental impact of greenhouse gases, Nottingham City Council, Newark and Sherwood District Council, Leicester City Council, Nottinghamshire County Council and the Borough Council have joined forces to support the international 'Cities for Climate Protection' campaign, and the UK Government's Climate Change strategy.

The scheme will provide the five local authorities with a framework to use to plan reductions in greenhouse gases and manage the energy usage of their communities.

Each of the local authorities will undertake and complete a five-step plan to:-

1. Establish a base year emissions analysis and a forecast of greenhouse gas emissions in a target year, for example 2001.
2. Set an emission reduction target.
3. Develop and adpot a Local Action Plan to reach that target.
4. Implement the Local Action Plan.
5. Monitor and report on greenhouse gas emissions and the success of the policy measures taken out to cut emissions.

Exam-style questions with answer comments for Chapter 11

Question 1 – F Tier

Study Figure 1, a newspaper report on acid rain written in 1986.

Figure 1

Disaster That Falls in the Rain

A new survey in Europe shows major destruction.

The report is packed with examples of deadly atmospheric corrosion: one inch of Portland stone eaten away from ST. PAUL'S CATHEDRAL, 4,000 dead Swedish lakes, and in Germany five million acres of woodland affected and 47,000 jobs have been lost.

Acid rain, the report says, has become one of the greatest environmental problems of our time.

Overall, power stations are said to be the main source of sulphur dioxide in the air and traffic the main source of nitrogen monoxide and dioxide.

The British power stations, said to be the main 'exporter' of sulphur dioxide pollution in Western Europe, have said it would cost £1,500 million to control the problem.

(a) (i) What chemicals cause rain to be acid? **(3)**

Sulphur dioxide; nitrogen monoxide; nitrogen dioxide.

(ii) Give **two** ways in which acid rain affects the *natural* environment. **(2)**

Kills lake life; damages woodland.

(iii) How does acid rain affect the *built* environment? **(1)**

Corrodes building stone.

(iv) What are **two** main sources of these chemicals? **(2)**

Power stations and traffic exhausts.

(v) Give **one** reason why the damage caused by acid rain was not being stopped. **(1)**

Costly to install cleaning technology.

(b) Study Figure 2 below and describe the **three** ways in which acid rain kills trees. **(3)**

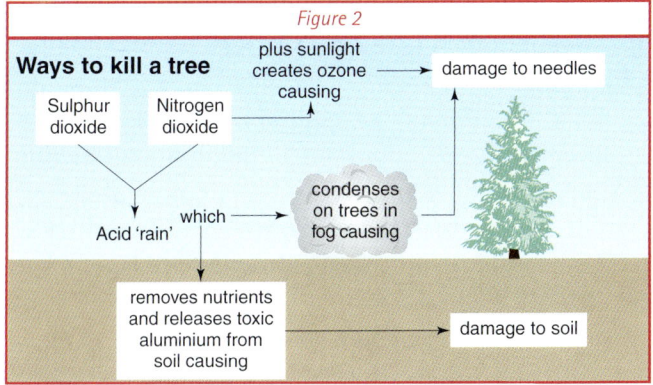

Figure 2

The three ways shown are though soil damage, through ozone damage to the trees' needles, and through acid 'fog' damaging the needles.

(c) Britain was blamed for damaging lakes and rivers in Norway and Sweden. Explain how pollution from Britain caused acid rain to fall in such countries. **(3)**

Pollution from industrialised and urban Britain is exported on the westerly winds across the North Sea.

(15 marks)

Question 2 – F Tier

Study Figure 3, a sketch which shows why a greenhouse gets hot.

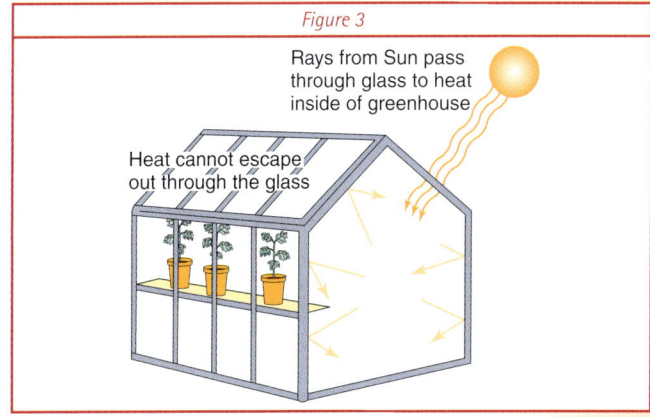

Figure 3

Some scientists think that an increase in carbon dioxide in the atmosphere may lead to the temperature of the Earth rising. They call this global warming due to a stronger 'Greenhouse Effect'.

(a) What is causing the increase in carbon dioxide in the atmosphere? **(3)**

The burning of fossil fuels as petrol in cars, coal and oil in power stations and factories, etc.

(b) Explain why scientists think that increased carbon dioxide will lead to a stronger Greenhouse Effect. **(3)**

Carbon dioxide acts in the atmosphere like the glass in a greenhouse, letting heat through and then trapping it in the lower atmosphere.

(c) Write about **three** effects on Britain if global warming does take place. **(3)**

A retreating coastline as sea levels rise, less energy needed for heating, and wetter, more violent weather is likely.

Chapter 11: Exam-style questions

(d) Describe **three** of the things that people and governments could do to try to cut down the problem of global warming. **(6)**

Find alternatives to fossil fuels, take public transport rather than the private car, and make preparations for living in a warmer, wetter world if developed. These could give maximum marks. Remember we can try to slow down the warming and/or accept it might happen and try to adjust.

(15 marks)

Question 3 – H Tier

Study Figure 4, a graph showing changes in atmospheric gases between 1965 and 2025 and their estimated effect on global temperatures.

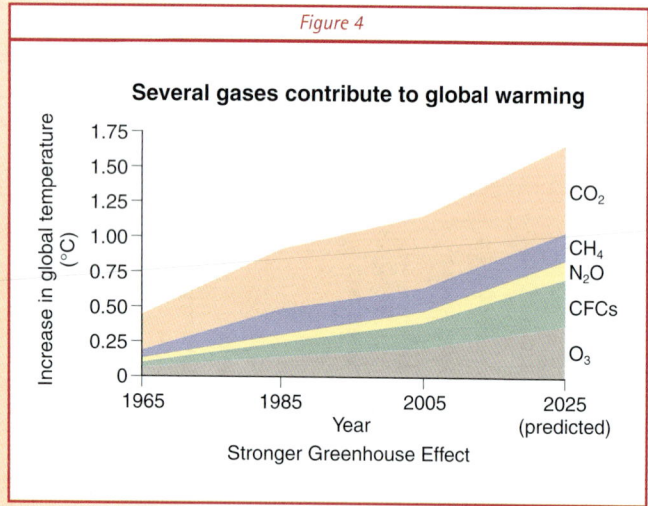

Figure 4

(a) (i) Work out the expected total increase in global temperature between 1965 and 2025. **(1)**

1.65–1.7°C

(ii) Which single gas of those shown in Figure 4 is expected to cause most global warming? **(1)**

CO_2 (carbon dioxide)

(iii) Give two reasons why this gas is increasing in amount. **(2)**

Refer to power stations and car exhausts.

(b) (i) Explain the term 'Greenhouse Effect'. **(2)**

This is the trapping of heat by gases in the lower atmosphere.

(ii) Explain how an increasing 'Greenhouse Effect' might be causing global warming. **(3)**

More CO_2 in the atmosphere strengthens the Greenhouse Effect which in turn might be causing world temperatures to rise.

(c) Increasing the amount of CFCs in the atmosphere has another effect other than increasing the 'Greenhouse Effect'.

(i) State this effect. **(1)**

Ozone layer depletion.

(ii) Briefly explain how CFCs have this effect. **(2)**

CFCs rise and react with sunlight in the upper atmosphere to produce chlorine monoxide which destroys the ozone at this level.

(iii) Explain why this effect is an issue of real concern to people and governments. **(3)**

The ozone layer acts as a shield against dangerous ultra-violet solar radiation. UV radiation is a threat to human health.

(15 marks)

Question 4 – H Tier

Study Figure 5, which shows various human activities which are believed to be responsible for an increase in the amount of certain gases in the atmosphere. These gases are thought to cause global warming, acid rain and ozone depletion.

Figure 5

(a) Name **three** activities shown and explain how each causes an increase in atmospheric gases. **(6)**

Landfill sites, cutting down trees and smoky chimneys all add gases to the atmosphere either by pollution or by not absorbing pollution.

(b) Describe how cars, factories and power stations can cause acid rain. **(2)**

If it rains through an atmosphere polluted by sulphur dioxide, the rain will fall as a weak acid.

(c) What is the link between the amount of these gases added to the atmosphere and a country's level of economic development? Use Figure 5 to briefly explain the link. **(2)**

MEDCs as shown on Figure 5 pollute more than LEDCs.

Study Figure 6, which shows average world air temperatures between 1880 and 2025.

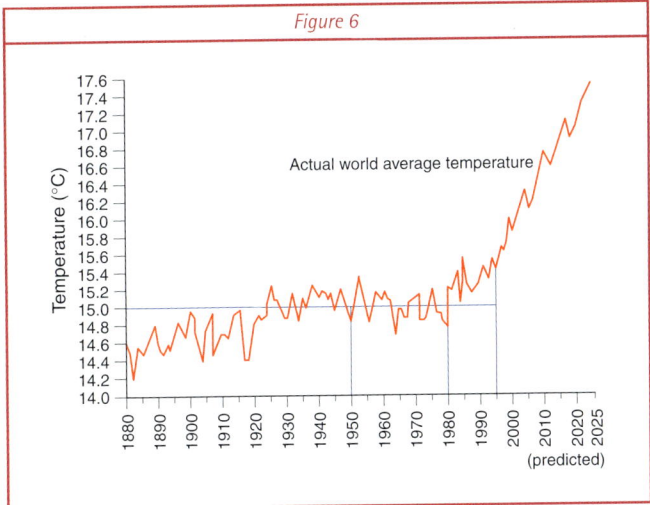

Figure 6

(d) Complete the sentences below by underlining the correct word.

The graph shows that, overall, average world temperatures have *risen* / *fallen* since 1880. It is forecast that from 1995 average world temperatures will rise more *rapidly* / *slowly*. 1883 was the coldest year with an average temperature of only *14.2°* / *15.2*°C. The hottest year up to 1990 was *1984* / *1986* when world temperatures averaged 15.6°C. Scientists forecast that by the year 2025 temperatures will be about *1.6°* / *2.6*°C higher than the average of 15°C recorded for the years between 1950 and 1980. **(5)**

Answers underlined.

(e) Suggest why building effective sea defences may be important in the future. **(1)**

An advancing sea as sea levels rise.

Study Figure 7, a recent newspaper headline.

Figure 7 © The Times

(f) Describe two ways in which the people and government of the UK could 'clean up' on acid rain and/or cut down their contribution to the Greenhouse Effect. **(4)**

One good answer would be to describe the technology now fitted to IK power stations, e.g. FGD and describe the Kyoto Conference CO_2 targets.

(20 marks)

Chapter 12: Tourism and the economy

Global tourism

Tourism is an activity that involves a visit, usually with at least an overnight stay, away from normal home. The activity will usually be of a leisure and recreational nature, and could include a day visit only. Taking holidays is the main tourist activity, though a day trip to a theme park or the seaside might also be one.

Tourism grew consistently through the 20th century, as Figure 12.1 shows, and is now the world's biggest industry.

Figure 12.1 Growth of tourism

Year	Number of international tourists
1950	25 million worldwide
1960	69 million worldwide
1970	160 million worldwide
1980	286 million worldwide
1990	425 million worldwide
2000	689 million worldwide
2010	1000 million (estimated) worldwide

By 2000 the industry worldwide:
- employed more people than any other (200 million people or 8% of all employment)
- created 6% of world economic output (GDP)
- was worth about $4 000 billion.

In the UK the average household spends around £1 500 a year on taking holidays, and is a major spending priority for most consumers. International tourism has shown tremendous growth, as shown by Figure 12.1. The foreign holidays taken by UK residents has risen six times since 1970; now approximately half of all holidays are foreign. A growing number of people now take two or more holidays a year. Many of these foreign holidays are to new locations. It is no longer just rich and famous UK residents who holiday in Australia, **Caribbean** Islands like Barbados and St Lucia, in African countries like Kenya and The Gambia, in the USA and in Pacific islands from Singapore to Fiji. **Long-haul flights** from UK airports to different continents have helped to turn the foreign holiday market into a global one. Around 35% of UK holiday-makers head for destinations outside Europe. France and Spain, with 50% of foreign holiday-makers, remain the most popular destinations for UK tourists. Long-haul holidays, even long-haul weekend breaks to, for example, Cape Town, South Africa, are growing rapidly in popularity. Tourism is a fashion industry. The most popular destinations are constantly changing as new locations become fashionable. Dubai, Nepal, Thailand, Mexico and Jamaica are currently popular with Europeans. China and Europe, including the UK, attract large numbers of American tourists. Wilderness destinations, including to as far as Antarctica are increasingly popular. There are few parts of the world now untouched by tourism. It has truly become a global industry though for various reasons, some countries have proved to be more popular with tourists than others. For example, the Middle East has not proved popular as a tourist destination, and Africa may attract tourists but is not a major source of tourists. However, it seems that more mass international tourism will become part of everyday life in the future. Global tourism will become even more globalised with fewer and fewer destinations remaining exotic and exclusive.

ACTIVITIES

1. Collect a range of holiday brochures (say, three to four) from a local travel agency. On a blank outline map of the world, plot the destinations from the UK marketed in the brochure.
2. Use an Internet website (e.g. World Tourism Organisation; World Travel and Tourism Council (WTTC); Office for National Statistics) to find out figures for a recent year which will show where UK tourists abroad visit, and where foreign tourists visiting the UK come from. Show the figures graphically (e.g. bar chart) and comment on how global this tourist pattern is. (IT1.1 – IT2.3, N1.3, N2.3)

The growth of international tourism, especially in LEDCs

The introduction of **package holidays** played a massive part in the growth of Mediterranean tourism in Spain, Italy, Greece and Turkey from the 1960s onwards. A package holiday arranged by a tour operating company (e.g. Airtours or Cosmos) combines travel and accommodation as a package with one price. A typical package holiday will be advertised in a brochure and include an air ticket, a hotel bedroom and meals, and a coach to transport you between aeroplane and hotel. Advances in air

Chapter 12: Tourism and the economy

travel during the past 50 years have been crucial to the package holiday boom. Improved airports and the introduction of faster, wide-bodied jet planes and charter flights, and now online bookings have cut the price of foreign travel. Bulk buying of these air tickets and hotel bedrooms abroad by tour operating companies has resulted in low-priced package holidays which sell well. D-I-Y (do-it-yourself) holidays where the Internet is used as a source of holiday information and a means of booking separate flights and hotels are becoming more common.

Other reasons often given for taking holidays are:

- relaxation
- broaden one's horizons
- adventure
- social life

The natural environment can influence a number of these reasons. The search for '**winter sun**' among west Europeans, including people from the UK, can be seen as part of the relaxation behind tourism. Figures 12.2(d) and (e) suggest why winter tourism to Barbados, an LEDC, is popular with those UK residents able to afford it. The **climate graph** (Figure 12.3) for the Greek island of Corfu, only three hours' flying time from London, suggests one reason why the island attracts tourists mainly between April and September – hot and dry weather!

Experiencing the climates of other countries also broadens one's experiences of life and can be an adventure. Recently, a World Tourism Organisation

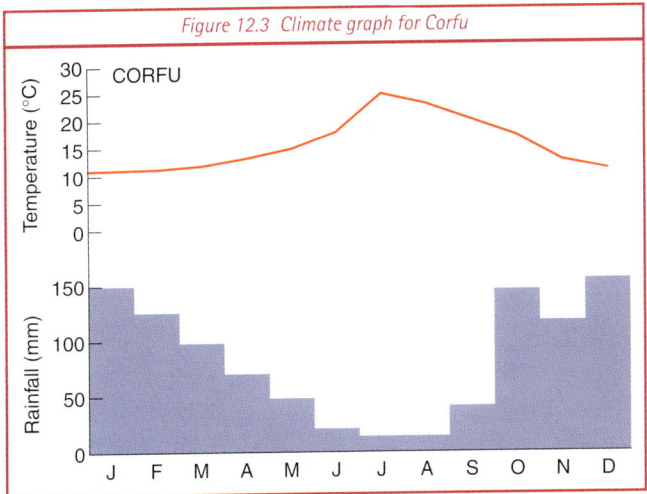

Figure 12.3 Climate graph for Corfu

Figure 12.2 Barbados – why it attracts winter tourism

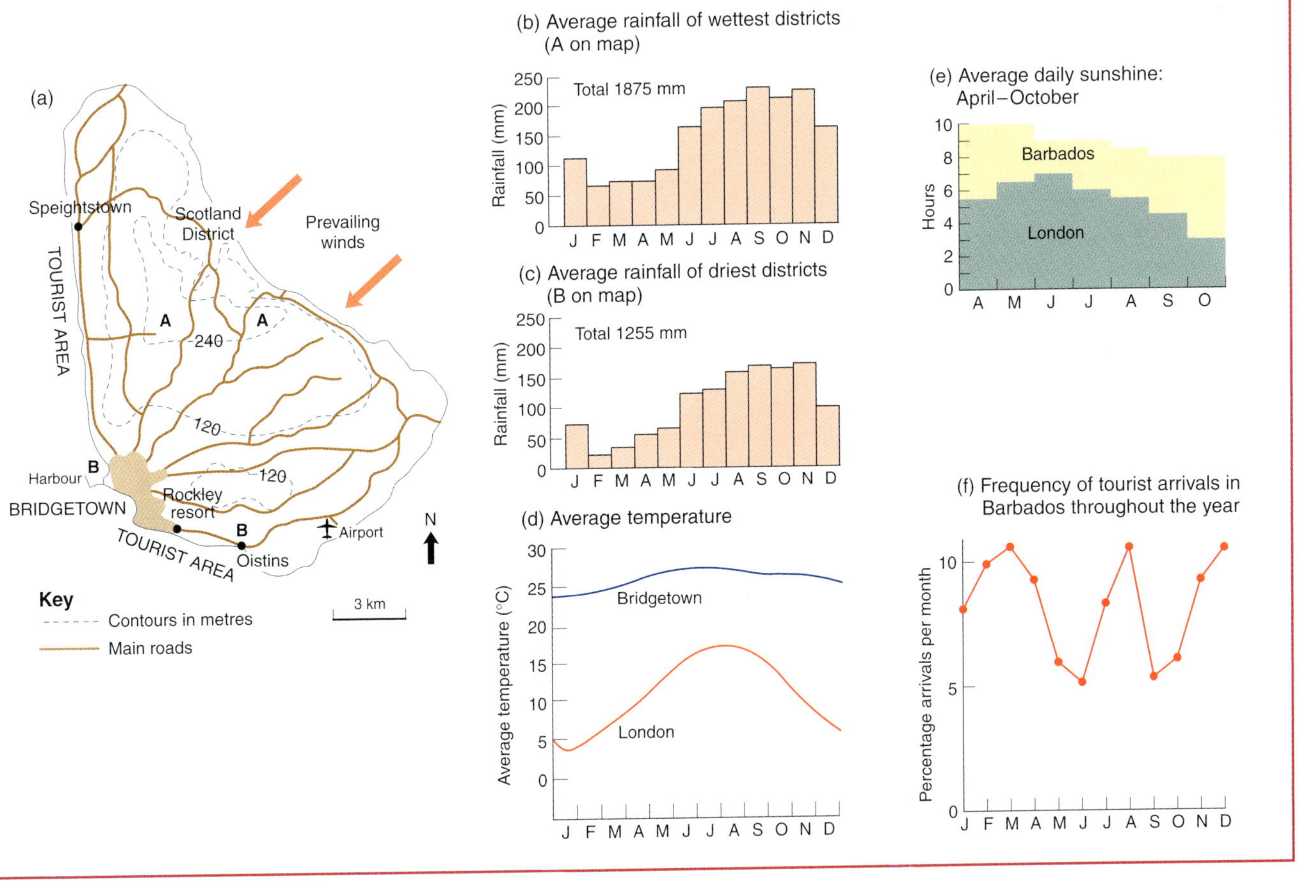

157

Section 3: Managing Economic Development

report predicted that the most rapidly growing types of tourism in the near future would be:

- **cultural tourism** – experiencing new cultures and lifestyles and visiting historic sites, e.g. Venice; the Egyptian Pyramids; stays in Hong Kong
- **eco-tourism** – seeing wildlife and experiencing new natural environments, e.g. safaris to African nature reserves
- **thematic tourism** – special interest holidays, e.g. golfing holidays to Portugal; night clubbing holidays to Cyprus; trips to Paris for shopping and Euro-Disney
- **adventure tourism** – activities in wilder exciting environments, e.g. whitewater rafting; mountain trekking in the Himalayas; diving expeditions to coral islands
- **cruises** – holidays aboard cruise ships, e.g. cruising the Caribbean islands

France, the world's most popular destination for international tourists, offers as well as seaside resorts and a Mediterranean climate on its south coast, cultural events, theme parks, high mountains and historic sites. Singapore attracts over 4 million tourists a year because of:

- its natural advantages, e.g. sunny equatorial climate; beaches; Bukit Timah rainforest nature reserve
- its culture, e.g. Chinatown with its cuisine, arts and crafts, music
- its artificial attractions, e.g. Singapore Zoological Gardens; Kranji War Memorial
- its organised events, e.g. the annual Great Singapore Sale for shoppers, the World Invitational Dragon Boat Race

People today have greater prosperity, especially in MEDCs, more leisure time than ever before, and face relatively lower travel costs than previous generations.

Greater prosperity as the income per head, especially in MEDCs, has risen has meant that more and more people can afford to travel abroad. As the length of paid **holiday entitlement** (30–35 days a year is no longer unusual) has grown for most workers, tourism has come to occupy more time in people's year and involve greater distances than ever before. We live in an age of **mass tourism**, ever more globalised, and made possible by the fall in air travel costs, greater incomes and longer holiday entitlements for working people. Greater prosperity has also seen the elderly retired holidaying abroad longer and more frequently.

ACTIVITIES

1. Use Figure 12.2. Suggest reasons for the pattern of tourist arrivals shown by Figure 12.2(f). Refer also to Figure 12.3. Suggest why Barbados is likely to attract more winter tourists than Corfu. (N1.1, N2.1)
2. Suggest reasons why the tourist industry has moved beyond providing only sun, sea and sand holidays. Why is Singapore's tourist industry well-placed to be successful in the future?

As we will see in the next section, LEDCs such as Barbados, St Lucia and other Caribbean islands have been particularly keen on developing their tourist industries. 25% of all international tourism is to LEDCs. Tourism is the major earner of foreign currency for many LEDCs (e.g. Kenya and St Lucia). It has been encouraged by the Governments of many LEDCs as the way forward for their economy. Concerns by foreign tourists over the level of security and safety are often quickly dealt with by Governments where the value of tourism is appreciated, e.g. Egypt, Sri Lanka.

The benefits and disadvantages of tourism

Tourist development brings both benefits and costs (disadvantages) for the local economy of areas receiving large numbers of tourists. It may also have important environmental effects, including on the natural environment. The benefits include these positive impacts:

- Strong economic gains from the income brought by tourists. Tourists spend money in the country, and other groups, e.g. foreign companies, governments, etc. invest in hotels, roads, airports etc. (i.e. direct tourist spending). Tourism creates employment. It creates jobs in hotel work, in coach travel, at the airport, in agriculture, in leisure services used by tourists (e.g. cafes, historic sites). Other parts of the economy can benefit too. The knock-on effects that spending in one sector of the economy has on other sectors is known as the **multiplier effect**. Tourist spending has a multiplier effect, e.g. the hotel buys supplies of locally produced drinks and food. Hotel waiters spend their earnings in local shops benefiting the economy (i.e. indirect spending on tourism). A third of St Lucia's working population are now employed in tourism and tourism-related jobs. It is now the most important part of their economy, and is seen by the government of St Lucia as the best way of boosting the economic development of the island. Tourism generates investment in the area's **infrastructure**, e.g. energy supplies, water supply, telecommunications cables, which can start to transform the whole area. For an LEDC, tourism can transfer resources to them from MEDCs, create jobs and help to ease poverty. Tourist spending and earnings from tourism are taxed by the government. This raises money that can pay for hospitals, schools and roads. For many LEDCs tourism is their biggest earner of foreign currency, which can be used to pay for imports or pay off debts.

- Improved local amenities; the preservation of heritage and some traditional culture. The money that tourists spend on local souvenirs can help prevent traditional arts and crafts skills from dying out (e.g. wood carving by hand in Kenya; lace making by hand in Cyprus). Many shops and services would go out of business were they not supported by tourists, for at least part of the year. Tourism is also credited with reducing the **depopulation** of some LEDC islands. The tourist boom in Cyprus over the past 20 years has encouraged the return of many British Cypriots to the island where they work in tourist and tourist-related jobs, and the regeneration of historic sites such as archaeological sites into tourist attractions.

- Greater environmental concern; the country becomes more international. Links with other countries are increased and the country increases its status and image in the world. Organisations to protect the natural environment that attracts tourists are set up and supported by money from tourists.

The opinion in most LEDCs in particular, remains that tourism provides a solution to their economic problems. Figure 12.4 summarises five positive impacts of tourism in LEDCs. The benefits brought to the UK economy by tourism are also considerable. It employs 1.7 million people in the UK and generates 8–9% of our GDP.

The costs or disadvantages of mass tourism include these negative impacts:

- Natural environments are spoilt and habitats threatened. Hotel, road and airport construction can destroy wildlife habitats and ruin scenery. Coastal environments can be particularly badly affected. Water pollution in the Mediterranean is at concerning levels, and the destruction of coral reefs in the Pacific from soil erosion, pollution, sewage and removal for tourist souvenirs is a major issue in the world. Honeypots where tourist activity is very concentrated, e.g. Benidorm and Miami, are particularly vulnerable to this unfortunate downside of tourism. Tourism can destroy the very quality of the environment that originally attracted tourists. Ayia Napa, Cyprus (see Figure 12.5) has changed beyond recognition, from the quiet fishing village of the 1970s to one of the liveliest resorts in the Mediterranean today, with rows of hotels, restaurants and night-clubs. Figure 12.6 shows an area of the resort in 1983 before the tourist building boom of the late 1980s/early 1990s, and the destruction of natural habitats and

Figure 12.4 Five positive impacts of tourism on LEDCs

Section 3: Managing Economic Development

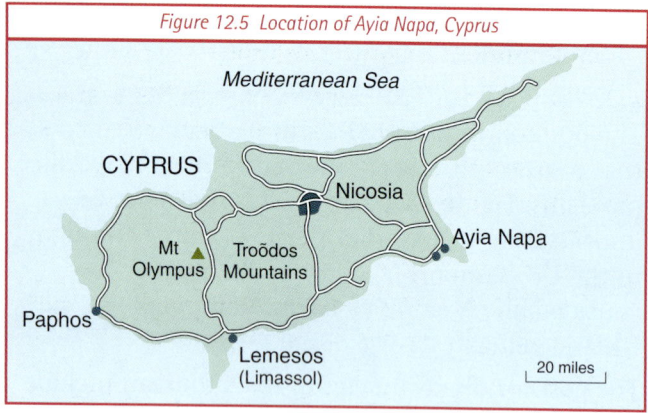

Figure 12.5 Location of Ayia Napa, Cyprus

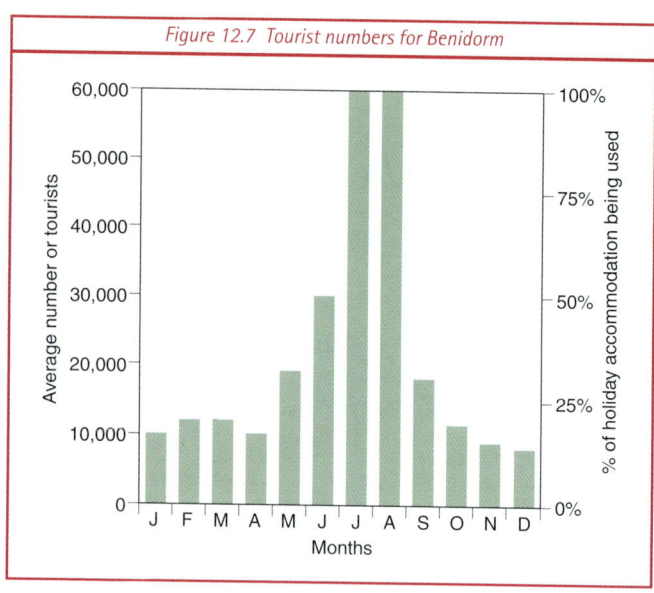

Figure 12.7 Tourist numbers for Benidorm

traditional livelihoods. Tourists also seem to be the worst offenders as beach litterers. Surveys show that they drop over a third of all beach litter, and that this proportion has risen over the past ten years. Tourism does not conserve natural systems or natural resources.

- The level of air and noise pollution also increases. Figure 12.7 shows the number of tourists staying in Benidorm, Spain through the year. Figure 12.8 shows this resort zoned according to noise levels in June. As Figure 12.7 shows, this is not the busiest time of the year yet the centre of Benidorm, and near the coast and roads are noisy. **Traffic congestion** and polluted air can be problems in heavily visited tourist areas.

- Local social, religious and cultural traditions can be swamped and suffer. The habits of tourists (e.g. drinking, fast food) can come to dominate the area receiving them. Increased contact with the MEDC tourists can lead to unrest and social problems among LEDC residents (e.g. drugs, crime, materialism). Local people often complain about poor treatment by tourists. The presence of richer tourists can encourage begging, and wheeling-and-dealing among the local population rather than traditional lifestyles and livelihoods. **Commercialisation** to meet the tourists' needs can change the culture of an LEDC.

- It can be wasteful of resources, e.g. encourages the over-use of water, especially in dry climates (e.g. in the Mediterranean summer) which leaves local people, including farmers, facing water shortages. Land is often sold to property developers and taken out of agricultural use. Land prices can rise beyond what local people can afford. Few local people can afford access to facilities catering for tourists.

- The country's import bill rises as tourists want home comforts and familiar foods, and local people's tastes change.

Figure 12.6 Unspoilt Ayia Napa

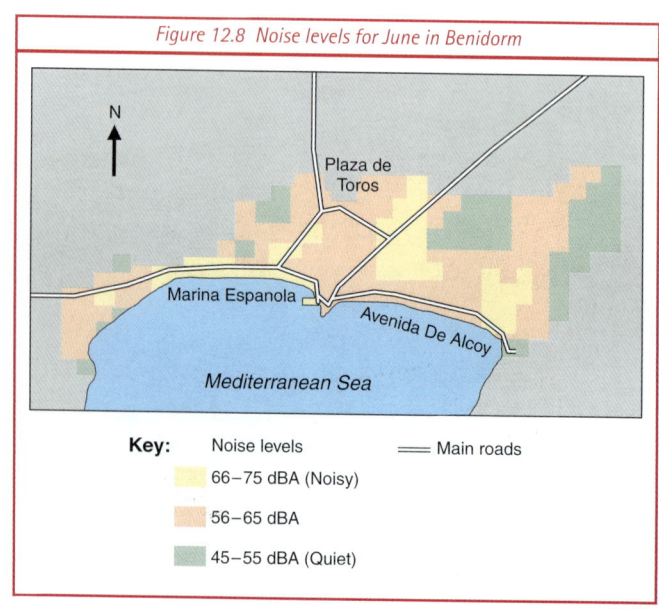

Figure 12.8 Noise levels for June in Benidorm

- Some of the human rights of local people can be threatened, e.g. some Caribbean beaches long used by local people are now only accessible to tourists; a feeling that local people always have to give way to visitors offends some local people.

- The start-up costs of a tourist area are very high, and in an LEDC are likely to need either foreign investment or government investment. Profits on investment in foreign-owned hotels leave the LEDC. Government investment in tourist areas drains resources from the rest of the country. A lot of tourist spending on a holiday benefits the companies in MEDCs (e.g. tour operators, travel agencies, airlines, foreign-owned hotels) rather than the local LEDC community where the holiday is taken.

- Many of the jobs for local people in tourist areas are only **seasonal** (e.g. summer) low-skilled and low-paid. Some are only informal sector jobs (e.g. roadside fruit and paintings sellers). The seasonal nature of tourism creates out-of-season unemployment and low activity problems for such resorts.

The benefits of tourism vary from time to time and from place to place. Surrounding areas can be left out of a tourist boom. All local people are not always helped by tourist income. Long term, an economy may not benefit very much. There are also the hidden costs, known as **externalities**, not included in the holiday price but picked up by the local population (e.g. pollution, damage to historic sites or rare wildlife habitats), perhaps when the tourists have gone home. The costs and benefits of tourism can conflict.

The general consequences for a growing tourist area and a declining tourist area

Tourist areas can go through a **life-cycle**, as Figure 12.9 shows.

To start with, there is an exploration stage when tourists first discover an area. There can follow a development stage in which the resort grows, a consolidation stage, in which it caters for a mass market, and then stagnation followed by decline. during the exploration to consolidation stages many of the impacts, positive and negative dealt with earlier may apply to the area. Figure 12.9 shows some of these consequences of tourist growth. Stagnation and decline can set in in a tourist area if, for example, it loses its holiday appeal because

externalities have been overlooked. Tourists stay away because of pollution and congestion.

It is not hard to find examples of stagnation and decline among British seaside resorts. Resorts faded as foreign tourism grew. Safety and security is another reason why tourists change their destinations, and leave some destinations facing decline. Internal conflicts in Zimbabwe since 1999 have 'killed' the country's important tourist industry. The Victoria Falls and the country's game reserves attracted two million tourists in 1999. Fuel shortages, violent protests over land ownership, mainly white-owned farms and other political uncertainties have resulted in only half a million tourists entering Zimbabwe in 2001. The **de-multiplier effects** on employment and the Zimbabwean economy have been devastating. There have been job losses in the tourist industry (e.g. hotels) and tourist related work (e.g. shops and markets struggling to make a living). Since 1999 the country's unemployment rate and population living below the poverty line have both risen. Recently, the tourism authorities in Zimbabwe have set about trying to halt the decline. They run campaigns to sell a more positive image of Zimbabwe abroad, and are trying to boost the number of tourists from within the country. Referring to the life-cycle model (Figure 12.9), they are looking to turn the decline stage into a rejuvenation or stabilisation stage. The September 11th 2001 attacks on the World Trade Towers in New York have not helped to achieve this recovery. Tourism to countries not seen as comfortable, safe and secure, especially by Americans, fell heavily in 2002.

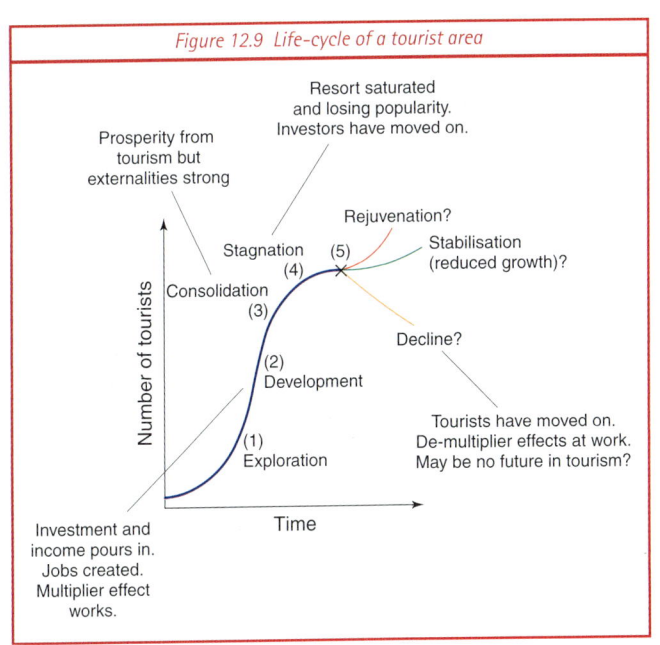

Figure 12.9 Life-cycle of a tourist area

Section 3: Managing Economic Development

ACTIVITIES

1. Tourist income accounts for about 40% of all Cyprus' export earnings. This has been achieved by increasing the number of tourists by almost four times since 1984. In this time the numbers employed by hotels has increased six times. Tourism has brought new prosperity to Cyprus but at a price. Figure 12.10, a cartoon, suggests some of these costs. Describe some of the conflicts that face the Cyprus Tourist Organisation when planning the island's tourist future. (PS1.1)

2. Study Figure 12.11 which suggests some of the advantages (benefits) and disadvantages (costs) which may affect a tourist area in an LEDC.

3. Explain why each of these comments is either an advantage or a disadvantage to the area concerned.

Figure 12.10

Figure 12.11

Sustainable tourism

It is clear that the impact of tourism on areas is not always positive, and that tourist developments may call for more careful management and planning than they have received in the past. Tourism needs to be sustainable. **Sustainability** is about seeing that developments which meet our present needs (e.g. holiday entertainments for us today) do not damage the ability of our children, grandchildren and future generations to meet their needs. Sustainable tourism is concerned about the well-being of local people now (e.g. local people have a say in developments; areas do not become totally dedicated to tourism), but also seeks to reduce the negative impacts of tourism on the environment (e.g. planners stop ugly buildings being a 'blot' on the landscape).

Eco-tourism

Eco-tourism is a form of sustainable tourism which aims to:

- conserve the natural environment (e.g. fragile ecosystems such as rainforests and coral reefs)
- market the appeal of these environments
- educate about nature
- cater for small numbers of tourists
- minimise the impact on environments and communities
- maximise local involvement and provide income for local people

The Government of Belize in Central America has made the sustainable development of its tourist industry a priority. After a surge in tourism during the 1990s, they are keen to prevent tourism from wrecking the country's natural resources and local ways of life for future generations. Eco-tourism is a key part of this sustainable tourism plan. Such trips for small groups of tourists are increasingly popular. Eco-tourists get the chance to see protected reserves for endangered species (e.g. manatees and jaguars) with their valuable natural habitats close up. Regulations apply to visiting tourists, e.g. must be accompanied by a guide; boats passing through the manatee reserve to switch off their engines. The incentive to preserve these natural resources comes from local management, local involvement in the projects, and benefits to the local community. Belizean law places a small conservation tax on all tourists to pay for environmental management.

Tropical rainforests are generally the best known examples of eco-tourist destinations. Conservation of the Kakum National Park in Ghana, West Africa, as well as some economic benefits from tourists have become the priorities. Tourism is Ghana's third largest earner of foreign currency. The idea is to maintain this income, some of it from rainforest visitors, but to do so without negative impacts on the natural environment and culture of Ghana. Guided tours, education centres and canopy walkways are ways in which it is hoped to gain income from the forest without destroying it or having tourists impact on it too negatively.

The Bukit Timah Nature Reserve is the only area of rainforest left on Singapore Island, and only 12 km from the city centre. It is an important attraction for tourists visiting Singapore as well as for day excursions by Singaporeans. It is an area protected, conserved and managed by the Singapore National Parks Board for:

- 'students as a place of discovery and experiencing a living ecosystem'
- 'visitors as a place for exploring and enjoying the varied equatorial plant and animal life'
- 'future generations of Singaporeans growing up in a highly urbanised environment'

It is not easy to develop a sustainable tourist industry. Small, local eco-tourist projects can work but they are never going to be a replacement for 'sun, sea and sand' mass tourism. Planning regulation have to be tight, and the negative impacts of tourists visiting sensitive environments need to be closely monitored. Local people rather

ACTIVITIES

1. What is a canopy walkway in a rainforest? How do they help to minimise the tourism impact on the rainforest ecosystem?
2. Visit the Iwokrama website at www.sdnp.org.gy/iwokrama. Iwokrama is an international centre for rainforest conservation and development in central Guyana, South America. Download the general rules in the visitor's guide. Explain how visitors following these rules are eco-tourists. (IT1.1 – IT2.3)
3. How should eco-tourists behave in Antarctica if they are to respect the natural environment? Surf the Internet to find out about the Lindblad Plan for Antarctica. (IT1.1 – IT2.3)

than foreign-owned tourism companies need to be the ones in charge, even though foreign expertise in managing fragile environments may be helpful. Local people should be receiving the benefits. Eco-tourism to Antarctica will require complete compliance with the international agreements about the protection of the Antarctic environment.

Other ways to ensure sustainable tourism

For tourism, especially mass tourism to have a long-term future, strict planning controls and conservation of natural resources and traditional cultures and livelihoods must be achieved. For example, planning policies which only allow low-rise building (Figure 12.12) may be more environmentally sustainable.

It may also be the case that some LEDCs need to diversify out of mass tourism. Over-reliance on income from the fickle tourist industry will never be a complete cure for the lack of economic development generally in most LEDCs. **Agro-tourism** is a new trend in some parts of the world. Here, tourists holiday in the countryside, living in traditional properties (e.g. farmhouses or restored cottages) and fitting into rural life and customs (e.g. traditional cooking and farming activities). The Cyprus Association of Agro-Tourism organises visits to the inland countryside of the island. These support established farming lifestyles and livelihoods.

CAMPFIRE (Communal Areas Management Programme for Indigenous Resources) tourism developments in Zimbabwe also help to improve the quality of life in poorer rural communities. CAMPFIRE is about diversifying into forms of tourism which sustain nature and traditions but help the rural poor. For example:

- selling licenses to foreign visitors to hunt wildlife up to set quota and so keep, for example, elephant numbers within sustainable limits.

- renting traditional round huts, sharing in local meals and going on wildlife-spotting hikes

CAMPFIRE profits have been spent on schools, wells and health centres, and this has helped to slow down rural-to-urban migration and rural de-population.

Pro-poor tourism projects are emerging throughout the LEDC world. They try to make tourism work for the poor, local people. The St Lucia Heritage Tourism Programme on the Caribbean Island of St Lucia is an example. New cultural tourism attractions based on the skills of poorer communities in farming, fishing, cooking and arts and crafts are being developed (e.g. seafood cuisine sampling centre). The idea again is to diversify St Lucian tourism out of 'sun, sea and sand' in resorts owned by large, often foreign companies. Such projects will take time to work but do offer a more sustainable future for tourism.

Tourism becomes more sustainable in the future if tourists shop in locally-owned shops and eat in local restaurants; if they buy items made using traditional skills and generally spend in a way which supports local livelihoods.

Finally, conserving natural resources, like beaches lies at the heart of sustainable tourism. The **Blue Flag scheme** is an attempt to improve the quality of beaches in the EU. A blue flag can be flown at beaches meeting high standards of:

- water quality, e.g. no sewage discharge
- environmental education and information, e.g. display boards about natural sensitive areas
- environmental management and appropriate facilities, e.g. coastal land use and development plans; litter bins, etc.
- safety services, e.g. beach guards and first aid facilities

The award is given by the foundation for Environmental Education in Europe, a non-government environmental organisation. The scheme acts as an incentive for local authorities to conserve their coastal environment and beaches and so sustains the flow of tourists. Spain and France have the most Blue Flag awards. Generally, the number of awards has increased each year since 1992, e.g. 55 British beaches were awarded Blue Flags in 2001, 17 in 1992. Awards for areas that make special efforts to keep their tourist environments, especially its natural resources, clean, can only help to make tourism more sustainable for future generations.

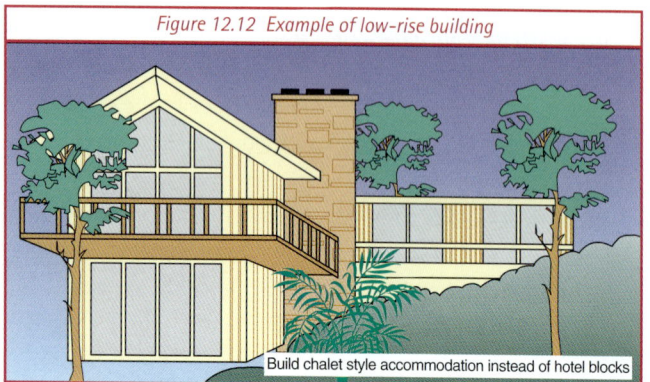

Figure 12.12 Example of low-rise building
Build chalet style accommodation instead of hotel blocks

Chapter 12: Exam-style questions

Exam-style questions with answer comments for Chapter 12

Question 1 – F Tier

Study Figure 1, two maps showing Benidorm in Spain in 1952 and 2000.

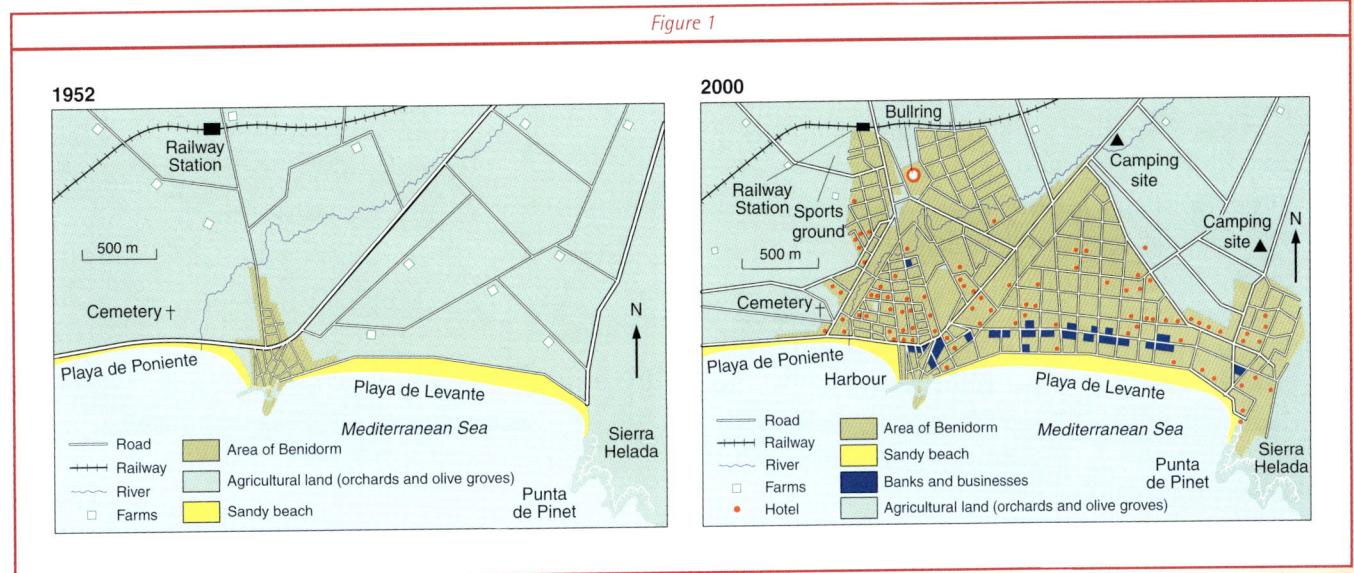

Figure 1

(a) The growth of package holidays contributed to the growth of Benidorm.

   (i) What is a package holiday? **(1)**

   Transport and accommodation combined into one product for sale.

   (ii) Give three land use changes that growth has brought to Benidorm. Use Figure 1. **(3)**

   Agricultural land now built on; land now used for hotels; new road network, etc. or any other changes to the way land is now used.

(b) The sandy beach shown on Figure 1 is a Blue Flag beach. Suggest how this indicates a willingness by the authorities to develop tourism sustainably. **(5)**

Blue flags are only awarded to beaches that are used sustainably. You need to explain the criteria for awarding Blue flags and then how this means that such beaches will have good environmental quality in the future.

(c) Explain why more people travel further to a greater variety of holiday destinations than ever before **(6)**

There are a variety of reasons but cheap and available flights are very important here. You should also write about people having higher incomes, longer retirements and more paid holidays than in the past.

(15 marks)

Chapter 12: Exam-style questions

Question 2 – F Tier

Study Figure 2, a newspaper article about the impact of tourism.

(a) Describe three problems resulting from the rapid increase in tourism in Cyprus during the 1980s. **(3)**

Unsightly coastal buildings; congested roads; insufficient water supplies, etc.

(b) What 'damage to the environment' might the Government have been concerned about? **(3)**

Disturbed wildlife habitats (e.g. turtle beaches); removal of natural vegetation from shorelines, etc.

(c) The article was written in 1989. Tourist development has continued in Cyprus during the 1990s, and to this day. What does this tell you about planning controls and attitudes to tourism in Cyprus? **(6)**

Commercialism and foreign exchange rule! Tourism seen as best way to develop island. Planning controls ineffective.

Figure 2

THE CYPRUS WEEKLY

June 2–8, 1989   25c   UK 60p   NO. 500

Government moves to limit development

ENOUGH!

The Government announced measures to control tourism. There is concern that the development of the coastline for tourism is out of control.

The measures include stopping permits for building new hotels for 10 months; making stricter financial controls; withdrawing incentives for the tourist industry.

The Government is concerned with the damage to the environment. Planning is required to solve problems such as overcrowded beaches, insufficient water supplies, traffic on the narrow coastal and hill roads, and unsightly building along the coastline.

However, there is another immediate problem. There is a shortage of labour. This will become worse as the tourist industry grows for there are 12 000 tourist beds under construction, e.g. new hotels, apartments, villas, etc.

(d) How can local people benefit from tourism? **(3)**

Jobs; greater awareness of life in other countries; better facilities bought with tourist earnings, e.g. airport.

(15 marks)

Question 3 – H Tier

Study Figure 3 which shows information about tourism in Europe.

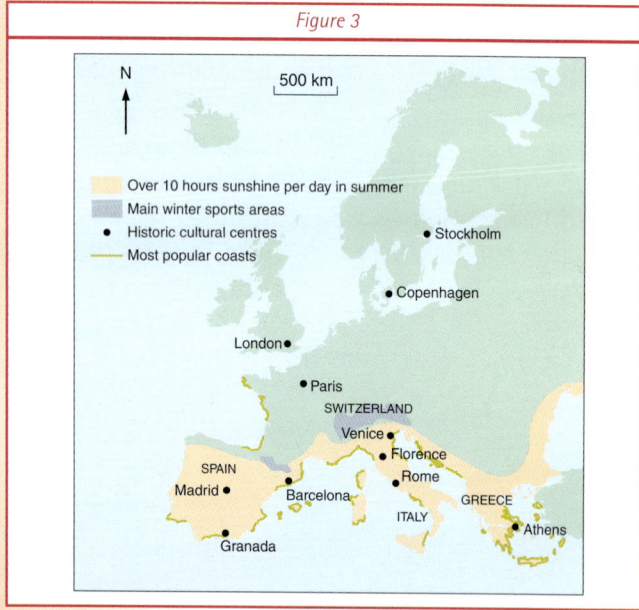

Figure 3

(a) Italy receives more international tourists than Spain. Use Figure 3 to suggest two reasons why. **(2)**

Greater winter-sports area, and more historic/cultural centres.

(b) Explain the attraction that Italy and Spain's weather/climate has for holiday-makers. **(3)**

A Mediterranean climate with over 10 hours of sunshine per day in summer; ideal for sea and sand holidays.

(c) Study Figure 4, two graphs showing monthly tourist patterns for two countries in Figure 3.

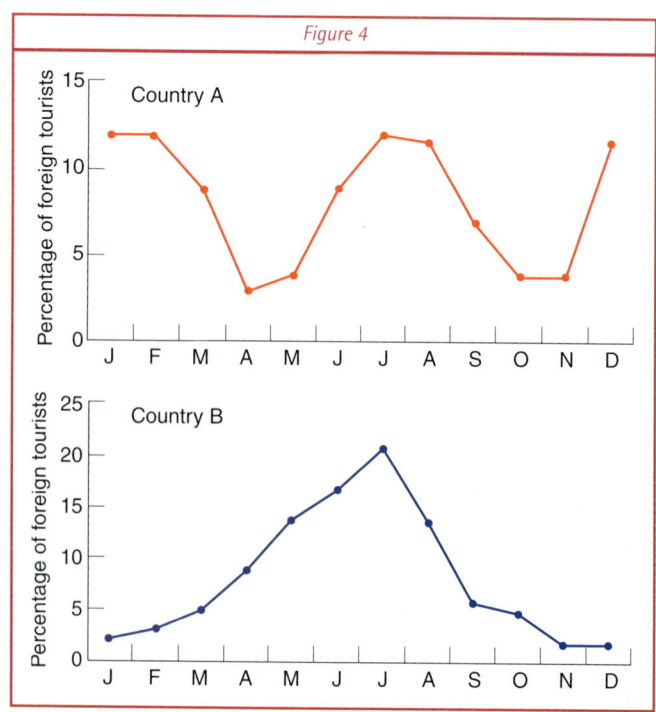

Figure 4

Chapter 12: Exam-style questions

For each graph, name a country which might fit the pattern, and give reasons for your choice. **(6)**

Switzerland is probably the best choice for Country A. The Swiss Alps attract winter sports visitors and summer tourists to the mountains. Greece or Spain for Country B – many sun-seeking summer visitors.

(d) Describe the problems caused by the seasonal nature of tourism at one named resort you have studied. **(4)**

Congestion and noise in the tourist season with unemployment and closed facilities in the off-peak season should be the basis of your answer.

(15 marks)

Question 4 – H Tier

Study Figure 5, information about an island in the Indian Ocean, which is part of an LEDC.

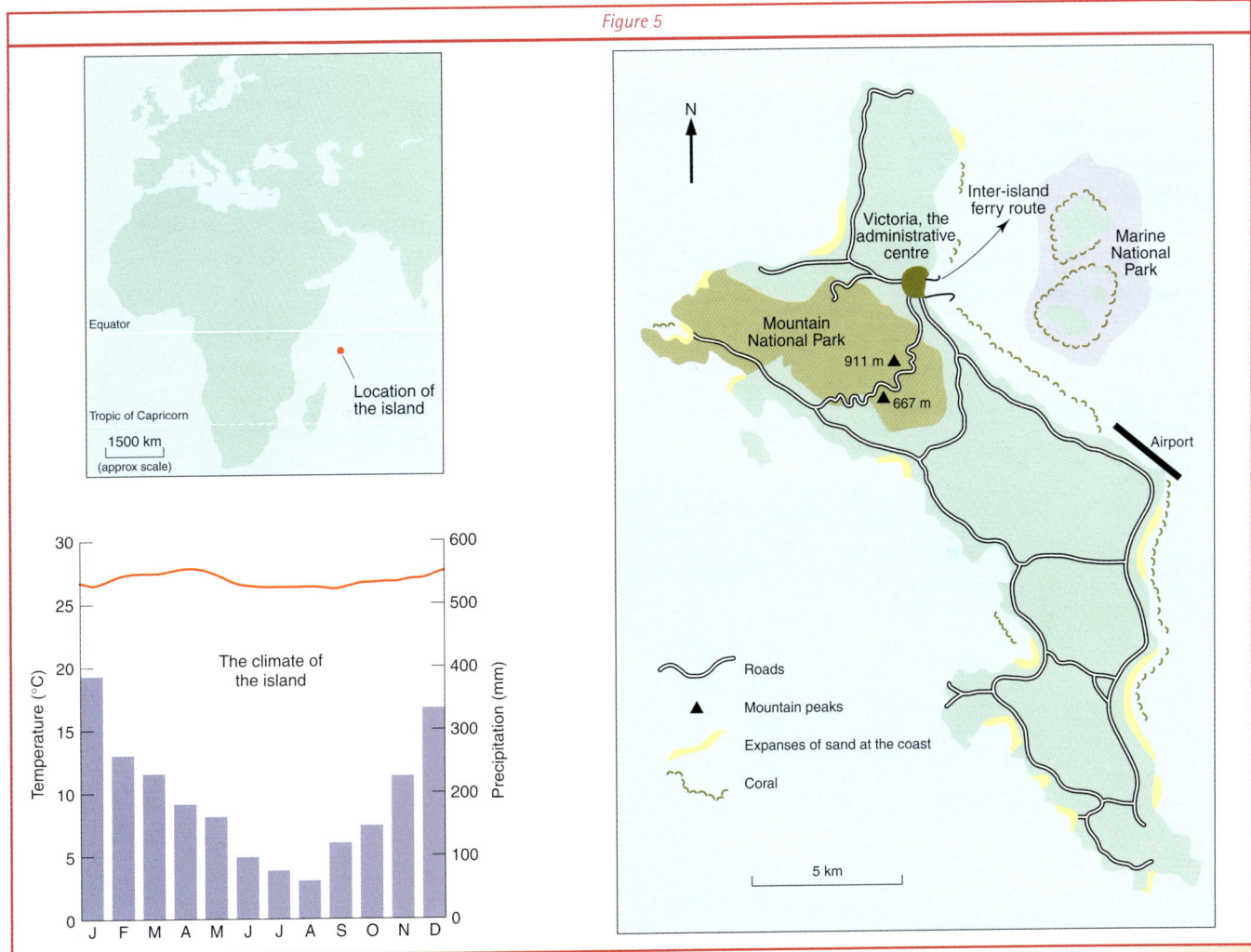

Figure 5

The traditional economy is the growing of coconut palms and tea on plantations and the cultivation of spices such as cinnamon and vanilla on small farms. There is a small island craft industry. Some fishing is done around the coast.

After much debate, with as many of the politicians and the public against the idea as for it, an international company has been given permission to build a resort complex on the island. The complex will include a hotel for 500 people, tourist apartments, tennis and squash courts, a golf course, restaurants, a shopping precinct and a workers' village.

Chapter 12: Exam-style questions

(a) What natural resources does the island possess that are likely to attract tourists? **(3)**

Hot all year; a dry season at European summer time; sandy beaches; coral reefs, etc.

(b) (i) What are likely to become the three most popular tourist months on the island? Give **one** reason for your answer.

June–August when precipitation is at its lowest.

(ii) What are the disadvantages of concentrating tourist activity in a few months of the year? **(2)**

An uneven flow gives crowds and labour shortages in one season, and unemployment in another season.

(c) Suggest a location on the island for the resort complex. Give **three** reasons for your choice. **(3)**

It is probably best to mark a location on the map where there is a sandy beach, with good road access, and perhaps not too far from the National Park. These would be your reasons. There is such a location in the north-west of the island. It is also quite close to Victoria.

(d) Explain the benefits and disadvantages of the resort complex for (1) the local peoples, (2) the environment? **(8)**

Jobs and income for environmental conservation scheme would be a benefit for (1) and (2) respectively. On the other hand, the building may ruin the natural environment, and tourists might introduce local people to bad habits.

(e) Those politicians and members of the public opposed to the complex prefer that the island adopts an eco-tourist strategy. Suggest why the island might suit the development of eco-tourism. **(4)**

With its marine and mountain National Parks eco-tourism could be developed on this island. You would need to explain what eco-tourism is about and how, for example, coral reefs can be used for tourism in this way.

(20 marks)

Coursework advice

Introduction

Coursework requires students to carry out a single geographical fieldwork investigation based upon a problem, hypothesis or issue. First hand data collection (e.g. a questionnaire, stream survey or traffic count) and data recording must be undertaken to support the investigation.

The fieldwork should be carried out at a local/small scale, and may be done individually or in groups. Any topic chosen should relate to some part of the specification content.

Four main areas provide obvious opportunities for fieldwork:

In section one Chapter 3: The changing town and city centre
Chapter 4: Pressure at the rural–urban fringe

In section two Chapter 6: Weather hazards
Chapter 8: Pressures on the physical environment

However, fieldwork may be derived from other areas. For example, water supply, reservoir sites, and agriculture could be investigated as part of Water and Food Supply, whilst recycling surveys and energy audits could support work on Resource Depletion. More traditional topics such as village studies, housing or out-of-town shopping centres can be related to other parts of the specification, and are acceptable lines of enquiry.

Rather than worry about this, you should plan tasks that produce sound results and which can be used to answer good geographical questions. An important point to realise is that titles which are 'on the edges' of the specification content are in effect 'extra work' which will not appear in the examination. You may well feel that these topics are worth the diversion involved.

The finished fieldwork report of less than 2 500 words should be written up and submitted for marking in an A4 folder by the date given to you by your teacher. It may be hand-written or printed using Information Technology: note that ICT skills are required in the mark scheme.

The following pages provide you with ideas and advice linked to chapters that appear in this book. When you write up your coursework report you will need the following items:

- **an introduction** – that sets out what you will investigate, where the study is and what ideas/theories are involved.
- **a method** – which explains what information you will collect, how to collect it, and how it will be used to test out your hypothesis/question.
- **presented data** – (some of which is ICT-based). This needs to be appropriate, varied in style and well-executed.
- **an analysis** – which interprets or explains your results, linking what you have found out to your original hypothesis or theories.
- **an evaluation** – which comments on how reliable your data is, how accurate you have been, and how secure your conclusions are.

The mark scheme on page 175 shows how marks will be awarded.

Coursework for Section 1: Managing Change in the Human Environment

Specification C requires that one geographical fieldwork investigation at a local scale should be produced. This fieldwork investigation (coursework) *must* involve first-hand data collection, i.e. information that you have collected on a school/college fieldtrip, or having undertaken an individual visit in order to collect information. The data that you collect should enable you to investigate an argument, problem or issue.

Therefore in this section of the specification, the key questions 'How do people in LEDCs and MEDCs meet the challenges of population change?' and 'How are the causes and consequences of rural–urban migration managed in LEDCs?' do not lend themselves to first-hand data collection. It may be possible to devise questionnaires and surveys on aspects of these areas of the specification, e.g. population age structure and migration into urban areas. However, these may be difficult to administer. The use of secondary data only in completing a coursework investigation is not acceptable. Although there is much valuable information on population change and the consequences of rural–urban migration in reference books, Internet web sites, computer simulations and videos, these alone cannot make up the basis of a piece of coursework. There must be first-hand collection of data.

The key questions from this section of the specification which do lend themselves very well to investigation are 'How are the changes occuring in and around MEDC city and town centres being managed?' and 'How can changes and pressures to develop rural–urban fringe locations in the EU or UK be managed more effectively?' Here is an example.

Investigating a proposed local development

This could be an actual development which is to take place in your local town or city, or it could be a hypothetical development, i.e. one which is made up for the purposes of your coursework.

Possible developments may include projects such as a new, large shopping centre, football/sports stadium, leisure park, retail park, business/science park, industrial estate, housing scheme, transport link (by pass, rail or LRT link), airport or airport extension etc. In class, you will have studied why many of these projects are needed and the possible impacts of them in areas such as the CBD, inner city, suburbs or rural–urban fringe. You will be able to *apply* this knowledge and understanding to your coursework.

Depending upon the way in which fieldvisits are organised in your school or college, visits by individual students or groups of students should be made to a site for a proposed development. Or, preferably, to a number of sites in contrasting locations within the urban area.

Some ideas for possible locations are shown in Figure 1 (page 31.)

Site A is an inner city location. A new development in a location such as this may be seen as part of attempts to regenerate the inner areas of towns and cities.

Site B is a location within the Central Business District. A new development in a location such as this may be seen as part of an attempt to reduce the impacts of competing developments on the edge of urban areas.

Site C is a location in an outer city estate. A new development in a location such as this may be seen as part of an attempt to solve the problems of above average unemployment and crime levels on some estates. Or, it may surround an issue such as the building of the new development in a park or other area of public open space.

Site D is a location at the rural–urban fringe. A new development in a location such as this may be seen as using up and damaging valuable greenbelt land as developers aim to gain maximum profit.

Site E is a located close to an outlying village. This would probably be a commuter village, with many of its residents travelling daily to work in the city.

Locations such as these offer opportunities for you to show in your coursework that you can apply understanding of geographical concepts, processes and issues. You should use the geographical terms that you have learnt in class. Through analysis of information that you have collected on your location(s), you should be able to draw valid conclusions which relate directly to the purpose of your investigation, i.e. which site is the best for the development? Should it be allowed to go ahead? Will the benefits of building a new development outweigh the costs?

Set up your hypothesis

For example: 'The new development should be built at site A.'

You then need to collect information which will help to support your hypothesis.

- How can I collect information?
- What should I collect information on?

You should collect information on the locational advantages offered by the site(s), i.e. features of the site and its surrounding area that may attract developers to it.

1. Is it accessible?

Road congestion levels at the site may be an important factor affecting the decision as to whether the new development should go ahead there. High levels of congestion may affect customer and worker access, deliveries etc. Traffic surveys can be done along roads close to the proposed site(s). Use a tally chart to record the type of vehicles which approach the site (inward traffic) and those which drive away from the site (outward traffic.) Do this for 15 minutes.

Figure 1 Tally chart for traffic survey				
HGV	LGV	Cars	Buses + coaches	Cycles + Motorcycles
III	IIII	IIII IIII	II	I

Back in class, you can convert the raw data collected into a **congestion index**. Each type of vehicle is given a score to convert it into a **passenger car unit (PCU)**:

- HGVs = 3 PCUs
- Buses and coaches = 2 PCUs
- LGVs and cars = 1 PCU
- cycles and motorcycles = 0.5 PCU.

You add together the inward and outward traffic PCU values and multiply this by 4 (to calculate the hourly total). This figure is the index of congestion. The point at which a two lane carriageway is said to be congested is 375 PCUs and a four lane carriageway, 1522 PCUs (per hour). The congestion index can be used to compare accessibility at different sites and gives a clear opportunity to explain why a method such as this was chosen.

Results can be presented as tables, graphs or flow maps.

If the development is a retail outlet or entertainment complex, similar pedestrian counts could be used. Secondary data may be used to support the first-hand data collected in the field. This may allow for greater depth of analysis. Bus and rail timetables can provide additional information and OS maps/bus or rail network maps can be used to create an accessibility matrix (table). From this, **an index of accessibility**, called a shimble number, can be calculated. This is a measure of how easy a site is to get to from a number of other places around it. Thus, access to a site from a wider area (as opposed to local access) can be calculated, and the index, or shimble number, can be used to compare the level of accessibility of different sites. The lower the shimble number, the more accessible the site. A simplified route map between four sites is shown in Figure 2.

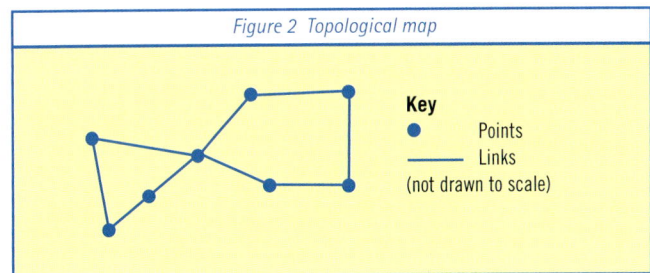

Figure 2 Topological map

Key
- Points
— Links
(not drawn to scale)

Main roads, rail lines, bus routes etc. are **links** between **points** (nodes) in the network. The number of links between sites can be counted and recorded in a matrix.

	A	B	C	D	Total
A	–	1	3	3	7
B	1	–	4	4	9
C	3	4	–	1	8
D	3	4	1	–	8

Figure 3 Accessibility map

Site A would be the most accessible. Techniques such as this offer opportunity for an evaluation of their limitations. They do not take account of actual distances or type of roads between the points.

2. What are the environmental factors?

The nature and quality of the environment at the site(s) may affect whether or not the new development should go ahead there. Factors such as the size of the site, how well drained it is, or the gradient of the land it is to be built upon, will affect decisions. Many developers now prefer an environment which reflects a modern or 'green' image. Some firms require similar types of business to be located nearby.

Coursework advice

Information on aspects of the environment can be collected using an **environmental perception survey**. Environmental perception means the way in which the environment is 'seen'. A number of factors, important to your choice of development, should be decided upon. These could include: noise level, nearness to housing, atmospheric/water pollution, vandalism and graffiti, quality of buildings, dereliction, land drainage, nearness to facilities (schools, leisure, shops, public open space etc.). Each factor should then be scored between 1 and 5, 1 being the worst score, 5 the best. A variety of scoring systems may be used. It may be left to the individual to assess the site by observation and allocate a score. A guidance sheet, giving examples of what to look/listen for in order to award a particular score, could be drawn up, e.g. noise level:

  1 = difficult to talk above the noise
  5 = very quiet, hear birds singing

If more scientific methods are available for certain factors, then these could be used. Clinometers can be used to measure gradient, noise level meters give a reading in decibels, atmospheric/water pollution testing kits are available.

The results of the whole group can be collated and an average score found. This can then be graphed/mapped. The subjective nature of this technique gives opportunities for evaluation of its limitations. What one person perceives as being attractive or noisy, is not necessarily the same as someone else. In addition, land use surveys, where the ways in which land surrounding the site(s) is used, can be recorded on a base map or transect. This can be useful as evidence of linkages or competition from similar firms as well as environmental quality. Annotated field sketches and/or photographs can also provide evidence of environmental quality and information on prominence, i.e. how clearly the site can be seen. This may be important to some businesses, who wish their logo/building to be clearly seen. Digital photographs provide an opportunity to use complex ICT techniques. Software is available that enables the user to superimpose a proposed development onto a photo of the site, in order to give some idea of its visual impact.

3. What will people think of a new development?

Questionnaires can be devised in order to assess the opinions of people who may be affected by a new development. A random sample of people could be interviewed at the site, or a specific group targeted, e.g. employees, local residents. A number of statements about a new development should be drawn up. Interviewees (people who are answering the questionnaire) can give one from a range of responses:

1. strong agreement
2. some agreement
3. no opinion
4. some disagreement
5. strong disagreement

Each interviewee's response to each statement is recorded on a tally sheet.

Figure 4 Opinion survey					
	Opinions				
Statement	1	2	3	4	5
A new development will bring much needed jobs		|||	||||	||||| ||	||||| |||||
A new development will not fit in with surroundings	||||| |||	|||||	|||				

The results of the opinion survey offer an opportunity to use a different type of graph in the presentation of data.

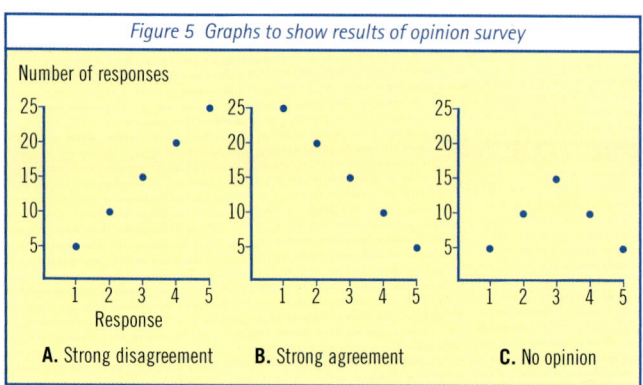

Figure 5 Graphs to show results of opinion survey

A. Strong disagreement   B. Strong agreement   C. No opinion

The information obtained from this number of surveys, mapping exercises etc. will provide sufficient evidence for analysis. On the basis of it, a decision can be made on whether to accept or reject your hypothesis. Evidence of originality and initiative may come from obtaining information from other sources, e.g. newspaper property adverts, interviews with members of planning departments, local authority structure plans (held in local libraries and web sites). Information on land values and availability of government/EU grants can be useful.

Coursework for Section 2: Managing the Physical Environment

Fieldwork suggestions

- **River processes** are an excellent topic to study, though it is important to do this safely.

  Measuring how rivers change downstream is interesting: looking at depths, speeds, channel shaped and load. You could also look at **river** landforms like meanders (see Figure 8.24, page 102). (N1.1, N1.2, IT1.2)

- **Coastal features** like cliffs and beaches can be studied similarly, and again safely. Try looking at the reasons for coastal defences like sea walls or groynes, and what effects they have. Their costs and benefits could also be studied. (C2.3, Ps)

- **Visitor impacts** in 'honeypot' locations could involve recording data about their effects on the environment – like footpath erosion. Alternatively, car parking problems or changes in local shops and businesses could mean **mapping** land use and using **questionnaires**. (Wo, Ps)

- **Weather data** recorded over a week locally could be compared with TV or newspaper reports and charts (rainfall, cloud cover and type, wind direction and strength, temperatures). Can you explain any patterns you find? (N1.1, N1.3, IT1.2)

- **Land use** near to rivers could be surveyed and mapped to look at the **risk of flooding**. Perhaps linked to river discharge measurements (see Research in chapter 8 page 102) or Environment Agency data. (Wo, C2.3)

- **A survey of flood** defences on a local river, looking at how effective they are, and a questionnaire to ask people about how safe they feel. (C2.1, Ps)

Coursework for Section 3: Managing Economic Development

Coursework ideas and advice

The focus in this section of the book is on international and global scale issues – varying levels of economic development; dwindling natural resources; atmospheric pollution; sustainable global tourism. Such large-scale, worldwide issues do not lend themselves well to small-scale field study; but it may be possible to devise an investigation around:

- development indicators and quality of life contrasts within an urban area

  e.g. how does environmental quality vary along a transect line between … and … within … ?

- quarrying and resource extraction, and environmental damage

  e.g. should quarrying for … be allowed close to the village of … ?

- recycling effectiveness within a local authority

  e.g. why are more recyclable materials collected in one area of the town than others?

- a renewable energy scheme

  e.g. how does a modern 'eco-building' with the latest design and construction encourage recycling and minimalise the use of natural resources?

- air pollution monitoring in an urban area

  e.g. where in … is the air most polluted? Why there?

- costs and benefits of tourism in a resort

  e.g. what are the environmental costs in … of large numbers of visitors during the school holidays?

- an eco-tourism project or sustainable tourism strategies in a particular place.

  e.g. in what ways is Centreparcs a sustainable tourist development?

Coursework advice

Any acceptable focus for fieldwork needs to allow you to have first-hand contact with the issue, so that specific information (e.g. data, opinions) can be collected. Once the focus has been agreed and information collected, it needs to be presented in maps and diagrams, then analysed, and finally conclusions drawn which relate to the original focus. Write it up as a series of 'chapters' in this sequence. This will correspond to the marking criteria in your syllabus. Check these criteria out and perhaps use them as 'chapter' headings.

(LP1.1 – 2.3).
(N1.2 & 2.2)
(WO1.1 – 2.3)
(PS1.1 – 2.3)
(C1.3 & C2.3)

Markscheme

Criteria for assessing coursework component.

Strand	Level 1 Marks 1–2	Level 2 Marks 3–4	Level 3 Marks 5–6
Applied Understanding	The candidate locates the study area in a basic manner and through description, using geographical terms, demonstrates some understanding of ONE idea or concept involved and can apply them in a simple manner to the geographical topic. Uses a limited range of geographical terminology.	The candidate locates the study area and through description and explanation, using a range of geographical terms, demonstrates and understanding of ONE idea or concept involved and can apply them to the geographical topic. Uses a range of geographical terminology.	The candidate locates the study area in detail and through description and explanation, using a wide range of geographical terms, demonstrates a thorough understanding of ONE idea, concept and process involved and can apply them constructively to the geographical topic. Uses a wide range of geographical terminology.
Methodology	The candidate identifies a question or issue and lists the methods used in obtaining the information. Selection observation, collection and recording uses ONE basic technique.	The candidate identifies a question or issue, the sequence of investigation and describes the methods used in obtaining the information. Selection, observation collection and recording uses TWO of appropriate techniques. The work is organised and planned and shows some evidence of the development of tasks.	The candidate identifies a question or issue explains why that particular question or issue was chosen. The candidate describes the sequence of investigation, the methods used in obtaining the information and explains why the methods selected are relevant to their investigation. Selection, observation, collection and recording uses THREE appropriate techniques. The work is well organised, planned and shows evidence or originally and initiative by the candidate.
Data Presentation	The candidate uses ONE basic technique, which is ICT based, to present the information and express simple ideas with some degree of accuracy.	The candidate uses accurately TWO techniques, ONE of which is ICT based, to present and develop the information; and express ideas with considerable accuracy in the use of English.	The candidate uses accurately THREE more complex techniques ONE of which is ICT based, to present and develop the information appropriate to their investigation; express ideas in a clear, fluent and logical form using precise and accurate English.
Data Interpretation	The candidate gives a brief description of the results and/or suggests basic reasons for the results.	The candidate makes valid statements about the results. Attempts are made to analyse the results. Conclusions are drawn that relate to the original purpose of the enquiry.	The candidate demonstrates links through a detailed analysis of the material. In referring specifically to the data valid conclusions are drawn that relate to the original purpose of the enquiry.
Evaluation	The candidate briefly describes how the enquiry process can be improved by questioning the reliability of the methods used to collect the data.	The candidate describes how the enquiry process can be improved by questioning the reliability of the methods used to collect the data and/or the accuracy of results.	The candidate describes how the enquiry process can be improved by questioning how the reliability of the methods used to collect the data have affected the accuracy of results and the validity of conclusions.

© AQA, 2000

Key Skills Guidance

Improving your key skills will enable you to perform better in your geography. Geography is also an excellent subject through which these skills can be improved. Most teachers would agree that it provides your best opportunity in the whole curriculum for developing a broad range of these skills. Coursework as geography students do it through fieldwork investigation is particularly good at providing key skills evidence.

As GCSE students you are looking to prove that you have these skills at either Level 1 or Level 2 standard. There are six Key Skills at each of these levels. Each Key Skill is divided into sub-skills, e.g. Communication is a Key Skill and its first sub-skill is to take part in discussions. This sub-skill is given the code C1.1. The other sub-skills and codes are given below.

Many of the activities in the chapters of this book can provide opportunities for you to develop key skills evidence. Those that do enable you to produce this evidence have been coded. Which sub-skill the codes stand for is given below. You can take responsibility for gathering some evidence for each of the sub-skills in a portfolio (file or folder). There are seventeen Level 1 sub-skills and nineteen Level 2 sub-skills. A grid which maps out where the evidence for achieving each sub-skill comes from say, within your geography programme of study, would be a very useful first page of your portfolio. The portfolio is assessed separately from your GCSE Geography, and needs to have separate sections for whichever Key Skill you are trying to prove you have. For example, it might have a Level 2 Communication section and a Level 1 Application of Number section. You will be able to negotiate this with your teacher. Here are examples of how you might produce evidence for each of the six Key Skills, Level 1 and Level 2 from the Managing Economic Development section of the book. Some activities/tasks can provide evidence for more than one Key Skill. The examples are based on Chapter 11 – Global Environmental Threats.

Key Skill – Communication

What you must do for Level 1		What you must do for Level 2		Possible tasks/activities	Form of evidence, e.g.
Code		Code		Causes, effects and management of acid rain issue. • Use textbook pages and Internet search to write notes • Prepare written work for class discussion and presentation as a small group	• Short written report in essay-style (with continuous prose sections) and word-processed • Illustrated poster or leaflet
C1.1	Take part in discussions	C2.1a	Contribute to discussions		
C1.2	Read and obtain information	C2.1b	Give a short talk		
C1.3	Write different types of documents	C2.2	Read and summarise information		
		C2.3	Write different types of documents		

Key Skill – Application of Number

What you must do for Level 1		What you must do for Level 2		Possible tasks/activities	Form of evidence, e.g.
Code		**Code**			
N1.1	Interpret information from different sources	N1.1	Interpret information from different sources	Global warming recent past and future • Presentation, interpretation and general handling of data in textbook • Internet search for more data	• Numerical data as tables/graphs with associated conclusions
N1.2	Carry out calculations	N1.2	Carry out calculations		
N1.3	Interpret results and present findings	N1.3	Interpret results and present findings		

Key Skill – Information Technology

What you must do for Level 1		What you must do for Level 2		Possible tasks/activities	Form of evidence, e.g.
Code		**Code**			
IT1.1	Find, explore and develop information	IT2.1	Search for and select information	• Internet search to research attainment of C1.2 and C2.2 (communication) Key Skill or for more global warming data • Word processing and image scanning	• Downloaded pages • PC – produced report on causes, effects and management of acid rain issue
IT1.2	Present information, including text, numbers and images	IT2.2	Explore and develop information and derive new information		
		IT2.3	Present combined information, including text, numbers and images		

Key Skills Guidane

Key Skill – Problem Solving

What you must do for Level 1		What you must do for Level 2		Possible tasks/activities	Form of evidence, e.g.
Code		Code			
PS1.1	Confirm understanding of given problems	PS2.1	Identify problems and come up with ways of solving them	• Issue-based geography involves you seeing acid rain, global warming and ozone depletion as management problems	• PC-produced report to include section on management strategies and early section on acid rain as problem
PS1.2	Plan and try out ways of solving problems	PS2.2	Plan and try out options		
PS1.3	Check if problems have been solved and describe the results	PS2.3	Apply given methods to check if problems have been solved and describe the results		

Key Skill – Improving own Learning and Performance

What you must do for Level 1		What you must do for Level 2		Possible tasks/activities	Form of evidence, e.g.
Code		Code			
LP1.1	Confirm short-term targets and plan how these will be met	LP2.1	Help set short-term targets and plan how these will be met	This is the progress review and general action planning between you and your geography teacher	• Action points and reviews in your tutorial records
LP1.2	Follow plan to meet targets and improve performance	LP2.2	Use plan and support from others, to meet targets		
LP1.3	Review progress and achievements	LP2.3	Review progress and identify evidence of achievements		

Key Skill – Working with others

What you must do for Level 1		What you must do for Level 2		Possible tasks/activities	Form of evidence, e.g.
Code		**Code**			
WO1.1	Confirm what needs to be done and who is to do it	WO2.1	Plan work and confirm working arrangements	● Acid rain work identified under communication key skill could be achieved as group work with each working as part of a small team	Action plan showing: ● steps ● dates ● individuals responsible ● planned discussions ● final review
WO1.2	Work towards agreed objectives	WO2.2	Work co-operatively towards achieving identified objectives		
WO1.3	Identify progress and suggest improvements	WO2.3	Exchange information on progress and agree ways of improving work with others		

©AQA, 2000

However, our advice is that these Key Skills are best demonstrated through the planning, researching, writing up and reviewing of your coursework for geography.

Appendix: Using Issues in Geography to develop moral, ethical, social and cultural awareness

This section is designed to help you develop your own values and attitudes in relation to some of the geographical issues contained within this book. Your values are the worth or respect that you give to issues about what is right or wrong. Your attitudes are how you feel in relation to an issue, or what you think about it. Two people may think that genetically modified crops may be of value to humankind, but their attitudes towards use of these crops may be very different. Throughout the book, information about important issues is given. These affect people today, both in the UK and across the world. Try to think about and discuss these in terms of their moral, ethical, social and cultural implications.

Spiritual issues may arise out of the earth's natural wonders, for example the awesome power of storms, floods and volcanic eruptions, or the uplifting effect of some of the beautiful natural landscapes around the world. Moral and ethical issues arise when decisions have to be made. These may have an impact upon people and the environment, for example should a new road be built through an area of attractive countryside? Should an MEDC provide financial aid for an LEDC? Social issues are those which directly affect people, for example what can be done to improve the quality of life for people in a shanty town? Cultural issues arise when the set of beliefs and traditions held by one group of people, are challenged by another group. For example, are tourists from the Western world damaging the local culture of areas in LEDCs?

Consider this examination question. Think of ways in which your own values and attitudes could be incorporated into answering the question.

'Explain why many people think that governments of MEDCs should provide disaster aid for LEDCs after earthquakes or volcanic eruptions.' (4 marks)

In addition to stating that MEDCs have the wealth to be able to afford to donate aid to LEDCs, there are opportunities to state the moral, social and cultural aspects of the issue. References could be made to the following:

- The idea of a 'global community' in which all fellow human beings help each other.
- All human beings having equal worth and equal rights.
- The responsibility to others through aid.
- The legacy of colonisation left by MEDCs, and the provision of debt relief in the light of this.

These points would make a convincing argument and help to achieve full marks.

Figure 1 The awesome scenery of the Grand Canyon, created by the powerful forces of nature

Figure 2 Three Gorges dam in China. A testimony to the technological achievements of humankind

Appendix

Opportunities to consider questions relating to spiritual, moral, ethical, social and cultural issues.

Chapter	Spiritual	Moral and Ethical	Social	Cultural
1. Population change		Can rapid population growth continue?	How will people adapt to a youthful or ageing population?	Is there a multicultural society?
2. Rural–urban migration in LEDCs.	The 'worth' of all people.	Why do people migrate? Can cities just keep on growing? Conflict or co-operation?	Can we improve living conditions in shanty towns? Can we improve living conditions in rural areas?	Are urban areas becoming more multicultural?
3. The changing town and city centre.	The wonder of human technological achievement and urban architecture	Should inner areas of cities be regenerated? Why do people migrate? Public transport or private car?	Are there varying social attitudes within urban areas? Will communities be broken up or created?	Is there a culture of city life and ethnic values? Are urban areas becoming more multicultural?
4. Pressure at the rural-urban fringe.		Should open land be developed? Why do people migrate?	Are there varying social attitudes to living in the urban fringe? Where will jobs be lost or gained?	Are we developing a 'leisure culture'?
5. Earthquakes and volcanoes (unstable plate margins).	The tremendous force that lies beneath the earth's crust.	Have we responsibilities to others through aid?	Do people's attitudes and perceptions of natural hazards differ?	Are there differences in cultural values regarding life and death? Are there cultural differences in attitudes to earthquake prediction?
6. Weather hazards.	The awesome effects of storms and floods and their impact upon people.	Have we responsibilities to others through aid? Is development sustainable?	Do people's attitudes and perceptions of natural hazards differ? Will managing the hazard affect people's lives?	Are there differences in cultural values regarding life and death?
7. Water and food supply.	The 'worth' of all people. The wonder of human technological achievement.	Have we responsibilities to others through aid? Do we use any means necessary to feed the world's population? Is development sustainable?	Is there a concept of 'world community'?	Are there cultural differences in attitudes to world trade?
8. Pressures on the physical environment.	The impact of beautiful scenery on the human spirit.	Are there conflicting attitudes to development and conservation?	Are social attitudes different in rural areas?	Is there a 'countryside culture'?
9. Contrasting levels of development.	The 'worth' of all people.	Have we responsibilities to others through aid? Should MEDCs provide debt relief? Do TNCs help LEDCs?	Are there social influences upon economic development? Is there a concept of a 'world community'?	Do MEDCs leave a colonial legacy?
10. Resource depletion.	The nature of humanity and threats to the planet.	Is development sustainable?	Will social attitudes to the use of resources have to change? Are there differing values towards waste disposal and recycling?	Are there differing cultural attitudes in MEDCs and LEDCs? They did it, why can't we?
11. Global environmental threats.	Our planet in the universe. The nature of humanity and threats to the planet.	Is development sustainable?	Will social attitudes to global problems have to change?	Are there differing cultural attitudes in MEDCs and LEDCs? They did it, why can't we?
12. Tourism and the economy.	Conflicting values of tourism and spirituality.	Is development sustainable? Are indigenous populations being exploited?	Is there an impact upon local customs and lifestyle?	Is local culture being eroded?

Glossary

Absolute poverty
people lacking the essential needs for a safe and healthy life

Accessibility
a measure of how easy it is to reach a location from elsewhere

Active
volcano which has erupted recently and seems likely to again

Adult literacy rate
the percentage of the population able to read and write

Aerosol
a pressurized spray

Afforestation
the planting of trees, perhaps where they did not grow before, to form a forest

Ageing population
a population in which the average age continues to rise over a period of time

Agro-forestry
farming which combines forestry with crop growing and/or animal rearing

Agro-tourism
a scheme where tourists take holidays on farms and in traditional villages

Appropriate (or intermediate) technology
technology given to a country as aid which is suitable for local conditions because it matches their level of skills and needs (e.g. not too advanced for an LEDC)

Arch
Distinctive coastal feature caused when waves are connected by wave erosion

Artificial substitutes
people-made replacements for natural resources

Attrition
how pebbles, etc. are rubbed together by water and get smaller and smaller

Bay
inlet where soft rocks have allowed coast to be eroded back

Bio-diesel
fuel produced from organic, once living matter e.g. agricultural waste or plant liquid

Biodiversity
the variety of plants, animals and other organisms in an area

Birth rate
the number of live births per 1 000 people in a year

Blue Flag scheme
a scheme to improve the quality of beaches in the EU

Braiding
how pebbles, etc. are deposited as small islands when water levels drop

Brownfield site
a location, often for industry, on land which: was once used for industry and/or housing; is now derelict; may be polluted by industrial waste; could be in an urban area, perhaps an inner city area

Caldera
a very large volcanic crater formed by major explosions

Capital-intensive
production which uses a lot of capital in relation to other factors of production

'Carbon sink'
the use of a forest or ocean to absorb excess carbon dioxide from the atmosphere

Car exhausts
the waste fumes from petrol combustion in an engine

Carrying Capacity
the number of people that an environment can reasonably cope with

Catalytic converter
technology fitted to car exhausts to remove the worst polluting gases

Central business district (CBD)
the major shopping and commercial area of a town or city. It is commonly known as the town or city centre

CFCs (chlorofluorocarbons)
chemicals used in foams, refrigeration, aerosols, etc. and thought to destroy the ozone layer

Climate graph
a graph showing average monthly temperatures and average monthly rainfall totals at a place

Commercial land use
land used for business, trade and commerce, e.g. shops and offices

Commercial logging
deforestation so that the timber can be cut and sold

Commercialisation
when business priorities like marketing, profit, etc. take over

Commute
to move, daily between a settlement in which you live and another in which you work

Comprehensive redevelopment
the complete demolition of a whole area of housing, industry, etc. and it's rebuilding

Cone
volcano shape formed by the build-up of layers of ash or lava around a crater

Conservation
the maintenance and protection of resources and environments OR action taken to protect and preserve the natural world, perhaps from pollution

Continental drift
the way in which the plates are slowly being moved by convection

Convection currents
powerful forces within the earth which can move the plates

Corrasion
the erosion of coasts and river channels by water and the pebbles, etc. in it

Counter-urbanisation
the movement of population away from large urban areas. It is the opposite process to urbanisation and has been occurring in most more economically developed countries, including Britain in recent times

Crater
hollow left when material is blown or forced out of the top of a volcano

Death rate
the number of deaths per 1 000 people in a year

Deforestation
the deliberate clearing of a forest. Large numbers of trees are burned or cut down in the tropical rainforests in order that the land can be used for other purposes such as farming or road building and/or to produce timber

Demographic transition
the population cycle. A model of the changes in the characteristics of birth rate and death rate as a country develops economically

Demography
the study of population

De-multiplier effects
declining economic activity in an area leads to unemployment , less spending and further economic decline in other activities

Depopulation
the loss of people from one area, often throughout migration

Deposition
the way in which material is left behind by rivers and glaciers

Depression
low pressure system that brings stormy weather to UK

Deprivation
refers to being poor or badly off by being deprived of important needs, e.g. decent housing

Desertification
the spread of a desert into a neighbouring area. This area must have previously received enough rain not to be a desert and support an almost continuous plant cover

Discharge
the amount of water in a river and how fast it travels (in cubic m per sec)

Dome
tall, steep-sided volcano shape formed by thicker (acid) lava

Dormant
volcano which has not erupted for a long time, but may do so again

Drought
a lack of rainfall over a long period of time

Dry deposition
when acids are deposited on the Earth's surface in a dry state from clouds and mist

Dry valleys
valleys where there are no permanent streams due to permeable rock like limestone

Dykes
either a ditch or drainage channel *or* an embankment built to prevent flooding

Earth Summit

a name for the various United Nations Conferences on Environment and Development held during and since the 1990s

Ecosystem

the links which exist in an area between living things and their environmental surroundings

'Energy crisis'

the threat that the supply of fossil fuels will run down while we still rely heavily on them

Entrepreneurship

innovation and risk-taking. Entrepreneurs possess the factor of production, enterprise

Epicentre

the point on the earth's surface directly above an earthquake focus

Erosion

the way in which water, ice and people wear away the ground

European Union (EU)

the fifteen countries, mainly in Western Europe, which are increasingly becoming ever more integrated and united

Export-led growth

economic development taking place because a lot of exports are bringing a lot of money into a country

Exposure (risk) maps

maps drawn to show the likely risks from volcanoes

Externalities

the wider effects of an action on third parties. Can be costs and benefits

Extinct

volcano which seems unlikely to erupt ever again

Factors of production

the resources used to produce goods and services, e.g. land, labour, capital and knowledge

Famine

a serious shortage of food, causing people to die of starvation

Fault

a line or zone where the earth has cracked or slipped, e.g. San Andreas

FGD (flue-gas desulphurisation)

technology fitted in power stations which burn fossil fuels to cut down on their sulphur dioxide emissions

Fissure

large opening that allows lava to spill out over a large area

Focus

the starting point or origin of an earthquake, below the ground

Food chain

a sequence showing the feeding relationships between organisms in an ecosystem

Foreign aid

the giving of resources by one country, usually an MEDC to another, generally an LEDC

Fossil fuels

the remains of past vegetation now used as an energy source

Freedom of speech

the ability to express opinions as long as libel and slander laws are not broken

Front

boundary between different types of air, bringing rainy conditions

Genetically modified (GM) crops

crops that have been deliberately altered by transferring genes from one organism to another

Global economy

production and markets are now worldwide and interlinked

Globalisation

an increasing number of geographical systems are worldwide ones with events in one place having effects in other parts of the globe

Government regulation

when the government intervene to influence the workings of a market so it is not just left to buyers and sellers to determine what it is like

Green belt

an area surrounding a large town/city which should not be built upon

Greenfield site

land that has previously not been used for industrial or urban development. It is likely to be in the countryside or on the edge of an urban area

Greenhouse Effect

the insulation of the Earth's surface by atmospheric gases

Greenhouse gases

gases like carbon dioxide and methane which insulate the Earth's surface and add to the Greenhouse Effect

Greenpeace

an international environmental and conservation organisation

Green revolution

an agricultural revolution in some *less economically developed countries*

Hard engineering

river and coastal management using engineering to fight floods and erosion

Hardwood

a tough, durable timber, mainly from tropical trees

Hazard maps

maps drawn to try to help predict where earthquakes might happen

HDI (Human Development Index)

a way of measuring people's quality of life and welfare by combining life expectancy, education levels and the value of currency

Headland

rocky outcrop where hard rock may be resisting erosion by waves

Hi-tech

the use of a lot of capital, knowledge and advanced technology in production

Holiday entitlement

the number of days of paid annual leave that a worker is entitled to

Honeypot

word used to describe places which attract large numbers of visitors

Human resources

the knowledge, capital and work that are inputs to production

Human welfare

the general well-being of people; their quality of life

Hurricane (or cyclone)

large seasonal tropical storm with very strong winds OR a violent, revolving tropical storm bringing high winds and heavy rainfall to the Atlantic coast of Central America and southern USA

Hydro-electric

electricity generated by moving water

Hydrological cycle

the movement of water between the sea, the land and the atmosphere

Immigration

the migration of people into a country from another. Those settling in the receiving country are known as immigrants

Impermeable

rocks and soils which prevent water entering the ground

Industrial Revolution

an enormous change in the amount and/or type of industry. Britain had its Industrial Revolution between 1760 and 1840

Industrialisation

when there is an enormous increase in the amount of manufacturing industry

Infant mortality rate

the number of deaths among under-one year olds in a year for every 1 000 live births in that year. It measures how many infants survive their first year of life as a key indicator of a country's level of development and the way in which it provides sanitation, nutrition and medical care for its people

Infiltration

the rate at which water soaks into the ground

Infra-red

long wave-length radiation

Inner city

the area of older residential and industrial development lying between the centre and suburbs of a city. In many cities this is where physical, social and economic problems – usually called urban deprivation – are concentrated

Interception

when water is held up by vegetation

Interlocking spurs

valley sides which have been cut back alternately by stream erosion

Intermediate technology

technology which is appropriately matched to the needs and skills of LEDCs

Irrigation

the watering of agricultural land by people rather than by rainfall. It is necessary for growing crops in areas of low rainfall

Isobar

line drawn on weather maps to show lines of air pressure (contour-like)

Keeling Curve
   the graph showing the increase in carbon dioxide in the atmosphere since 1958

Knowledge-based
   industries like pharmaceutical manufacturing that need a lot of knowledge in the production process

Kyoto Protocol
   the proposal to cut carbon dioxide emissions so as to reduce the risk of global warming

Lag time
   the delay (in hours) between peak rainfall and peak discharge

Lahar
   mudflows triggered by rainfall and volcanic ash during an eruption

Landslides
   slopes formed when water saturates (clay) cliffs, which then slide seawards

Licences
   an official form that gives legal permission for something

Life-cycle
   the successive stages in the life of an object or organism

Life expectancy
   the average number of years a new-born baby can expect to live

Living standards
   usually taken to mean the GNP/GDP per person. They refer to material or physical well-being, e.g. number of cars per 1 000 people

Local authorities
   in the UK the town, city, district or county councils

Long-haul flights
   flights lasting many hours and covering thousands of miles, e.g. to other continents, the other hemisphere

Longshore drift
   coastal currents which move beach material along the coast

Long-term aid
   assistance with development programmes

Low-tech
   simple, traditional technology

Magma
   molten rock below ground

Malnutrition
   a condition in people which occurs when their diet is insufficient and/or unhealthy

Mass tourism
   the majority of the population taking at least one holiday a year, perhaps abroad

Meanders
   a pair of bends that are a distinctive feature of many rivers

Mercalli Scale
   used to indicate earthquake intensity based on the damage caused

Migration
   the movement of people to change their home

Mobile geographically
   movement from place to place

Montreal Protocol
   the proposal to stop the use of CFCs

Multilateral aid
   aid that goes through international organisations, e.g. the UN

Multiplier effect
   an increase in investment in an area leads to an increase in incomes greater than the size of the investment

National Park
   management authority charged with looking after areas of outstanding natural beauty

Natural environment
   those aspects of surroundings that are down to Nature, e.g. weather

Natural resources
   those gifts of Nature which are of use to people, e.g. minerals

Newly Industrialising Countries (NICs)
   countries which until quite recently were LEDCs but are now becoming wealthier because of industrialisation

North Atlantic Drift (Gulf Stream)
   a warm ocean current that slows north of the Gulf of Mexico and close to western Britain

Notch
   overhang or small cave cut by waves at high tide

Nuclear power
electricity generated by splitting the nucleus of uranium/plutonium atoms

Occluded front
boundary where cold front lifts warm air clear of the ground

Ocean trenches
deep areas of the ocean where the earth's crust is being recycled

OPEC
Organization of Petroleum Exporting Countries. An organisation formed in the 1970s to protect the oil trade interests of Venezuela and Middle East countries

Organic farming
farming without the use of artificial chemicals such as fertilisers, pesticides or growth stimulants. Instead naturally occurring materials and traditional labour-intensive approaches to farming are used

Overcultivation
the growing of too many crops on an area of land. Continual cultivation, without allowing the land time to replenish itself

Over-felling
cutting down trees at too high a rate

Overgrazing
the loss of grass and other vegetation from an area because too many animals, especially cattle and sheep, are grazed on it. The exposure of the soil will lead to it being eroded, and can result in the area becoming a desert

Package holidays
a holiday where travel and accommodation are sold as one

Pavements
flat areas of limestone rocks that have been weathered

Peak flow
the highest discharge in storm hydrograph

Petro-chemical smogs
a mixture of fog and smoke (pollution) occurring in large cities

pH value
measures the acidity of a substance, e.g. soil, rainwater

Plate margins
the edges of tectonic plates (can be constructive or destructive)

Plunge pool
river bed eroded to form deeper area below a waterfall

Point bars
low areas of sandy deposition found on the inner side of meander

Population age structure
the way in which a population is broken down into the various age-groups

Population explosion
the rapid increase in the world's population that has occurred during the 20th century. Most of this explosion is now taking place in the LEDCs

Population growth rate
a measure of how quickly the number of people in an area increases

Population pyramid
a diagram to show the age and sex structure of a population. This data is represented as horizontal bars, with males on the right and females on the left, in intervals of five years, e.g. 0-4, 4-9 year olds, with the youngest at the bottom and oldest at the top

Population structure
the make-up of a population by age and by sex (gender). Population pyramids are drawn to show population structure

Pothole
vertical limestone cave, often where there is a fault, into which water sinks

Power stations
a plant/station where primary energy is used to generate electricity

PQLI (Physical Quality of Life Index)
a way of measuring people's quality of life and welfare by combining life expectancy, and the rates of literacy and infant mortality

Primary sector
includes all industries producing raw materials – mining, farming, fishing

Pro-poor tourism
projects in LEDCs that try to make tourism help poor, local people

Quality of life
the general well-being of people

Quotas
a fixed amount of a good or service that can be produced or traded

Recycling
processing waste so that it can be re-used

Regeneration
  to give new life to a part of a town or city by attempting to improve housing, community amenities and attracting new industries to improve employment prospects

Renewable resources
  one which will not run out because it can be grown or because the supply is endless

Residential land use
  land used for housing

Richter Scale
  used to indicate earthquake intensity using a numeric scale

Rift valley
  area where faults are causing land to drop downwards making valleys

River cliffs
  steep banks caused by erosion on the outside of meanders

Rural
  pastoral or agricultural area; relating to the countryside

Rural–urban fringe
  the area of transition between town and countryside. In this area there are both urban (e.g. housing and supermarkets) and rural (e.g. riding schools and garden centres) features, often side by side

Rural–urban migration
  the movement of population from the countryside to the city

Satellite photography
  images taken from space that show cloud patterns, etc.

Scars
  steep cliffs found in limestone areas

Screes
  groups of boulders at the foot of slopes, often falling after frost action

Seasonal
  changes between winter or summer, the dry time or wet time of the year

Self-sufficient
  able to meet your own needs, especially for foods

Seismograph
  an instrument used to record shock waves during earthquakes

Shanty town
  an area of unlawful, makeshift housing. Housing is made from any available cheap material (e.g. packing cases, cardboard); sanitation and supplies of water, gas and electricity often do not exist; overcrowding is common; and roads generally unmade

Shield
  low, gently-sloping volcano shape formed by fluid (basic) lava

Short-term aid
  immediate relief for disasters and emergencies

Skylark Index
  a recent way of measuring quality of life which combines 13 indicators

Slumping
  like landslides only much more rapid and unpredictable

Soft engineering
  more natural approach to floods, etc. that helps protect the environment

Soil erosion
  the removal of topsoil by water, wind and gravity (i.e. mass movement)

Solution
  how rocks like limestone dissolve in rainwater

Spit
  distinctive coastal feature formed when longshore drift leads to deposition

Stack
  Isolated rock pillar formed when sea caves and arches collapse

Storm hydrographs
  specialised graphs that show how river discharge changes

Storm surge
  local increases in sea level caused by low atmospheric pressure

Stratosphere
  a layer in the upper atmosphere. It contains the ozone layer but is above the weather layer just above the Earth's surface

Subduction zone
  zone where the crust under pressure becomes molten again

Suburbs
the outlying districts of a town/city. They are: mainly residential, usually with a low-housing density per hectare; grow outwards and extend the urban/rural fringe; increasingly becoming the location for industrial developments, e.g. hi-tech industries and superstores

Surface run-off
water flowing above ground in rivers or as overland flow

Sustainability
the meeting of today's needs as well as those in the future

Sustainable development
development which combines meeting today's needs for progress with conservation for the future

Sustainable strategies
schemes that will not cause worse problems in the future

Swash and backwash
how waves move pebbles, etc. up and down beaches

Synoptic chart
specialised chart that shows pressures and weather patterns

Tectonic plates
parts of the earth's crust which fit together like pieces of a jigsaw

Tertiary/quaternary sectors
the services sector. Services to the public are called tertiary; those to other service industries are called quaternary

Tidal barrages
a large dam wall built across an estuary so that electricity can be generated from the rise and fall of a tidal sea

Tourist honeypot
a site which attracts large numbers of tourists

Traffic congestion
when there are too many cars for the road space available

Transnational companies (TNCs)
large internal companies operating across national boundaries, usually in many countries. Another name for multi-national companies

Transportation
the way in which material is carried downstream by rivers in Traction (rolled), Saltation (bounced) or Suspension (floated)

Tsunami
massive waves triggered by earthquakes

Ultra-violet (UV)
short wave-length radiation

United Nations (UN)
the international organisation whose main objective is maintaining peace in the world

Unleaded and lead-free petrol
petrol that has all or some of its lead removed before use

Urban
a built-up area; relating to towns and cities

Urban sprawl
the outward spread of the urban area of towns and cities into the countryside

Urbanisation
an increase in the percentage of the population living in urban areas (i.e. towns and cities)

Vent
opening in a volcano, through which magma escapes

Watertable
level of water within permeable rocks, reflects recent rainfall amounts

Wave-cut platform
Area of rock planed off by the effects of waves and tides

Weathering
the way in which rocks, etc. are broken down by rain, wind and temperature changes

Wet deposition
when acids are deposited on the Earth's surface and mix with precipitation, e.g. rainfall

'Winter sun'
holiday-makers leaving in the winter for sunny destinations abroad

Youthful population
a population with a young average age and a high proportion of children and young people in it

Index

absolute poverty 107–8
adult literacy rate 106
adventure tourism 158
Afghanistan 68
agro-forestry 133–4
agro-tourism 164
aid 80–2
    foreign 117–18
    long-term 117
    multilateral 117
    short-term 117
Alaska 128
appropriate technology 118, 131
Atacama desert 71

Bangladesh 5, 49, 57, 61
Bangkok 120
Belize 163
biodiversity 134
Blue Flag scheme 164
Borneo 126
brownfield site 17
Burgess, E.W. 14
Burkino Faso 76

California 34, 43, 46–7
carbon sink 151
Central Business District 14–17, 20–2, 26–7, 170
CFCs 139, 143
China 6, 180
coasts
    features of 90
    management of 94
    processes 93–5
commercial logging 126
commuting 20
conflicts 89, 98, 100, 102
conservation 101, 134
counter-urbanisation 27
cultural tourism 158
cycle of growth and poverty 114
Cyprus 159–60

dams 62
deforestation 55–6, 73, 127–8, 151
de-multiplier 161
depopulation 159
depressions 50, 53–4
deprivation 107

Derby 16
desertification 72–4
disease 67
diversification 116
drought 68–79, 72
dry deposition 141

Earth Summit 150
earthquakes
    management of 43–5
    occurrence of 36
    risks of 34
    strength of 35–6
    terms 35
ecosystem 128
eco-tourism 158, 163
El Niño 72
employment 98
energy crisis 125
entrepreneurship 113
environment
    and climate change 72, 74
    and farming 73–4
    impacts of 99
    population pressure (LEDCs) 73
    urban development (MEDCs) 28–30
Environment Agency 59 Etna 33
European Union (EU) 22–3
export-led growth 119
externalities 161

factors of production 131
farming
    and environment 73–7
    changes 78–80
FEMA 59
FGD 142
floods
    and engineering 60
    impacts of 55–8
    management of 60–1
    risks of 60
    'flood watch' 60
food chain 128
food supply 66–9, 78–82
foreign aid 45
fossil fuels 125, 139
freedom of speech 116
fronts 53

Galapagos Islands 128–9
Gambia 119
Gateshead 24
GDP 105–7, 113
Genetically modified (GM) crops 79–80
geographically mobile 116
Ghana 163
global economy 119
globalisation 119
GNP 105–7, 113
government regulation 116
Grand Canyon 180
green belt 28–30
greenfield site 29–30
greenhouse gases 140
green revolution 78
Gujurat 33

hardwood 126
HDI 107, 113
hi-tech 113
Holderness coast 94–5
Honduras 69
honeypots (tourist) 89, 92, 143
hot deserts 70–1
human resources 125
human welfare 107
hurricanes 51–2, 63, 146
    India 13, 76, 78
    Mitch 51–2, 63
hydro-electric 126
hydrographs 55–6
hydrological cycle 55

industrialisation 115–16
Industrial Revolution 139
infant mortality rate 106
infra-red 147
infrastructure 159
Ingleborough 90–2
intermediate technology 76–7, 80–5, 114, 118, 131
irrigation 75–6, 83–4

Jamaica 13
Japan 106

Keeling Curve 147
Kenya 68, 105

Kobe 33, 37
Kyoto Protocol 150

life expectancy 106-7
limestone scenery 91
living standards 108
London 21
low-tech 131

Malaysia 134
Mali 74
Malthus, Thomas 66
Manchester 18-19, 21, 24-6, 30
manufacture
    capital-intensive 115
    knowledge-based 115
mass tourism 158
Meteorological Office 59
migration
    rural-urban 9
Montreal Protocol 144-5
Montserrat 38, 40
multiplier effect 158

National Parks 97, 101, 104
natural environment 125
natural resources 125
Newly Industrialising Countries 108, 117
NGOs 130
NHWC 59
Nicaragua 51-2
Nile, River 75-6
Non-Governmental Organisations (NGOs) 80

OPEC 108, 115

package holiday 155
Patagonian desert 71
Peru 12
petro-chemical smog 143
plates
    boundaries 41-2
    tectonic 41-2

population
    changes 1-3, 6
    growth rates 1-2
    structure 4-5, 106
PQLI 107
primary sector 106

quality of life 107-8, 113
quarternary sector 109

recycling 129-31
renewable resources 125
residential environments
    city centre 17
    inner city 15, 17-19, 27, 170
    regeneration 17-19
    shanty towns 10-13
    suburbanised village 28, 170
    suburbs 28, 170
rivers
    features 90
    processes 90, 102
rural enterprise 13
rural-urban fringe 24-30, 170

Sahara desert 70, 72
Sahel 72, 75
Saudi Arabia 115-16, 127, 130
self-sufficient 115
settlement
    function of 14-15
    shanty towns 10-13
Sheffield 112
shopping
    Central Business District 16, 26
    out of town 24-5
Sicily 34
Singapore 163
Skylark Index 107, 113
soil erosion 72-4
soil management 76-7, 83-4
Spain 76
sustainable development 75-7, 79-80, 163
synoptic charts 53-4

Taiwan 111, 116
Tasmania 126-7
tertiary sector 109
Thailand 119-20
thematic tourism 158
tidal barrages 130
TNCs 118
transport
    commuting 20
    management 21-2
    problems 20
    light rapid transit 21

ultra-violet 143
United Kingdom 5, 6
Urbanisation 55-6
    LEDCs 7-8, 10-13
    MEDCs 7-8
    problems of 10-13

volcanoes
    features 39
    hazards 38
    occurrence 39

water supply 65-8, 81-4
'weather watch' 59
wet deposition 141

York 58, 60
Yorkshire dales 90-2, 98

Zambia 79, 128-9
Zimbabwe 161